Topics in Mining, Metallurgy and Materials Engineering

Series editor

Carlos P. Bergmann, Porto Alegre, Brazil

More information about this series at http://www.springer.com/series/11054

R. Ganesh Narayanan · Uday Shanker Dixit
Editors

Advances in Material Forming and Joining

5th International and 26th All India Manufacturing Technology, Design and Research Conference, AIMTDR 2014

Springer

Editors
R. Ganesh Narayanan
Mechanical Engineering
Indian Institute of Technology Guwahati
Guwahati, Assam
India

Uday Shanker Dixit
Mechanical Engineering
Indian Institute of Technology Guwahati
Guwahati, Assam
India

ISSN 2364-3293 ISSN 2364-3307 (electronic)
Topics in Mining, Metallurgy and Materials Engineering
ISBN 978-81-322-2354-2 ISBN 978-81-322-2355-9 (eBook)
DOI 10.1007/978-81-322-2355-9

Library of Congress Control Number: 2015934677

Springer New Delhi Heidelberg New York Dordrecht London

Printed on acid-free paper

Springer (India) Pvt. Ltd. is part of Springer Science+Business Media (www.springer.com)

About the Conference

AIMTDR series of conferences is a highly prestigious biennial event in the field of mechanical engineering. It has a glorious history. The first conference entitled All India Machine Tools Design and Research Conference was held at Jadavpur University, Kolkata. Later on it was renamed as All India Manufacturing Technology, Design and Research Conference in order to widen its scope. It became an international event in 2006 with the first international conference held at IIT Roorkee. The conference aims to bring together the academicians and industry professionals working in the field of manufacturing for the exchange of ideas.

Indian Institute of Technology Guwahati organized 5th International and 26th All India Manufacturing, Technology, Design and Research (AIMTDR) conference during 12–14 December 2014. The conference was attended by more than 400 delegates from Academia and Industry. Apart from the addresses of Chief Guest Professor Amitabha Ghosh, former Director of IIT Kharagpur and Mr. P.K. Borthakur, Former Director (offshore), ONGC, a total of 22 invited lectures were presented in the conference. Some of the notable speakers from academia were Prof. Kornel Ehmann, Prof. Shiv G. Kapoor and Prof. Marc Madou from USA, Prof. Hocheng from Taiwan and Prof. V. Radhakrishnan from India. Some of the invited speakers from industry/R&D organizations were Dr. V.K. Suri from BARC, Mr. Khongwir from IOCL Bongaigaon, Dr. Sumitesh Das from Tata Steel, Dr. Sanyal from CSIR and Dr. K. Ramesh from DRDL Hyderabad. Some manufacturing organizations also gave presentations on their products. Total 423 papers were published in the proceedings of the conference by young and senior engineers/ scientists.

Prof. M. Rahman of National University of Singapore delivered Prof. Amitabha Bhattachraya Memorial Lecture. He advocated sustainable manufacturing that should not harm the environment. He emphasized the role of revolutionary and evolutionary innovation to achieve sustainable manufacturing. Overall, The conference was a huge success in terms of the participation of delegates from the premier educational and research institutions of India and abroad.

Preface

Material forming and Joining are amongst the oldest manufacturing processes, that are in use till today to fabricate micro to macro components. While material forming involves mainly elastic and plastic deformation of materials to obtain desired change in shape, joining essentially is an assembly process to make large-sized or complex parts. The material and metallurgical properties undergo substantial change during these processes. Researchers try to understand these processes by experimental and computational means. Although experiment-based research is inevitable, there is a trend to supplement it with computational techniques based on sound theoretical principles.

The chapters in this book discuss some of the basic and advanced topics in material forming and joining, including problems on rolling, sheet metal forming, adhesive bonding, welding operations like submerged arc welding, friction stir welding, accumulative roll bonding, and magnetic pulse welding. These chapters are extended versions of peer reviewed manuscripts presented in 5th International and 26th All India Manufacturing Technology, Design and Research (AIMTDR) conference held during December 12–14, 2014 at Indian Institute of Technology Guwahati, India. In total, 53 authors from academia and industry have contributed to this book. The book will be useful for researchers and practicing engineers.

The book contains 18 chapters. They are on varied topics. Chapter "An Efficient Inverse Method for Determining the Material Parameters and Coefficient of Friction in Warm Rolling Process" deals with numerical modelling of warm rolling process for predicting the materials behaviour using an efficient inverse method. Chapter "Incremental Sheet Forming (ISF)" discusses incremental sheet forming. Chapter "Effect of Yield Strength, Pre-strain and Curvature on Stiffness and Static Dent Resistance of Formed Panel" discusses the influence of selected process parameters on the dent resistance of sheets. Chapter "Formability Prediction and Springback Evaluation of Adhesive Bonded Steel Sheets" describes how the springback and formability of base sheets can be controlled by adhesive bonding. Chapter "Feasibility Study of Thermal Autofrettage Process" proposes a method for achieving autofrettage that involves creating thermal gradient in the wall of the cylinder. Chapters "Numerical Analysis of Warm Deep Drawing for Ti–6Al–4V

Alloy", "Springback and Formability Studies on Friction Stir Welded Sheets" and "Prediction of Formability of Bi-axial Pre-strained Dual Phase Steel Sheets Using Stress-based Forming Limit Diagram" present some of the recent research activities in sheet metal forming. The authors of chapter "Numerical Analysis of Warm Deep Drawing for Ti–6Al–4V Alloy" have predicted the warm deep drawing behaviour of a titanium alloy. In chapter "Springback and Formability Studies on Friction Stir Welded Sheets" controlling springback and formability of sheets using friction stir welding has been discussed. The stress based forming limit diagram has been used to predict the formability of dual phase steel in chapter "Prediction of Formability of Bi-axial Pre-strained Dual Phase Steel Sheets Using Stress Based Forming Limit Diagram". The microforming of a particular sheet material has been discussed in chapter "Size Effect on Mechanical Behaviour of SS304".

Chapters "Prediction of Weld Induced Angular Distortion of Single Sided and Double Sided Fillet Joints by SAW Process" and "Optimization of Process Parameters in Submerged Arc Welding Using Multi-objectives Taguchi Method" deal with the role of submerged arc welding process on the joint distortion and weld joint quality. Chapter "Monitoring of Weld Quality in Friction Stir Welding based on Spindle Speed and Motor Current Signals" presents real-time weld quality monitoring during friction stir welding using spindle speed and main motor current signals. The finite element modelling simulation of magnetic pulse welding of tubes is presented in chapter "Coupled Electromagnetic–Structural Simulation of Magnetic Pulse Welding". In chapter "Numerical Analysis of Heat Transfer of Arc Welded Plate", finite difference method has been used to provide solution for moving heat source with Gaussian distribution of heat flux density during submerged arc welding. The micro-friction stir welding of aluminium sheets and the joint strength and hardness characteristics are described in chapter "Development and Analysis of Butt and Lap Welds in Micro-friction Stir Welding (μFSW)". Chapter "Numerical Modeling of Impact and Solidification of a Molten Alloy Droplet on a Substrate" discusses the solidification characteristics of an alloy droplet impacting on a substrate using a two dimensional axisymmetric model. Plasma assisted friction stir welding, a hybrid system, is described in chapter "Hybrid Friction Stir Welding of Similar and Dissimilar Materials" with the aim of improving the weld joint quality and joint efficiency. Finally, the basic concepts of roll bonding and accumulative roll bonding as applicable to light weight ductile metals and the influence of various parameters on the joint strength are described in chapter "Solid-State Joining by Roll Bonding and Accumulative Roll Bonding".

Authors and editors have taken utmost care in presenting the information and acknowledging the original source wherever necessary. Editors are thankful to authors, organizers of AIMTDR and staff of AIMTDR (India) for making the publication of this type of research book possible. Readers are requested to provide their feedbacks and constructive suggestions to us. We hope that manufacturing fraternity will welcome this book.

R. Ganesh Narayanan
Uday Shanker Dixit

Editorial Acknowledgments

The editors thank a number of reviewers involved in reviewing the papers submitted to the conference. We acknowledge all the authors, invited speakers, sponsors, sub-committee members, members of national advisory committee and members of scientific advisory committee for successful conduct of the conference. We are grateful to Prof. Gautam Biswas, Director, Prof. Gautam Barua, former director, Prof. Anoop K. Dass, HOD Mechanical Engineering, and Prof. P. Mahanta, former HOD Mechanical Engineering for their continual motivation to organize such a huge event, that paved way for this book issue. The assistance extended by our colleagues and students, beyond their work schedule, needs a special mention and we thank them for active participation. Finally, we acknowledge the staff members of Springer.

Contents

About the Editors

Dr. R. Ganesh Narayanan is currently Associate Professor at the Department of Mechanical Engineering at Indian Institute of Technology Guwahati. He obtained doctoral degree from Department of Metallurgical Engineering and Materials Science, IIT Bombay and Master's from Department of Manufacturing Technology, Anna University, Chennai. He has published several papers in international journals and conferences and contributed a number of book chapters. He has edited as book on Metal forming. His areas of research include sheet metal forming, forging, joining and forming, adhesive bonded sheets, formability and forming limit and computational applications in metal forming.

Dr. Uday Shanker Dixit obtained a bachelor's degree in Mechanical Engineering from the University of Roorkee (now Indian Institute of Technology Roorkee) in 1987, an M.Tech. in Mechanical Engineering from Indian Institute of Technology (IIT) Kanpur in 1993, and a Ph.D. in Mechanical Engineering from IIT Kanpur in 1998. A Professor in the Department of Mechanical Engineering, Indian Institute of Technology Guwahati, Dr. Dixit has published numerous papers and five books. He has also edited a book on Metal Forming, guest-edited a number of special journal issues, and is an associate editor for the Journal of Institution of Engineers Series C.

An Efficient Inverse Method for Determining the Material Parameters and Coefficient of Friction in Warm Rolling Process

V. Yadav, A.K. Singh and Uday Shanker Dixit

Abstract In the present work, the material parameters for power law and the coefficient of friction are obtained using inverse analysis by measuring exit strip temperature and slip. The procedure makes use of a finite element model for deformation and an analytical method for the estimation of temperature. A heuristic optimization algorithm is used for this purpose that minimizes the error between the measured and estimated flow stresses. The method is verified by conducting some numerical experiments.

Keywords Flow stress · Temperature distribution · Finite element method · Inverse analysis · Warm rolling

1 Introduction

Rolling is one of the oldest and most important metal forming processes. It plays an important role due to its versatility and high production rate in industry. It is an economical method for the manufacturing of metal strip (Kalpakjian 2008). In general, the rolling can be considered into three categories, viz. cold, warm, and hot rolling. It has been suggested (Hirschvogel 1979; Hawkins 1981, 1985; Pietrzyk and Lenard 1990) that for hot rolling, the lower limit of the temperature range

V. Yadav · U.S. Dixit
Department of Mechanical Engineering, Indian Institute of Technology Guwahati, Guwahati 781039, Assam, India
e-mail: v.yadav@iitg.ac.in

U.S. Dixit
e-mail: uday@iitg.ac.in

A.K. Singh (✉)
Department of Civil Engineering, Indian Institute of Technology Guwahati, Guwahati 781039, Assam, India
e-mail: arvind@iitg.ac.in

R.G. Narayanan and U.S. Dixit (eds.), *Advances in Material Forming and Joining*, Topics in Mining, Metallurgy and Materials Engineering,
DOI 10.1007/978-81-322-2355-9_1

should be taken as 0.6 T_m, where T_m is the melting point of the metal in K. In hot rolling, the effect of strain hardening is removed by the relevant active dynamic and/or static softening mechanisms. In cold rolling, the upper limit of working temperature is 0.3 T_m and strain hardening is not relieved. If the working temperature for the metal is 0.3–0.6 T_m during rolling, then it defined as warm rolling process. In this range of temperature, the degree of strain hardening that occurs is less as compared to cold rolling.

The proper estimation of roll torque, roll force, and temperature distribution in strip as well as rolls is important for warm rolling because it greatly influences the lubricant behavior, roll wear characteristics, and quality of the rolled metal. Hawkins and Shuttleworth (1979) studied the effect of deformation temperature on the structure and properties of a low-carbon steel. They observed that by rolling a 0.017 % plain carbon steel to natural strains of up to 1.6 at 923 K, the rolled metal strength can be increased up to 60 %. The longitudinal and transverse tensile tests indicated that for both the test directions, warm rolling increases the strength, reduces the ductility, and decreases the ratio of the tensile strength to the proof stress. Reducing the rolling temperature to 723 K provides greater strength, but ductility and toughness decreases in comparison with rolling at 923 K. Hum et al. (1996) measured the roll separating forces, roll torques, pressure, and interfacial shear stresses as a function of the reduction and rolling speed of commercially pure Al strip at 500 °C. Serajzadeh (2004, 2006) developed the rigid viscoplastic finite element model considering the occurrence of dynamic strain aging and its effect on strain distribution during warm rolling.

The focus of this present work is to develop an inverse procedure for the estimation of material properties and friction coefficient in a warm rolling process. A few researchers have carried out inverse analysis to estimate parameters in rolling process. Kusiak et al. (1996) proposed a methodology for the evaluation of material parameters in hot forming of metals by an inverse analysis. The methodology was tested for hot compression of medium-carbon tool steel. Huang et al. (1995) presented an inverse analysis based on the conjugate gradient method to minimize heat flux and temperature distribution in roll considering the effect of water cooling on the surface of the roll. The inverse model is capable of obtaining the surface heat transfer coefficient if the surface heat flux is known. Chen and Yang (2010) used the conjugate gradient method to estimate the heat generation at the roll–strip interface by temperature measurement at some locations in the roll. For validation of the proposed method, simulated temperatures were used instead of the real experiments. Cho and Ngaile (2003) proposed an inverse method to determine the flow stresses and friction coefficient. The inverse method was based on the minimization of the difference between the experimental loads and the corresponding finite element method (FEM) predictions. In their work, the rigid viscoplastic finite element formulation was used to obtain the flow stress, while optimization algorithm adjusts the parameters used in the simulation until the calculated response matches the experimental measurements within a specified tolerance. Han (2005) applied a modified two-specimen method (MTSM) for the online determination of flow stresses and coefficient of friction in rolling. In their method, the strip is rolled twice with two different sets of roll radii. Instead of real experiments, ABAQUS FEM simulations were used for validating the methodology. Cho and Altan (2005) proposed an inverse method to determine flow stress and friction factor of the bulk and sheet materials under isothermal condition. First, the material parameters are determined by minimizing the difference between the experimental and calculated loads. Afterward, the shape of the deformed specimen is compared with the computed shape. If they do not match, the friction factor is adjusted in order to reduce the difference in the next iteration. The methodology is applied to the ring compression and the modified limiting dome height test.

Byon et al. (2008) proposed inverse method for the prediction of flow stress–strain curve and coefficient of friction using actual mill data. The roll force and forward slip are taken as the basis for inverse estimation. In their work, forward slip is decided to find the coefficient of friction and the material parameters are subsequently obtained based on roll force matching. Lenard and Zhang (1997) estimated the friction through an inverse technique by matching the measured and computed roll force, roll torque, and forward slip. Lenard and Nad (2002) estimated the coefficient of friction in rolling as a function of process and material parameters by an inverse analysis. The basis was matching of the measured and calculated roll force and roll torque. Recently, Weisz-Patrault et al. (2011) proposed an inverse method to determine the temperature field at the surface of the roll by measuring the temperature using thermocouple at only one point of the roll. Further, they extended their work (Weisz-Patrault et al. 2012, 2013) to predict the contact stress in the roll gap by measuring the stress tensor with fiber optics at only one point of the roll. In their recent paper (Weisz-Patrault et al. 2014), they presented a fully three-dimensional inverse analytical method for determining the temperature field and heat fluxes by measuring the temperatures at several locations in the rolls. Hsu et al. (2000) predicted the thermal behavior of work-roll in rolling by three-dimensional inverse model. The temperature measurements are carried out at several locations inside the roll.

In the present work, an inverse analysis is carried out to predict the material parameters and coefficient of friction during warm rolling. The heuristic algorithm is used to identify the unknown material parameters and the friction. The main contributions of this chapter are as follows. Firstly, a simple computationally and experimentally efficient algorithm has been adopted for minimization of the objective function. Secondly, a method is proposed for estimating process parameter-dependent coefficient of friction as well as material parameters based on the measurement of slip and temperature of strip at exit.

The rest of the chapter is organized as follows. Section 2 describes the modeling of warm rolling. Section 3 deals with the inverse estimation of parameters methodology. Results comprising validation and parametric study are discussed in Sect. 4. Section 5 presents the sensitivity study on the material parameters based on exit strip temperature measurement. Section 6 concludes the chapter.

2 Modeling of Warm Rolling

A two-dimensional thermo-mechanical analysis based on Eulerian flow formulation has been employed to simulate the strip rolling process. The analysis is divided into two modules—deformation module and thermal module. Deformation module is based on the flow formulation of FEM. Thermal module uses an approximate analytical method. Initially, the deformation module carries out the analysis at inlet temperature of the strip. The output of deformation module is used in thermal module to estimate the average temperature of the deformation zone. Now, deformation module is activated again to carry out the analysis at this average temperature. This iterative procedure continues till convergence. A brief description of deformation and thermal module is provided in following subsections. Figure 1 shows the flowchart of the procedure.

2.1 Deformation Module

Eulerian flow formulation FEM is employed to analyze stresses, roll torque, roll force, and forward slip considering plane strain condition. This formulation is based on steady-state deformation of the strip. Due to symmetry, only half domain is considered. The FEM-meshed domain of strip along with boundary conditions is shown in Fig. 2. In this work, elasticity is neglected. Hence, the plastic strain is the same as the total strain.

The rigid plastic material behavior is assumed where the deviatoric part S_{ij} of the stress tensor σ_{ij} is related to the symmetric part of velocity gradient ($\dot{\varepsilon}_{ij}$) as

$$S_{ij} = 2\eta\dot{\varepsilon}_{ij}, \tag{1}$$

where η is the proportionality factor.

Note that the plastic incompressibility condition has been assumed in Eq. (1). The use of von Mises yield criterion provides

$$\eta = \frac{\sigma_y}{3\sqrt{\frac{2}{3}\dot{\varepsilon}_{ij}\dot{\varepsilon}_{ij}}} = \frac{\sigma_y}{3\dot{\varepsilon}_{eq}}, \tag{2}$$

Two models of flow stress are used in this work. One is the Johnson–Cook (J–C) model (Johnson and Cook 1983) given by

$$\sigma_{eq} = \left(A + B\varepsilon_{eq}^{n}\right)\left[1 + C\ln\frac{\dot{\varepsilon}_{eq}}{\dot{\varepsilon}_0}\right]\left[1 - \left(\frac{T - T_{\mathrm{amb}}}{T_{\mathrm{m}} - T_{\mathrm{amb}}}\right)^{m}\right], \tag{3}$$

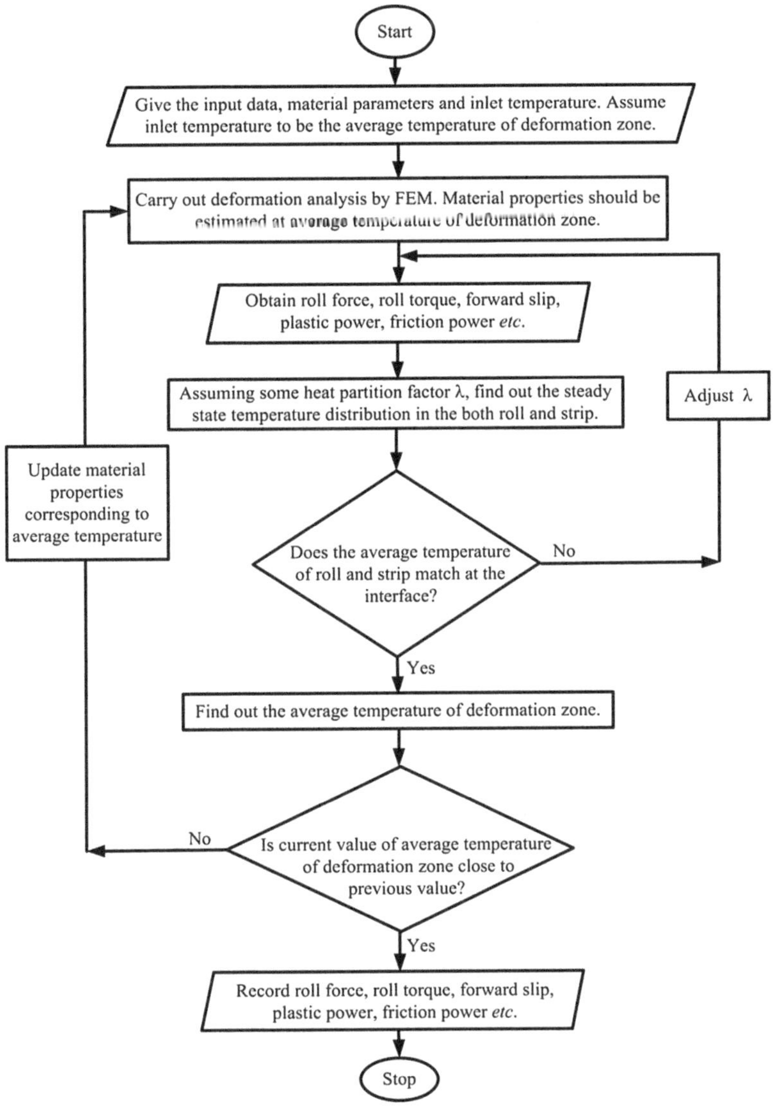

Fig. 1 Flowchart illustrating the methodology of direct model

where $\dot{\varepsilon}_0$, T_{amb}, and T_m are the reference strain-rate, the ambient temperature, and the melting temperature, respectively. A, B, C, m, and n are the material parameters. The other is a power law given by

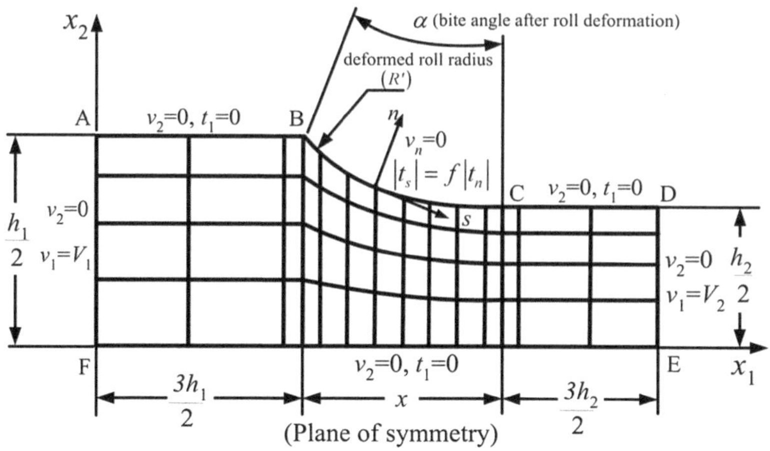

Fig. 2 A meshed domain of strip with boundary conditions

$$\sigma_{eq} = \sigma_0 \varepsilon_{eq}^n \left(\frac{\dot{\varepsilon}_{eq}}{\dot{\varepsilon}_0}\right)^{\beta} \left(\frac{T}{T_m}\right)^{-\gamma}, \tag{4}$$

where σ_0, β, and γ are the material parameters. The J–C model has a fairly wide range of applicability. Hence, it is used for carrying out the master simulations of warm rolling, and these simulations are taken in lieu of real shop floor experiments. The power law model is used for the inverse estimation of material parameters in somewhat restricted range of parameters.

In this formulation, two-dimensional continuity and momentum equation for the steady-state process are solved, which are given in index notation as

$$\frac{\partial v_i}{\partial x_i} = 0 \tag{5}$$

$$\frac{\partial \sigma_{ij}}{\partial x_j} \equiv -\frac{\partial p}{\partial x_i} + \frac{\partial S_{ij}}{\partial x_j} = 0 \tag{6}$$

where p is the hydrostatic pressure. The FEM procedure is similar to the cold rolling model of Dixit and Dixit (1996) except that the material models contain the effect of strain-rate and temperature. Roll deformation is taken into account using Hitchcock's formula. Wanheim and Bay (1978) friction model is used instead of Coulomb's model. In the notations of this chapter, f is the ratio of the magnitudes of normal and tangential tractions at the roll–work interface and μ is the equivalent Coulomb coefficient of friction.

2.2 Thermal Module

A methodology of Yadav et al. (2014) has been employed. Figure 3 shows upper half geometric arrangement of roll and strip. The plastic deformation is assumed to occur in the domain BGHC. It is assumed that 90 % of the work in plastic deformation gets converted into heat and all the frictional work gets converted into heat input at the roll–strip interface. At the interface of the roll and strip, heat transfer takes place. In this work, two sub-modules are developed for obtaining the temperature distribution. One sub-module finds the temperature distribution in the strip when the heat generation due to plastic deformation and due to friction is known. The other sub-module estimates the temperature distribution in rolls, when the heat transfer through the roll–strip interface is known. Actual heat transfer through the roll–strip interface is not known. Here, it is obtained by an inverse procedure such that both the sub-modules predict the same average temperature at the interface of roll and strip. If Q is the total friction plus plastic deformation generated heat, the quantity λQ is assumed to pass to the rolls and $(1 - \lambda)Q$ remains in strip. The value of partition factor λ is obtained to match the temperatures from both the modules.

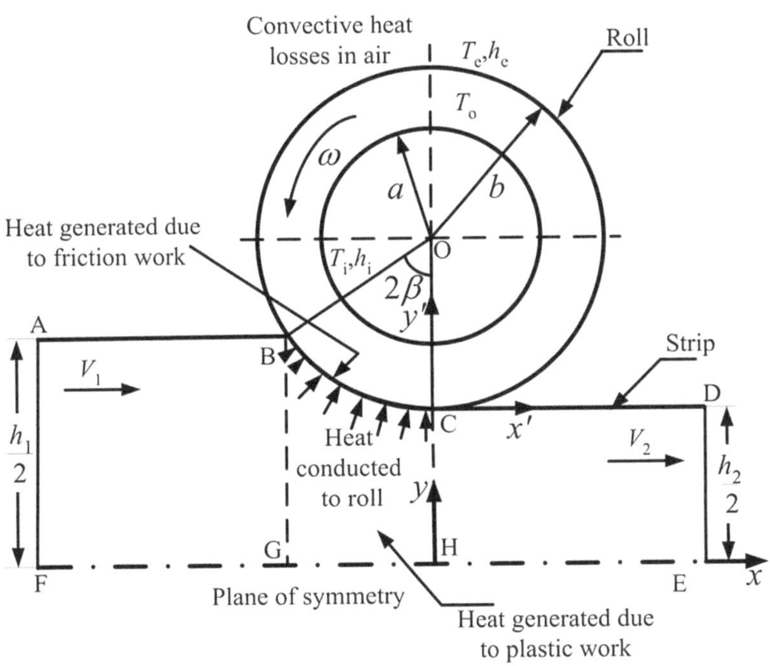

Fig. 3 Geometric arrangement of the roll and upper half of the strip

2.2.1 Module for the Estimation of Temperature Distribution in Rolls

Two different coordinate systems (x–y and x'–y') employed for the roll and strip are shown in Fig. 3. Considering non-dimensional coordinates ξ and ζ given by $x' = 2\beta b\xi$, $y' = \delta\zeta$, where δ is heat penetration depth given by

$$\delta = \sqrt{\frac{2b\beta\alpha}{\omega R}}. \tag{7}$$

The temperature distribution for zero initial temperature in the roll has been obtained by Fischer et al. (2004) as

$$T(\xi,\zeta) = T_{\max}\left[\sqrt{\xi}\exp\left(-\frac{\zeta^2}{4\xi}\right) - \frac{\sqrt{\pi}}{2}\zeta\left\{1-\operatorname{erf}\left(\frac{\zeta}{2\sqrt{\xi}}\right)\right\}\right], \quad \text{for } 0 \le \xi \le 1, \tag{8}$$

$$T(\xi,\zeta) = T_{\max}\left[\begin{array}{c}\sqrt{\xi}\exp\left(-\frac{\zeta^2}{4\xi}\right) - \frac{\sqrt{\pi}}{2}\zeta\left\{1-\operatorname{erf}\left(\frac{\zeta}{2\sqrt{\xi}}\right)\right\} \\ -\left(\sqrt{\xi-1}\exp\left(-\frac{\zeta^2}{4(\xi-1)}\right) - \frac{\sqrt{\pi}}{2}\zeta\left\{1-\operatorname{erf}\left(\frac{\zeta}{2\sqrt{\xi-1}}\right)\right\}\right)\end{array}\right], \quad \text{for } \xi > 1 \tag{9}$$

where $T_{\max}$ is given as

$$T_{\max} = \sqrt{\frac{8}{\pi}\frac{b\beta\alpha}{\omega R}}\frac{\dot{q}}{k}, \tag{10}$$

and $\dot{q}$ is the constant heat flux (bite angle) in the interval [0, 2β], k is the thermal conductivity of the roll, α is the thermal diffusivity of the roll, and ω is the angular velocity of the roll. Equations (8) and (9) provide the temperature distribution for initial condition of zero temperature distribution. For the steady-state case, a constant temperature field T_0 is imposed on the surface such that

$$T_0 + \frac{\int_0^{(\pi/\beta)} T(\xi,0)\mathrm{d}\xi}{(\pi/\beta)} = T_b, \tag{11}$$

where T_b is the average outer surface temperature of the roll which is obtained as described in the following paragraph. Similar equations can be written for other radial distances.

At steady state, the heat balance provides

$$2\pi a h_i(T_a - T_i) + (2\pi - 2\beta)bh_e(T_b - T_e) = 2\beta b\dot{q}, \tag{12}$$

where $\dot{q}$, a, b, T_a, T_b, T_i, T_e, h_i, and h_e are the heat input, inner radius, outer radius, average temperature of inner radius, average temperature of outer radius, ambient temperature at the inner radius side, ambient temperature at the outer radius side, convective heat transfer coefficient at the inner radius, and convective heat transfer coefficient at the outer radius of the roll, respectively. The validity of this equation is better for high-speed rolling, where the temperature is mainly a function of radial distance. As there are two unknowns, viz. T_a and T_b, one additional equation is provided by the following approximate boundary condition:

$$h_i(T_a - T_i) = 2bk\frac{T_b - T_a}{b^2 - a^2}, \tag{13}$$

Now solving Eqs. (12) and (13), the average temperatures T_a and T_b along the roll periphery are obtained. It can be assumed that average temperatures are the linear function of radial distance, and thus knowing the outer and inner average temperatures, the average temperature at any radial distance can be obtained. Equation (11) can now provide temperature T_0 to be superimposed on the temperature distribution of outer periphery. Similar modifications can be made for the temperature distributions of other radial locations.

2.2.2 Module for the Estimation of Temperature Distribution in Strip

The plastic deformation is assumed to occur in the domain BCHG as shown in Fig. 3. Kim et al. (2009) solved one-dimensional equation by eigen-function method. The temperature distribution as per the coordinate system of Fig. 3 is given by

$$\begin{aligned} T(y,t) = {} & \frac{1}{h}\int_0^h T_0(y)\mathrm{d}y + \frac{1}{h\rho c_p}\left(\dot{q}_s + \int_0^h \dot{q}_v \mathrm{d}y\right)t \\ & + \sum_{n=1}^{\infty}\left[\left\{\exp\left(-\lambda_n^2\frac{k_s}{\rho c_p}t\right)\left\{a_n(0) + \frac{2(-1)^n}{h\rho c_p}\int_0^t\left(\dot{q}_s\exp\left(\lambda_n^2\frac{k_s}{\rho c_p}t\right)\right)\mathrm{d}t\right\}\right\}\right. \\ & \left. + \frac{2}{\lambda_n^2 k_s h}\left(1 - \exp\left(-\lambda_n^2\frac{k_s}{\rho c_p}t\right)\right)\int_0^h \dot{q}_v\cos(\lambda_n y)\mathrm{d}y\right]\cos(\lambda_n y), \end{aligned} \tag{14}$$

where $\lambda_n = \frac{n\pi}{h}$ and $a_n(0) = \frac{2}{h}\int_0^h T_0(y)\cos(\lambda_n y)\mathrm{d}y$, $T_0(y)$ is the initial strip temperature, h is the average thickness of strip at deformation zone, k_s is the thermal conductivity of strip, ρ is the density of strip, c_p is the specific heat of strip, $\dot{q}_v$ is the heat generated per unit of volume, and $\dot{q}_s$ is the average heat flow rate to the strip at the roll and strip interface. It is assumed that there is no variation of temperature along longitudinal direction in the deformation zone. The approximate time t_a spent by the strip in the roll bite can be calculated by dividing L with the average velocity of the strip. The time t_a is put in Eq. (14) to obtain the temperature in the roll bite. The rate of heat generated per unit volume $\dot{q}_v$ and the average heat flow rate to the strip at the roll–strip interface $\dot{q}_s$ are given by

$$\dot{q}_v = \frac{0.9P_P}{0.5(h_1 + h_2)Lw}, \quad (15)$$

$$\dot{q}_s = \frac{P_f}{wl_d}, \quad (16)$$

where P_p, P_f, h_1, h_2, L, w, and l_d are the plastic deformation power, friction power, sheet thickness at inlet, sheet thickness at outlet, length of plastic deformation zone (=GH in Fig. 3), width of the sheet, and contact length of roll and strip, respectively. The plastic deformation power P_p and friction power P_f are obtained from the deformation module.

2.3 Validation

To test the validity of the model developed, the results obtained by the model are compared with experimental data of Shirizly and Lenard (2000). They performed rolling experiments on low-carbon steel AISI 1018 with steel rolls. In their experimental work, the specimens were preheated in air at 850 °C for 60 min to ensure a homogeneous deformation temperature. The material used in the rolling is cold-rolled, low-carbon AISI 1018 steel. The rolling speed is 157 mm/s, and the coefficient of friction is 0.25. The strip dimension is taken as (305 × 44 × 4.71) mm^3. The outer diameter of the work-roll is 150 mm. The inner diameter of the roll is not mentioned by authors. Here, it is assumed as 130 mm, considering typical practical value. (Later on, the inner diameter of 120 mm was also chosen and it was found that results differ by an amount less than 1 %. Thus, the inner diameter has an insignificant effect on roll torque and roll force in this case.) The convective heat losses at the outer and inner periphery roll are assumed as 260 and 2.6 W/m^2 °C, respectively. The thermal properties of strip and roll are taken from Khalili et al. (2012). The flow stress of the strip material is governed by modified J–C model. The modified J–C model material parameters are taken from Vural et al. (2003) for AISI 1018 cold rolled. In the modified J–C model, two separate J–C models are modeled for low and high strain-rate cases. Vural et al. (2003) have suggested the smooth transition between two models over a narrow strain-rate zone. However, here, discontinuous model is taken, as it does not pose any problem in FEM code. The parameters of J–C model given by Eq. (3) are $A = 560$ MPa, $B = 300$ MPa, $n = 0.32$, $m = 0.55$, and $T_{\mathrm{m}} = 1773$ K. For strain-rate of less than $\dot{\varepsilon}_t$, C is given by

$$C = n_1 \left(\frac{\dot{\varepsilon}}{\dot{\varepsilon}_{01}} \right)^{n_1}, \quad (17)$$

and for strain-rate of greater than $\dot{\varepsilon}_t$, C is given by

$$C = n_2\left(\frac{\dot{\varepsilon}}{\dot{\varepsilon}_{02}}\right)^{n_2}, \tag{18}$$

where $\dot{\varepsilon}_{02}$ is calculated as

$$\dot{\varepsilon}_{02} = \left(\dot{\varepsilon}_{01}\right)^{n_1/n_2}\left(\dot{\varepsilon}_t\right)^{(n_2-n_1)/n_2}, \tag{19}$$

For AISI 1018 cold-rolled steel (Vural et al. 2003), $n_1 = 0.007$, $n_2 = 0.075$, $\dot{\varepsilon}_{01} = 5\times10^{-6}\,\text{s}^{-1}$, and $\dot{\varepsilon}_t = 96\,\text{s}^{-1}$.

Figures 4 and 5 compare experimental and simulated roll forces and roll torques, respectively, (Shang et al. 2012). The input process parameters are given in the figures. It is seen from Fig. 5 that the roll torque gives the maximum 12.6 % error with experimental results of Shirizly and Lenard (2000). However, the roll forces almost fully match with the experimental results as shown in Fig. 4.

The second example is considered in which the roll surface temperature is matched with the average temperature at the interfacing surface of the strip. The average temperatures at the deformation zone are compared with the experimental results of Serajzadeh and Mohammadzadeh (2007). Serajzadeh and Mohammadzadeh (2007) performed the warm rolling experiments of low-carbon steel with steel rolls. They conducted the experiments with inlet strip temperature in the range of 500–750 °C for different rolling speed and percentage reduction. The dimension of the strip was (100 × 5 × 3) mm^3, and the work-roll diameter was 150 mm.

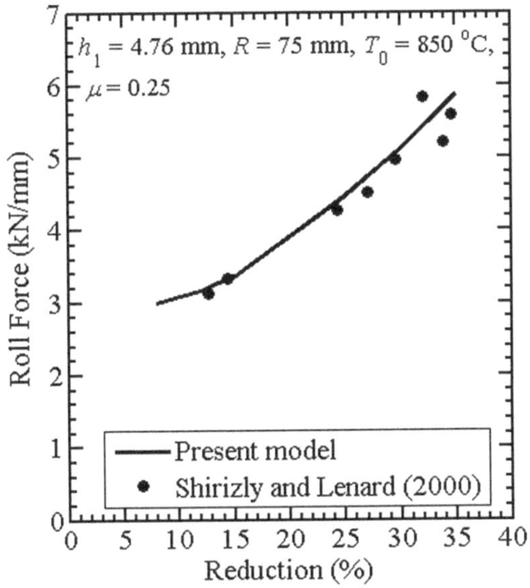

Fig. 4 Comparison of present model with experimental results of Shirizly and Lenard (2000) of roll force per unit width

Fig. 5 Comparison of present model with experimental results of Shirizly and Lenard (2000) of roll torque per unit width

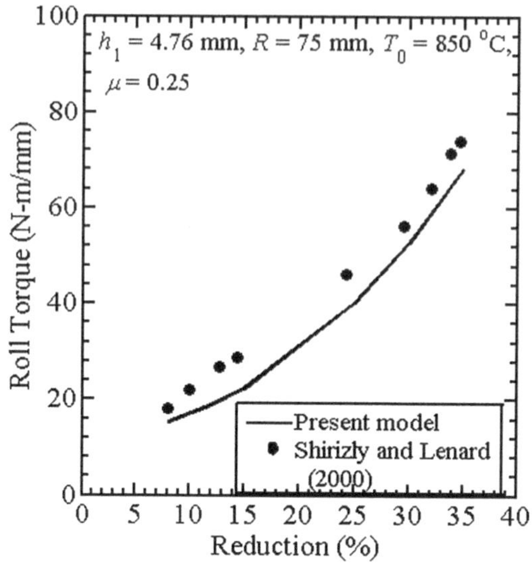

The flow stress of the strip material is governed by the following power law given by Serajzadeh (2004):

$$\sigma_{\mathrm{eq}} = B\dot{\varepsilon}^m \varepsilon^n, \tag{20}$$

where B, m, and n are the temperature-dependent material constants expressed by

$$B = 1198 - 0.084\,T - 0.002\,T^2, \tag{21}$$

$$n = 0.36 - 2.96 \times 10^{-4}\,T, \tag{22}$$

$$m = 0.788 - 0.0046\,T + 9.11 \times 10^{-6}\,T^2 - 7.04 \times 10^{-9}\,T^3 + 1.9 \times 10^{-12}\,T^4, \tag{23}$$

where T is the temperature in °C. The inner diameter of the roll and convective heat transfer coefficients are assumed as in the previous example.

For different values of the inlet temperature of strip, inlet thickness of strip, percentage reduction and angular velocity of roll, results of the present work and Serajzadeh and Mohammadzadeh (2007) are given in Table 1. It is observed that the error in the prediction of temperature is less than 9 %. Thus, the model is validated.

Table 1 Comparison of present model with experimental results [Serajzadeh and Mohammadzadeh (2007)]

Sample no.	Initial temperature of strip (°C)	Inlet thickness of strip h_1 (mm)	Percentage reduction (% r)	Revolution per minute (r.p.m) of the roll ω	Average temperature at deformation zone (°C)	Expt. [Serajzadeh and Mohammadzadeh (2007)] temperature (°C)	% error
1	500	3	30	50	426.95	441	−3.19
2	650	3	15	50	599.62	595	0.78
3	650	3	30	50	577.16	530	8.89
4	750	3	15	65	692.52	700	3.28

3 Inverse Estimation of Parameter

It is assumed that the material behaves as per the power law material model given in Eq. (4). This model may not be valid over wide ranges of strain, strain-rate, and temperature. However, it is always possible to select smaller domains for fitting this model. A number of such models in various smaller domains collectively can model the behavior in a wider zone.

The root-mean-squared (RMS) fractional error, E of the estimated and the measured quantity (exit temperature, roll force, or roll torque), is taken as the objective function, which is minimized with respect to the decision variables σ_0, n, β, γ, and μ. Mathematically,

$$E = \sqrt{\frac{1}{n} \times \sum_{i=1}^{n} \left(\frac{\text{measured parameter} - \text{estimated parameter}}{\text{measured parameter}} \right)^2}, \tag{24}$$

where n is the number of experimental observations. It was found that any one parameter among exit temperature, roll force, and roll torque can be the basis for calculating E, although the exit temperature measurement is the most convenient from a practical point of view. Figure 6 shows the schematic representation of arrangement for temperature measurement. The strip temperature is estimated as proposed in Kim et al. (2009) and summarized in Appendix. Alternatively, one can measure as many parameters as convenient for better reliability. Minimization of

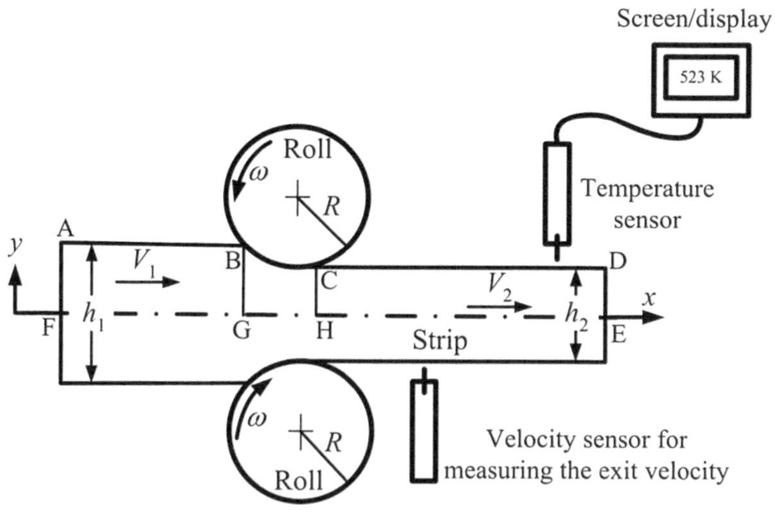

Fig. 6 Schematic arrangements of forward slip and temperature measurement during rolling process

the objective function is carried out by a heuristic method inspired by the similar method used by Chandrasekeran et al. (2012) and Eideh and Dixit (2013).

3.1 Details of the Algorithm

The methodology for finding out the material parameter and friction during rolling is as follows:

Step 1: Decide the ranges of strain, strain-rate, and temperature in which the power law model has to be fitted. The total domain can be represented as a rectangular parallelepiped in a three-dimensional space with strain, strain-rate, and temperature as axes. Carry out 8 experiments approximately corresponding to eight corners of rectangular parallelepiped. It is not possible to carry out the experiments at exact desired values of strain, strain-rate, and temperature. Actual controllable parameters are percentage reduction, rolling speed, and initial temperature. The computational model of the rolling guides in the proper selection of controllable parameters.

Step 2: Choose suitable ranges for material parameters, σ_0, n, β, and γ of power law, and friction coefficient μ.

Step 3: For each parameter, the range is divided into three linguistic zones, viz. low (L), medium (M), and high (H) as shown in Fig. 7 for two parameters. Thus, the entire domain gets divided into $3^5 = 243$ cells.

Step 4: Select the middle (M) values of all the parameters as initial guess parameters. Calculate E using (Eq. 24). The E is first calculated only on the basis of two measurements—one corresponding to high reduction, low temperature, and high strain-rate and other corresponding to low reduction, high temperature, and low strain-rate. If $E \geq 0.1$, the error is considered very high and there is no need to calculate errors for other experimental points. If $E < 0.1$, then gradually other experimental points are included in the error calculation for finding out E subject to maximum of 8 experimental points. This strategy helps in reducing the computational time involved in function evaluation.

Step 5: Keeping 4 other parameters (n, β, γ and μ) constant, carry out one-dimensional search for optimum σ_0 in the following manner:

- If the estimated temperature at the current point is greater than the measured temperature, then decrease the value of σ_0 by jumping to the center of adjacent cell. If reduction in E is significant based on test of significance, then the center of adjacent cell becomes the current point. If the reduction in E is insignificant, both points are taken as current point and further exploration is carried out in a parallel manner from both points. If E increases, then σ_0 is not changed.
- If the estimated temperature at the current point is lesser than the measured temperature, then increase the value of σ_0 by jumping to the

center of adjacent cell. If reduction in E is significant based on test of significance, then the center of adjacent cell becomes the current point. If the reduction in E is insignificant, both points are taken as current point and further exploration is carried out in a parallel manner from both points. If E increases, then σ_0 is not changed.

- Else do not change the value of σ_0.

Step 6: The similar methodology as discussed in Step 5 is repeated for optimizing other parameters, i.e., one parameter at a time is changed keeping other four parameters constant. After completing an iteration consisting of five one-dimensional searches, the search domain gets reduced to one cell.

Step 7: For the further refinement, the optimum cell is further divided as in Step 3. Repeat the procedure of Steps 4–6. A graphical representation for reducing sizes of the search domain toward optimum is shown in Fig. 7.

After carrying out this procedure, if the E could not be reduced significantly, then the ranges of strain-rate, temperature, and strain need to be reduced.

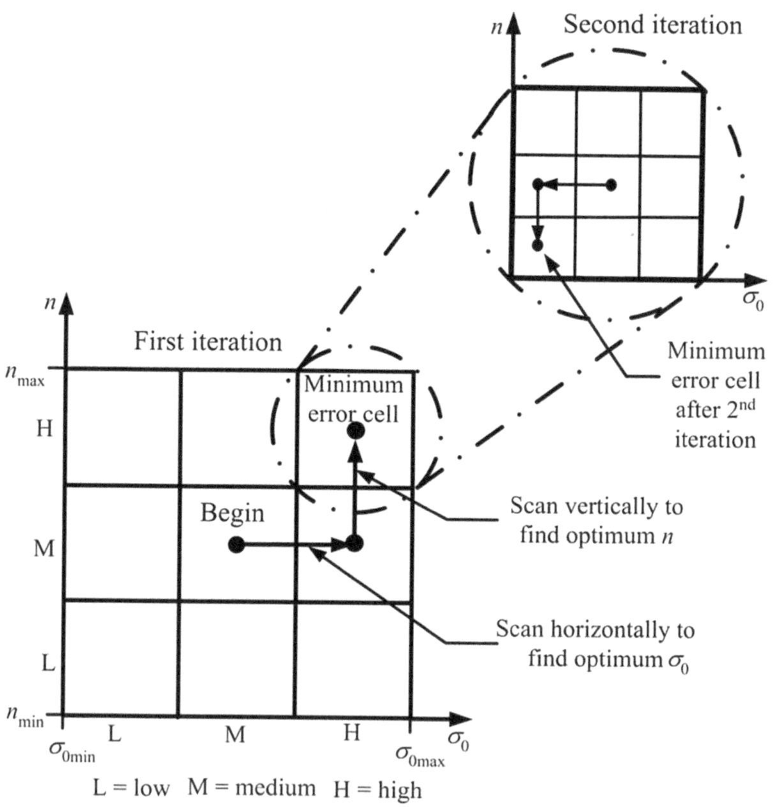

Fig. 7 Two-dimensional graphical representation of search procedure

3.2 Modification of the Algorithm for Variable Friction Case

In many cases, the friction is a function of process parameters and it cannot be considered constant. Most predominantly, it depends on the temperature. The friction can be measured by measuring the slip at each experimental point. For each experimental point, the coefficient of friction may be different. The algorithm developed in Sect. 3.1 can be used for this case also. Now, the number of decision variables reduces from 5 to 4. However, in each cell, while estimating the observable parameter (temperature, roll force, and roll torque), the proper value of coefficient of friction as estimated based on slip measurement needs to be considered.

4 Results and Discussion

In the following subsections, inverse estimation of material parameters and coefficient of friction and inverse estimation of material parameters with considering the coefficient of friction are variable. The inverse estimation has been carried out for different strip materials, namely AISI 4142 steel, copper, and aluminum alloy considering roll material to be steel.

4.1 Inverse Estimation of Material Parameters and Coefficient of Friction

In the present work, actual shop floor experiments have not been carried out. For validating the proposed procedure, numerical experiments have been carried out. It is assumed that the actual flow stress is governed by the well-known J–C relation. The constants of different material of J–C model are given in Table 2.

The actual friction coefficient is assumed to be 0.20. The roll radius (R) is 65 mm, and inlet thickness of the strip (h_1) is 1 mm. With these data, virtual simulations are carried out using the FEM analysis for different initial temperature

Table 2 Constants of J–C model for different material

Material	A (MPa)	B (MPa)	n	C	m	$\dot{\varepsilon}_0$	T_m (K)	References
AISI 4142 steel	598	768	0.2092	0.0137	0.807	0.001	1768	Boisse et al. (2007)
Copper	90	292	0.31	0.025	1.09	1	1338	Raczy et al. (2007)
Aluminum alloy AI6061-T6	293	121.3	0.23	0.002	1.34	1	925	Shang et al. (2012)

Table 3 Thermal properties of work material

Material	Density (kg/m^3)	Thermal conductivity (W/mK)	Heat capacity (J/kgK)	Heat transfer coefficient of air (W/mK^2)	References
Steel (AISI 4140)	7800	43	470	10	Yadav et al. (2011a)
Copper	8960	400	385	10	Zhang et al. (2011)
Aluminum alloy	2700	237	900	8	Yadav et al. (2011a)

of strip, percentage reduction, and exit velocity of strip. Forward slip is recorded. The mechanical properties are obtained by the inverse analysis using the thermal properties of material as given in Table 3. The work-roll material is considered as steel only for the entire analysis. The mechanical and thermal properties of roll are given in Khalili et al. (2012).

Inverse analysis provides the material parameters σ_0, n, m, and γ and the coefficient of friction μ of different materials, viz. AISI steel, copper, and aluminum alloy AI6061-T6 as given in Table 4. The convergence is obtained in 32–42 function evaluation, i.e., in 3–4 iterations for all three different materials of strip. The convergence depends on the initial hypercube size of the unknown parameters. Starting with a large range of domain is better, but the number of iterations required for convergence will be large.

Figures 8 and 9 show the variation of flow stresses with strain at the estimated and the actual material properties for work material (steel) of strip. It is assumed that the power law gives the better results for smaller range of process parameter. Hence, temperature is defined in the two ranges, i.e., 450–650 K and 650–850 K. Later on, the larger range is also considered. Figure 8a, b shows the variation of flow stresses with strain for temperature range (450–650 K) at different strain-rates. Similar graphs have been plotted for the temperature range (650–850 K) in Fig. 9a, b. It is observed that the estimated flow stresses are in good agreement with actual material flow stresses; the maximum error is found to be less than 1 %.

Table 4 Optimum values of material parameters by inverse analysis

Obtained material and process parameters	Steel (AISI 4142)		Copper		Aluminum alloy AI6061-T6
Temperature range (K)	450–650	650–850	450–650	650–850	400–550
σ_0 (MPa)	461.1	408	117	124	202
n	0.085	0.0966	0.155	0.174	0.0567
β	0.013	0.0122	0.0468	0.0065	0.0032
γ	0.565	0.83	0.8	0.967	0.75
μ	0.2005	0.1995	0.2005	0.2005	0.1875

Range Strain-rate: 10–100 s^{-1} and strain: 0.02–0.4

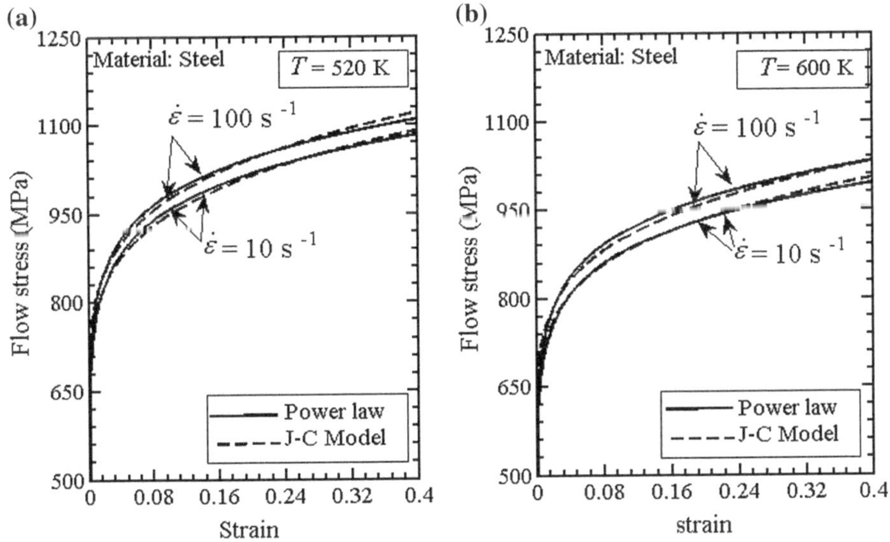

Fig. 8 Variation of flow stresses with strain for steel (strip) at actual and estimated material properties **a** 520 K, **b** 600 K

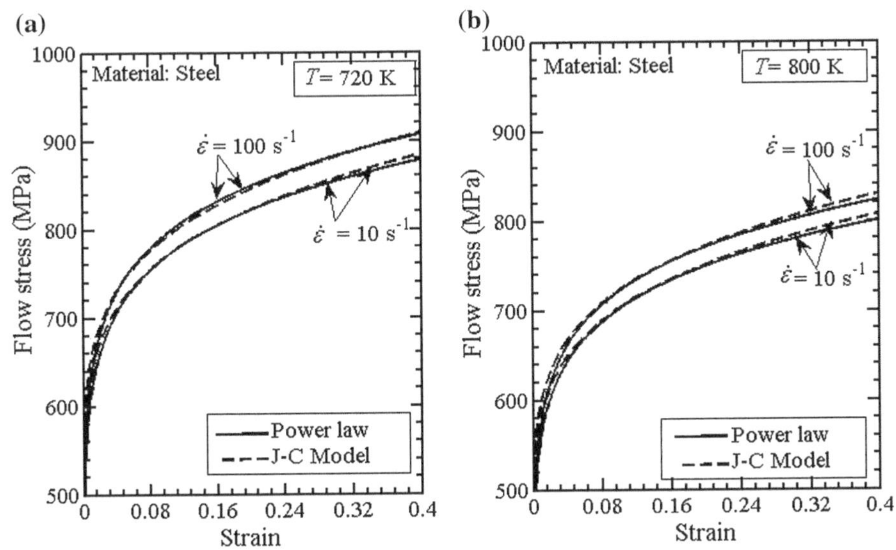

Fig. 9 Variation of flow stresses with strain for steel (strip) at actual and estimated material properties for temperature **a** 720 K, **b** 800 K

Figure 10 shows the variation of flow stress with strain at different strain-rates for the temperature range of 450–850 K. Here, the inverse estimation of material parameters has been carried out for larger range of temperature. The same inverse methodology is used to obtain the following material parameters: σ_0 = 443 MPa, n = 0.0835, β = 0.0105, γ = 0.701, and friction coefficient μ = 0.1995. It is observed that in this case, there is somewhat larger deviation with J–C model.

For the work material, copper, the same inverse methodology is implemented to obtain the optimum material and process parameters. Figures 11 and 12 show the variation of flow stress with strain for different temperatures at different strain-rates. It is seen from Figs. 11a, b and 12a, b that the flow stresses are in good agreement with actual material parameters. The maximum error between actual flow stress and estimated flow stresses is lesser than 1 %.

Figure 13 shows the variation of flow stress with strain at different strain-rates for the temperature range of 450–850 K of work material, copper. The inverse methodology is used to obtain the unknown material parameters as σ_0 = 128.14 MPa, n = 0.172, β = 0.0233, γ = 0.837, and friction coefficient μ = 0.1895. Although larger error compared to smaller temperature ranges is obtained, the error is lesser than 5 %.

A similar methodology of inverse analysis has been used for aluminum Al6061-T6. Figure 14 shows the variation of flow stress with strain for actual and estimated material parameters for aluminum strip. Here, the temperature range considered is

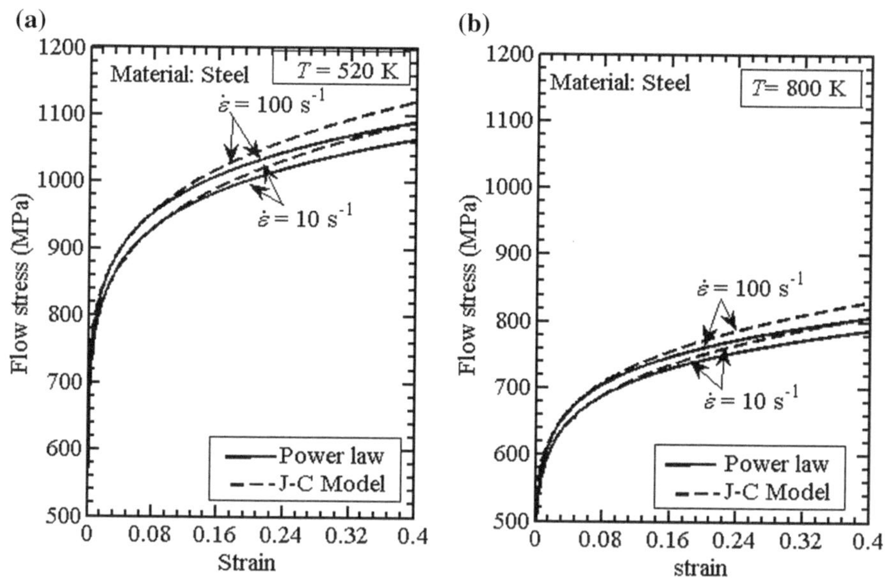

Fig. 10 Variation of flow stresses with strain for steel (strip) at actual and estimated material properties for temperature **a** 520 K, **b** 800 K

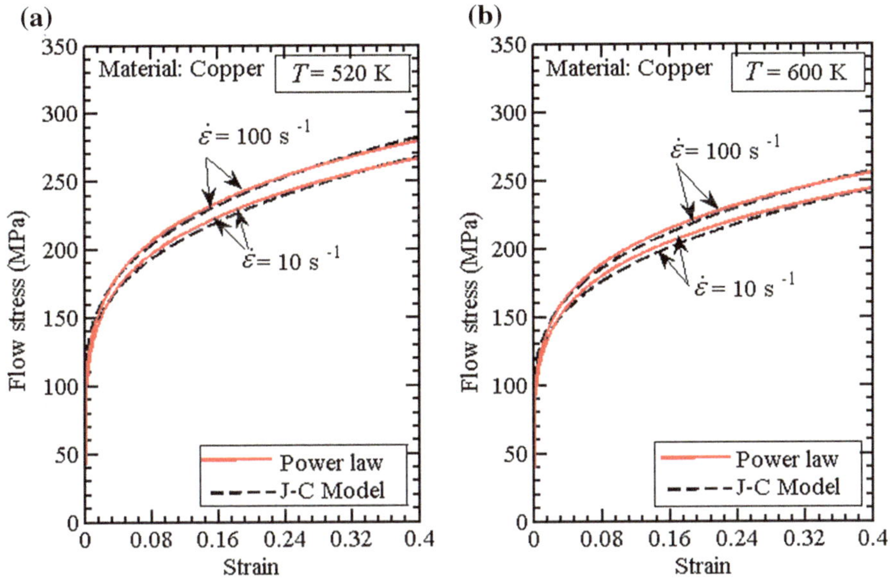

Fig. 11 Variation of flow stresses with strain for copper (strip) at actual and estimated material properties for temperature **a** 520 K, **b** 600 K

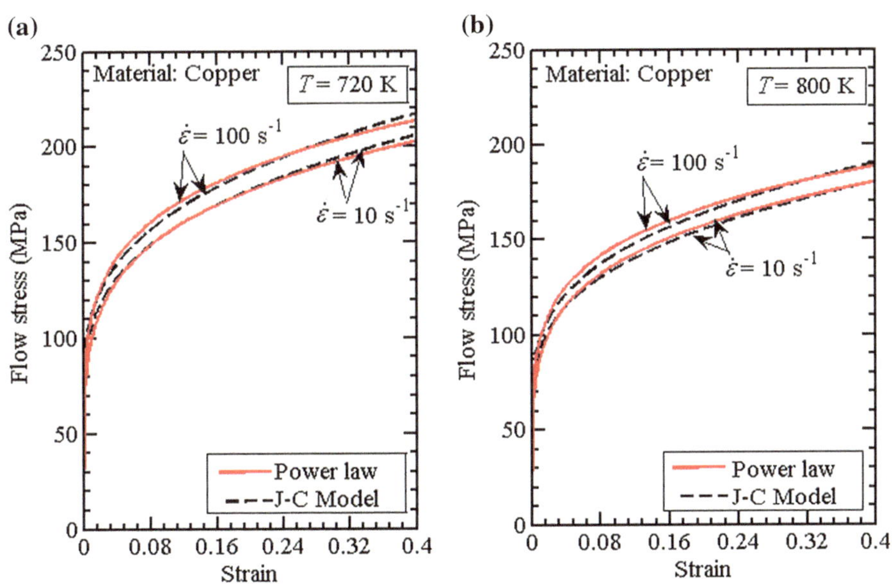

Fig. 12 Variation of flow stresses with strain for copper (strip) at actual and estimated material properties for temperature **a** 720 K, **b** 800 K

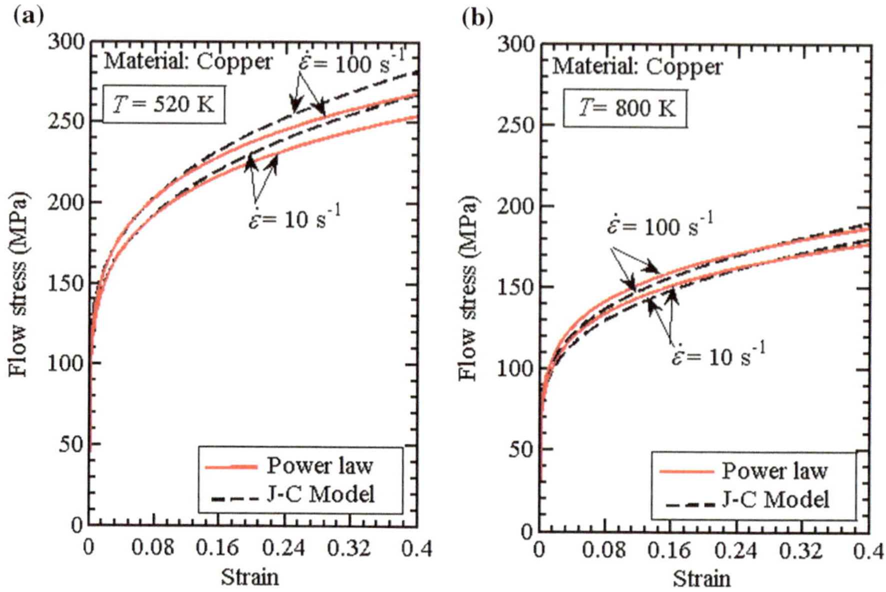

Fig. 13 Variation of flow stresses with strain for copper (strip) at actual and estimated material properties for temperature **a** 520 K, **b** 600 K

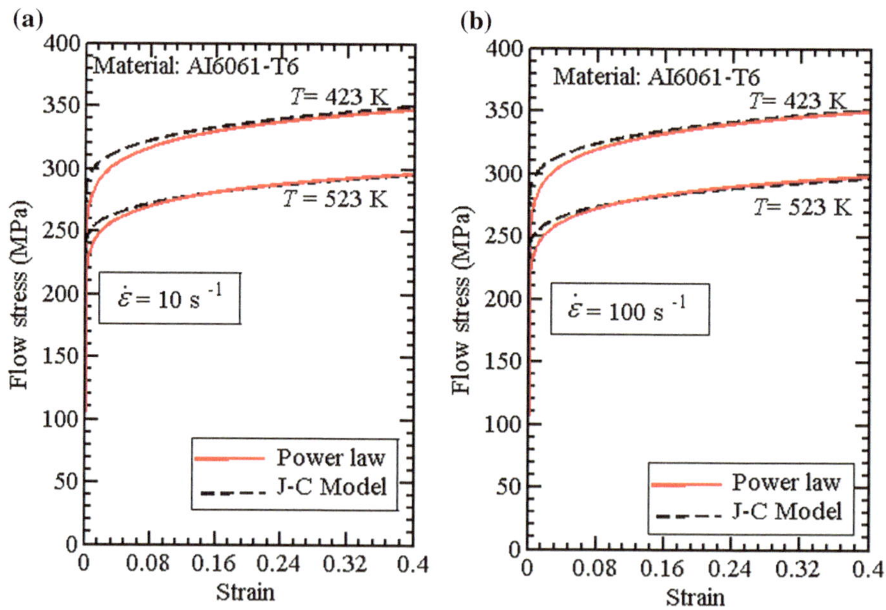

Fig. 14 Variation of flow stresses with strain for aluminum (strip) at actual and estimated material properties for temperature **a** 423 K, **b** 523 K

400–550 K. For the case of aluminum strip material, the inverse estimated coefficient of friction and parameters of power law show the good agreement with J–C model at high strain compared to the low strain as shown in Fig. 14.

4.2 Inverse Estimation of Material Parameters and Coefficient of Friction for Variable Friction Case

In this section, the coefficient of friction is assumed to vary with temperature. The friction coefficient is decided on the basis of forward slip measurement. The exit speed of strip is higher than the speed roll (peripheral speed of work-roll) during rolling due to reduction in strip at deformation zone. The relative difference of the these speeds is defined as forward slip (f_s) as follows

$$f_s = \frac{V_2 - V_r}{V_r} \times 100 \tag{25}$$

where V_2 is the exit speed of strip, and V_r is the roll speed. The forward slip is calculated by deformation FEM module. The roll speed may be measured by a tachometer and velocity of exit speed may be measured by velocity sensor as shown in Fig. 6. Yadav et al. (2011b) also used the velocity sensor and the temperature sensor for estimating the coefficient of friction and flow stress of the strip material by measuring the forward slip and the exit strip temperature for cold rolling.

The friction coefficient is assumed to follow the following relation:

$$\mu = 0.41 - 0.00025\,T - 0.028\,V \tag{26}$$

where T is the average temperature at the deformation zone in °C, and V is the roll speed in m/s. Table 5 shows the estimated friction coefficients for different cases.

The optimum material parameters for power law are obtained as $\sigma_0 = 456$ MPa, $n = 0.077$, $\beta = 0.0035$, and $\gamma = 0.735$. Figure 15 shows the variation of flow stresses

Table 5 Optimum friction coefficient

Case	Inlet temperature T_0 (K)	Exit velocity V_2 (m/s)	% reduction r	Optimum friction coefficient μ
1	600	0.5	4	0.243
2	600	0.5	28	0.232
3	600	3	4	0.235
4	600	3	28	0.227
5	520	0.5	4	0.252
6	520	0.5	28	0.263
7	520	3	4	0.241
8	520	3	28	0.234

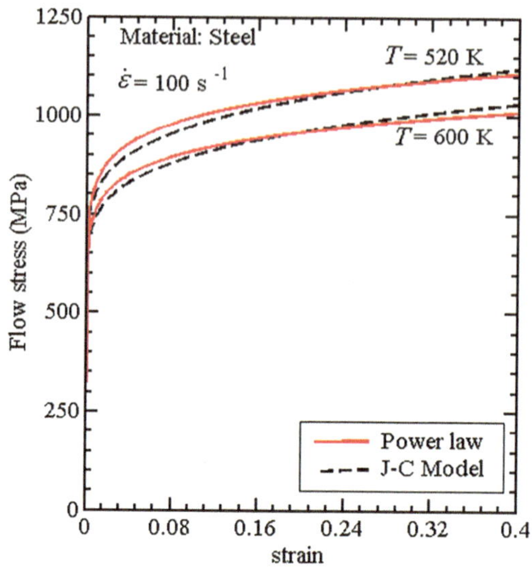

Fig. 15 Variation of flow stresses with strain at actual and estimated material properties of steel for strain-rate 100 s^{-1} (450–650 K)

with strain for actual and estimated material parameters. The difference between the actual and estimated flow stresses is observed to be less than ±3 %.

5 Sensitivity Analyses

A sensitivity study is carried out of the material parameters, and the coefficient of friction is obtained by an inverse analysis for steel (strip) at 520 and 600 K inlet temperature of strip. Table 6 shows the errors in exit strip temperature computed for various friction coefficients and material parameters of power law coefficient. Here, the actual temperature was obtained using J–C model. Computational results have been obtained by taking inverse estimated parameters as well as perturbing them one-by-one. The error in temperature at 150 mm from the exit of roll bite has been calculated for the two cases, i.e., E_1 for high inlet temperature, low reduction, and low exit velocity and E_2 for low inlet temperature, high reduction, and high exit velocity.

The sensitivity of error in temperature with respect to error in the estimation of material parameters, viz. $\sigma_0, n, \beta, \gamma$ and equivalent Coulomb coefficient μ, is estimated. This information helps in deciding the permissible errors in $\sigma_0, n, \beta, \gamma$ and μ. The convergence criteria can be chosen accordingly. E_1 and E_2 for the inversely determined parameter along with ±5 % material and process parameters variation are listed. The exit strip temperature measurement is more sensitive for low inlet temperature, high reduction, and high exit velocity as shown in Table 6. It is observed that temperature estimates are very sensitive to γ.

Sensitivity derivative is calculated using the central difference method. Table 6 can be used to estimate partial derivatives of E with respect to parameters. For

Table 6 Error in the temperature estimation at optimum values and at ±5 % variation in the material parameters and friction coefficient for work-roll material is steel

Material parameters and friction coefficient used to estimate the temperature	Error in exit strip temperature	
	Case I E_1 (T = 600 K, V_2 = 0.5 m/s, r = 4 %) K	Case II E_2 (T = 520 K, V_2 = 3 m/s, r = 28 %) K
$(\sigma_0, n, \beta, \gamma, \mu)$ = (461.11 MPa, 0.085, 0.013, 0.565, 0.2005)	+0.98	−1.51
$(1.05\sigma_0, n, \beta, \gamma, \mu)$	+1.27	+30.3
$(0.95\sigma_0, n, \beta, \gamma, \mu)$	−0.68	−4.48
$(\sigma_0, 1.05n, \beta, \gamma, \mu)$	−0.84	−32.65
$(\sigma_0, 0.95n, \beta, \gamma, \mu)$	+1.1	+19.35
$(\sigma_0, n, 1.05\beta, \gamma, \mu)$	+0.96	+16.19
$(\sigma_0, n, 0.95\beta, \gamma, \mu)$	−0.90	−40.68
$(\sigma_0, n, \beta, 1.05\gamma, \mu)$	+1.07	+58.61
$(\sigma_0, n, \beta, 0.95\gamma, \mu)$	−0.77	−62.08
$(\sigma_0, n, \beta, \gamma, 1.05\mu)$	+0.89	+35.26
$(\sigma_0, n, \beta, \gamma, 0.95\mu)$	−1.25	−6.05

The exit temperature of strip is obtained by J–C model: T_{exit} = 523.85 K for Case I and T_{exit} = 569.93 K for Case II

example, $\partial E_2/\partial\sigma_0$ comes out to be 0.754 K/MPa by central difference method. Similarly, $\partial E_2/\partial n = -5.20$ K, $\partial E_2/\partial m = 5.68$ K, $\partial E_2/\partial\gamma$ =12.06 K, and $\partial E_2/\partial\mu = 4.13$ K. It is noted that the derivatives of error in temperature at actual and estimated parameters with respect to γ are high. Hence, the most sensitive parameter is γ. This type of analysis helps in deciding the convergence criteria.

6 Conclusion

In the present work, an inverse method is proposed for the estimation of material parameters of power law and coefficient of friction. The methodology requires the estimation of the exit temperature of the strip and slip. The slip measurement gives the proper estimation of coefficient of friction. If the coefficient of friction is assumed constant for different process conditions, then only the measurement of temperature can be carried out. In place of the exit temperature, the roll torque or roll force can also be measured. A heuristic method is used for the minimization of the error between actual and estimated temperature of exit strip. A sensitivity analysis is also carried out.

Appendix

Estimation of Temperature Distribution in the Strip at Exit Side

The temperature distribution of strip at the exit side is obtained as (Kim et al. 2009)

$$T_e(y,t) = T_0 + \sum_{n=1}^{\infty} \exp\left(-\frac{k_s}{\rho c_p}\lambda_n^2 t\right)\left(\frac{4\lambda_n \int_0^{\frac{h_2}{2}} T(y,0)\cos(\lambda_n y)\mathrm{d}y - 4T_0 \sin\left(\lambda_n \frac{h_2}{2}\right)}{\lambda_n h_2 + \sin(\lambda_n h_2)}\right)\cos(\lambda_n y), \tag{27}$$

where $T(y, 0)$ is the temperature at the exit of deformation zone that corresponds to time $t = 0$, T_0 is the temperature of the coolant, h_a is the convective heat transfer coefficient at the strip surface, k_s is the thermal conductivity of strip, ρ is the density of strip, c_p is the specific heat, and λ_n are obtained by solving the following equation:

$$k_s \lambda_n \sin\left(\lambda_n \frac{h_2}{2}\right) - h_a \cos\left(\lambda_n \frac{h_2}{2}\right) = 0. \tag{28}$$

The temperature below the sensor is obtained by substituting

$$y = \frac{h_2}{2} \quad \text{and} \quad t = \frac{\text{distance from the exit of roll bite to sensor}}{\text{exit velocity of the strip}} \tag{29}$$

in Eq. (27).

References

Boisse P, Altan T, Luttervelt KV (2007) Friction and flow stress in forming and cutting. 1st south asian edition, London, pp 125–126

Byon SM, Kim SI, Lee Y (2008) A numerical approach to determine flow stress-strain curve of strip and friction coefficient in actual cold rolling mill. J Mater Process Technol 201:106–111

Chandrasekaran M, Muralidhar M, Murali Krishna C, Dixit US (2012) Online machining optimization with continuous learning. In: Paulo Davim J (ed) Computational methods for optimizing manufacturing technology: models and techniques. IGI Global, Hershey

Chen WL, Yang YC (2010) Inverse problem of estimating the heat flux at the roller/workpiece interface during a rolling process. Appl Therm Eng 30:1247–1254

Cho H, Altan T (2005) Determination of flow stress and interface friction at elevated temperatures by inverse analysis technique. J Mater Process Technol 170:64–70

Cho H, Ngaile G (2003) Simultaneous determination of flow stress and interface friction by finite element based inverse analysis technique. CIRP Ann Manuf Technol 52:221–224

Dixit US, Dixit PM (1996) A finite element analysis of flat rolling and application of fuzzy set theory. Int J Mach Tools Manuf 36:947–969

Eideh A, Dixit US (2013) A robust and efficient inverse method for determining the thermal parameters during laser forming. In: Proceedings of national conference on recent advancements in mechanical engineering, NERIST, Nirjuli, India, 8–9 Nov 2013

Fischer FD, Schreiner WE, Werner EA, Sun CG (2004) The temperature and stress fields developing in rolls during hot rolling. J Mater Process Technol 150:263–269

Han H (2005) Determination of mean flow stress and friction coefficient by modified two-specimen method in cold rolling. J Mater Process Technol 159:404–408

Hawkins DN (1981) An etchant for revealing the substructure in low-carbon steels. Metallography 14:61–68

Hawkins DN (1985) Warm working of steels. J Mech Working Technol 11:5–21

Hawkins DN, Shuttleworth AA (1979) The effect of warm rolling on the structure and properties of a low carbon steel. J Mech Working Technol 2:333–345

Hirschvogel M (1979) Recent developments in industrial practice of warm working. J Mech Working Technol 2:317–332

Hsu PT, Yang YT, Chen CK (2000) A three dimensional inverse problem of estimating the surface thermal behavior of the working roll in rolling process. J Manuf Sci Eng ASME 122:76–82

Huang CH, Ju TM, Tseng AA (1995) The estimation of surface thermal behavior of the working roll in hot rolling process. Int J Heat Mass Transf 38:1019–1031

Hum B, Colquhoun HW, Lenard JG (1996) Measurements of friction during hot rolling of aluminum strips. J Mater Process Technol 60:331–338

Johnson GR, Cook WH (1983) A constitutive model and data for metals subjected to large strains, high strain rates and high temperatures. In: Proceedings of the 7th international symposium on ballistics (vol 21), International Ballistics Committee, The Hague, Netherlands, 19–21 Apr 1983, pp 541–547

Kalpakjian S (2008) Manufacturing engineering and technology, 5th edn. Addison-Wesley, London

Khalili I, Serajzadeh S, Koohbor B (2012) Thermomechanical behavior of work rolls during warm strip rolling. Metall Mater Trans B 43:1638–1648

Kim J, Lee J, Hwang AM (2009) An analytical model for the prediction of strip temperatures in hot strip rolling. Int J Heat Mass Transf 52:1864–1874

Kusiak J, Kawalla R, Pietrzyk M, Pircher H (1996) Inverse analysis applied to the evaluation of material parameters in the history dependent flow stress equation in hot forming of metals. J Mater Process Technol 60:455–461

Lenard JG, Nad LB (2002) The coefficient of friction during hot rolling of low carbon steel strips. Trans ASME J Tribol 124:840–846

Lenard JG, Zhang S (1997) A study of friction during the lubricated cold rolling of an aluminum alloy. J Mater Process Technol 72:293–301

Pietrzyk M, Lenard JG (1990) The effect of the temperature rise of the roll on the simulation of the flat rolling process. J Mater Process Technol 22:177–190

Raczy A, Altenhof WJ, Alpas AT (2007) An Eulerian finite element model of the metal cutting process. In: Proceedings 8th international LS-DYNA users conference, 2–4 May 2004, pp 9–11

Serajzadeh S (2004) Modelling the warm rolling of a low carbon steel. Mater Sci Eng A 371: 318–323

Serajzadeh S (2006) A model for prediction of flow behavior and temperature distribution during rolling of a low carbon steel. Mater Des 27:529–534

Serajzadeh S, Mohammadzadeh M (2007) Effects of deformation parameters on the final microstructure and mechanical properties in warm rolling of a low-carbon steel. Int J Adv Manuf Technol 34:262–269

Shang J, Hatkevich S, Wilkerson L (2012) Experimental study and numerical simulation of electromagnetic tube expansion. In: Proceedings of the 5th international conference on high speed forming, 24–26 Apr 2012, pp 83–92

Shirizly A, Lenard JG (2000) The effect of lubrication on mill loads during hot rolling of low carbon steel strips. J Mater Process Technol 97:61–68

Vural M, Rittel D, Ravichandran G (2003) Large strain mechanical behavior of 1018 cold-rolled steel over a wide range of strain rates. Metall Mater Trans A 34:2873–2885

Wanhiem T, Bay N (1978) A model for friction in metal forming processes. Ann CIRP 27: 189–194

Weisz-Patrault D, Ehrlacher A, Legrand N (2011) A new sensor for the evaluation of contact stress by inverse analysis during steel strip rolling. J Mater Process Technol 211:1500–1509

Weisz-Patrault D, Ehrlacher A, Legrand N (2012) Evaluation of temperature field and heat flux by inverse analysis during steel strip rolling. Int J Heat Mass Transf 55:629–641

Weisz-Patrault D, Ehrlacher A, Legrand N (2013) Analytical inverse solution for coupled thermoelastic problem for the evaluation of contact stress during steel strip rolling. Appl Math Model 37:2212–2229

Weisz-Patrault D, Ehrlacher A, Legrand N (2014) Temperature and heat flux fast estimation during rolling process. Int J Therm Sci 75:1–20

Yadav V, Singh AK, Joshi SN, Dixit US (2011a) Comparison of the performance of lubricants in rolling based on temperature measurement. In: Proceedings of the 14th international conference on material forming-ESAFORM2011, 1353, Belfast, United Kingdom, 27–29, pp 357–361

Yadav V, Singh AK, Dixit US (2011b) Online determination of material parameters and coefficient of friction in cold flat rolling process. In: Proceedings of the international conference on computational methods in manufacturing (ICCMM2011), IIT Guwahati, India, 15–16 Dec 2011, pp 35–42

Yadav V, Singh AK, Dixit US (2014) An approximate method for computing the temperature distributions in roll and strip during rolling process. Proc IMECHE Part B J Eng Manuf 228:1118–1130

Zhang Y, Zhang HL, Wu JH, Wang XT (2011) Enhanced thermal conductivity in copper matrix composites reinforced with titanium-coated diamond particles. Scr Mater 65:1097–1100

Incremental Sheet Forming (ISF)

Yogesh Kumar and Santosh Kumar

Abstract Incremental sheet forming (ISF) process is an advanced flexible manufacturing process to produce complex 3D products. In conventional forming process, the products are produced by dedicated tools, i.e. die and punch. However, ISF does not require any dedicated tool. A laboratory setup of single-point incremental forming machine has been developed using a motion card to control the 3 servo motors for controlling individual 3-axis and spindle movement at IIT (BHU), Varanasi. The strain distribution on the sheet over the length of deformation has been computed. The surface quality of the products is found to be good. Simple simulation has also been carried out.

Keywords Metal forming · Incremental forming · Dieless forming · Deforming tool

1 Introduction

Metal forming is the backbone of modern manufacturing industry besides being a major industry in itself. Throughout the world, hundreds of million tons of metals go through metal forming processes every year. As much as 15–20 % GDP of industrialized nations comes from metal forming industry. Besides, it fulfills a social cause by providing job opportunities to the millions of workers. Metal forming industry, in general, is a bulk producer of semi-finished and finished goods

Y. Kumar (✉) · S. Kumar (✉)
Mechanical Engineering Department, Indian Institute of Technology (BHU), Varanasi, Uttar Pradesh 221005, India
e-mail: yogeshiitbhu@gmail.com

S. Kumar
e-mail: santoshkr.mec@gmail.com

R.G. Narayanan and U.S. Dixit (eds.), *Advances in Material Forming and Joining*, Topics in Mining, Metallurgy and Materials Engineering, DOI 10.1007/978-81-322-2355-9_2

and this is one reason that it is viable to undertake large-scale research and development projects because even a small saving per ton adds up to huge sums (Juneja 2013).

The incremental forming process is originated from stretch forming and metal spinning process. The ISF has been originated with partial hybridization of stretch forming process and metal spinning processes. Thus, ISF process has combined advantages of stretch forming processes and metal spinning process. In the later developments in ISF process, the process is found to be capable of producing 3D complex shapes with multiple features on it.

1.1 Elements of Incremental Forming Process

Incremental forming is a new technique for deforming sheet metals by the application of step-by-step incremental feed to a deforming tool (DT). For the production of parts by conventional sheet forming techniques, dedicated tools are required. The dedicated tools are complex 3D design and thus are expensive. The design of tooling (die and punch) for complicated shapes is very difficult and costly. In incremental forming technique, only a DT is required for deforming the sheet metals. There are four basic elements of an ISF process as shown in Fig. 1—a sheet metal, a blank holder, a single-point forming tool or DT, and CNC machine.

The shape of DT may be hemispherical. The design of DT is quite simple and economical. A CNC controller having capability for controlling individual 3 axes can be used to control the movements of the DT. The tool path needs to be optimized and a free run is taken before performing the actual manufacturing on the sheet metal in order to ensure the proper functioning of the machine.

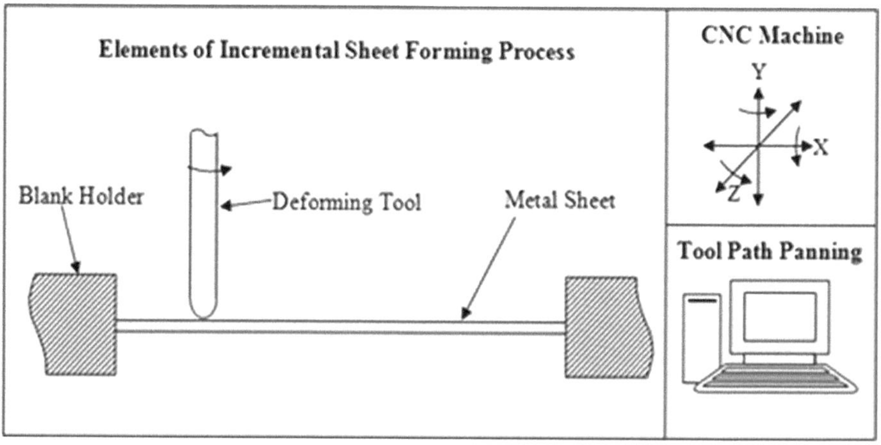

Fig. 1 Basic elements of ISF

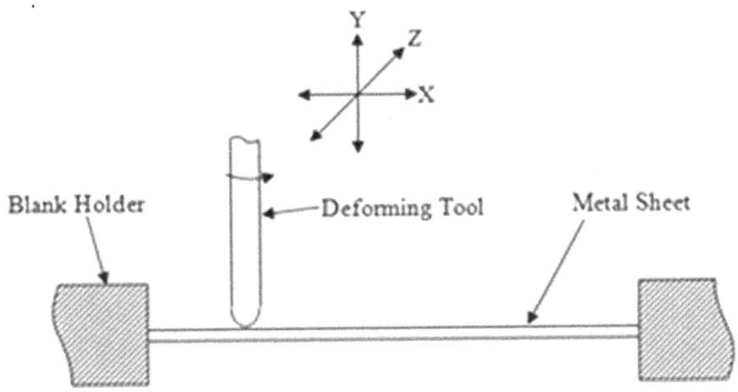

Fig. 2 Single-point incremental forming (SPIF)

1.1.1 Classification of Incremental Sheet Forming

The Incremental forming processes are broadly classified into two categories:

(i) **Conventional Incremental Sheet Forming (CISF)**:
In CISF process, generally a sheet of metal is deformed by progressive and localized plastic deformation using a simple hemispherical/ballpoint tool, and this path of the DT is controlled by a CNC machine. The DT moves over the surface of the sheet and results the final shape. There is no other tool or external pressure applied for deforming the sheets into the desired shape. The conventional incremental forming process can be further classified as follows:

(a) **Single-Point Incremental Forming (SPIF) also known as Negative Dieless Forming**:
In SPIF, only one tool moves over the surface of the sheet as shown in Fig. 2.

(b) **Two-Point Incremental Forming (TPIF) or Positive Dieless Forming**:
In TPIF, two tools, one called DT and another one supporting tool, move over the surface of the sheet as shown in Fig. 3.

(ii) **Hybrid Incremental Sheet Forming (HISF)**:
HISF processes are the modified forms of conventional incremental forming. In these processes, DT moves over the surface of sheet metals, while the another side of surface of sheet metals is supported by pressurized hydraulic fluid, partial die or full die to get the desire shape and size. Hybrid incremental forming processes are further classified as follows:

(a) ***SPIF with Hydraulic Fluid also known as Single-Point Incremental Hydro-Forming***:
In this type of hybrid incremental forming process is different from conventional SPIF process that a single tool moves over one side of the surface of the sheet metals and other side of surface of sheet metals is supported by the pressurized hydraulic fluid as shown in Fig. 4.

(b) ***TPIF with Partial Die***:
In this type of hybrid incremental forming process, a single tool moves over one side of the surface of the sheet metals and other side of surface of sheet metals is supported by a partial die to get the desire impression as shown in Fig. 5.

(c) ***TPIF with Full Die***:
In this type of hybrid incremental forming process, a single tool moves over one side of the surface of the sheet metals and other side of surface of sheet metals is supported by a full die to get the desire shape and size as shown in Fig. 6.

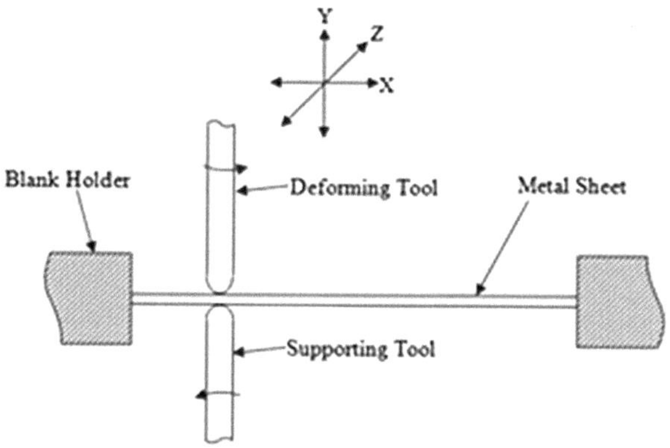

Fig. 3 Two-point incremental forming (TPIF)

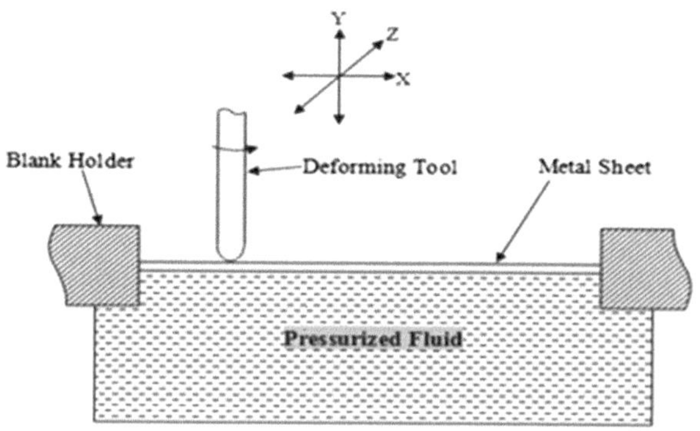

Fig. 4 Single-point incremental hydro-forming (SPIHF)

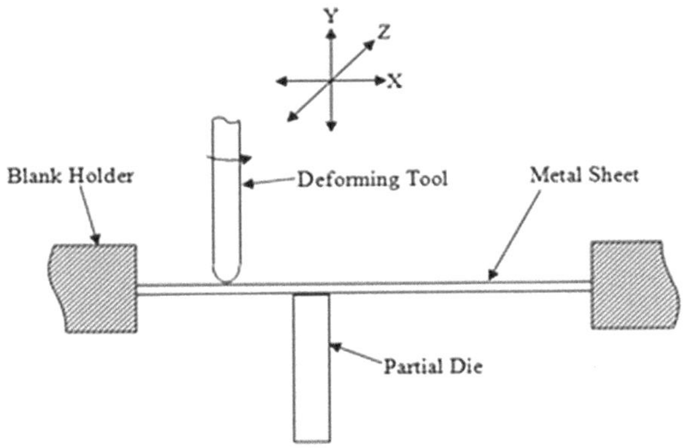

Fig. 5 Two-point incremental forming with partial die (TPIFPD)

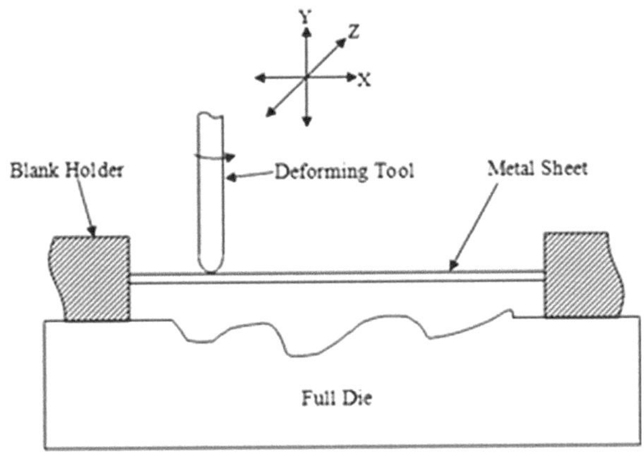

Fig. 6 Two-point incremental forming with full die (TPIFFD)

1.1.2 Advantages and Disadvantages of ISF

The advantages and disadvantages of SPIF are as follows

A. **Advantages**:

- Useable parts can be formed directly from CAD data with a minimum of specialized tooling. These can be either rapid prototypes or small volume production runs.
- The process does not require either positive or negative dies; hence, it is dieless. However, it does need a backing plate to create a clear change of angle at the sheet surface.
- Changes in part design sizes can be easily and quickly accommodated, giving a high degree of flexibility.
- Making metal rapid prototypes is normally difficult, but easy with this process.
- The small plastic zone and incremental nature of the process contribute to increased formability, making it easier to deform low formability sheet.

- A conventional CNC milling machine or lathe can be used for this process.
- The size of the part is limited only by the size of the machine. Forces do not increase because the contact zone and incremental step size remain small.
- The surface finish of the part can be improved.
- The operation is quiet and relatively noise free.

B. **Disadvantages**:

- The major disadvantage is the forming time is much longer than competitive processes such as deep drawing.
- As a result, the process is limited to small-size batch production.
- The forming of right angles cannot be done in one step, but requires a multi-step process.

1.1.3 Applications of ISF

There are a number of areas, where high precision of the products is required for the accuracy of the performance. Areas of products can manufacture by ISF are as follows:

- **Aerospace Industry**: Instrument panel, body panel, passenger seat cover, etc.;
- **Automobile**: Door inner/outer panel, hood panel, engine cover, etc.;
- **High customized products**: Denture plate, ankle support, metal helmet, etc.;
- Cellular phones;
- IC lead frames;
- Electronics;
- Health care;
- Miniature fasteners;
- Hard disk drives;
- Products of national security and defense;
- Automobiles; and
- Sensors.

1.1.4 Basic Terminology in Incremental Sheet Forming

The typical terminology in ISF can be understood from the terminologies used in the incremental sheet hydro-forming (ISHF) as shown in Fig. 7, with notations below:

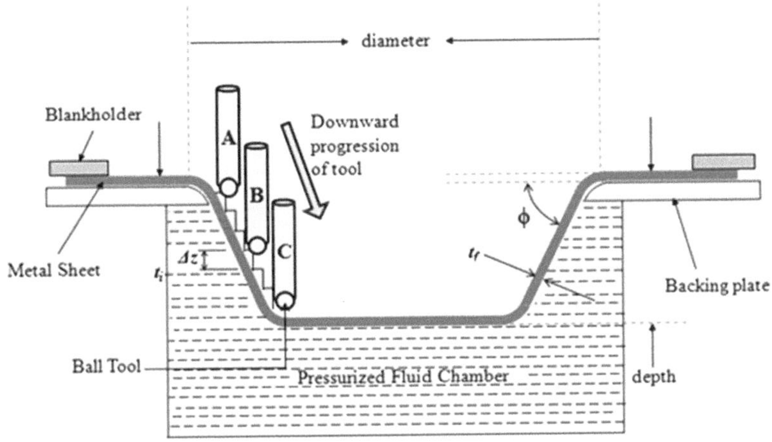

Fig. 7 Basic terminology in deformed part in ISHF

$$t_i = \text{Initial sheet thickness.}$$
$$t_f = \text{Final sheet thickness.}$$
$$\Delta z = \text{Incremental step} - \text{down size.}$$
$$\varphi = \text{Draw angle or forming angle.}$$

The incremental step-down size (step size, Δz) affects the machine time and the surface quality. Feed rate is the speed the forming tool moves around the mill bed. The angle between the un-deformed sheet metal and the deformed sheet metal is defined as forming angle (φ) as shown in Fig. 7. The forming angle can be used as a measure of material formability. The maximum angle ($\varphi_{\max}$) is the greatest angle formed in a shape without any failures. The forming angle is set within CAD software (Ham and Jeswiet 2007).

1.2 Literature Review

Single-point incremental forming (SPIF) is a sheet metal forming process that allows manufacturing components without development of complex tools in comparison with stamping process (Thibaud et al. 2012). ISF process depends strongly on the forming tool path which influences greatly the part geometry and sheet thickness distribution (Azaouzi and Lebaal 2012).

There are several issues that need separate discussion as peoples around the world are trying to optimize the ISF process the governing process parameters. The concerned issues (formability, surface quality, geometric accuracy, forming forces, etc.) are discussed briefly.

Kim and Park (2002) identified that the deformation pattern in ISF is very similar to spinning process and the sheet metals deform by shear-dominant deformation like shear spinning. They also found that the formability is improved when a ball tool of a particular size is used with a small feed rate and a little friction. The formability differs according to the direction of the tool movement because of the plane anisotropy (Kim and Park 2002). Several investigations have been done with emphasis on assessing and improving the formability in this forming method. Kim and Yang proposed the double-forming technique to improve formability, assuming that only shear deformation occurs in the material (Kim and Yang 2000). The ball tool is more effective than the hemispherical head tool in terms of formability. Little friction at the tool and sheet interface helps to improve the formability. Also, the formability increases as the feed rate decreases (Kim and Park 2002).

Filice et al. proposed the two characteristics of deformation in this forming method. One is the deformation pattern. While the tool moves straight on a horizontal plane, the deformation that occurs at the starting and ending points of the straight line is biaxial stretching. The deformation that occurs between these points is plane-strain stretching. As the curvature of the tool movement increases, the deformation turns more into biaxial stretching. The other characteristic is the formability of the deformation. As shown in Fig. 8, the forming limit curve, which depicts the formability in the major and minor strain space, is expressed as a straight line with a negative slope. Especially, for an aluminum sheet, the formability can be quantified as a scalar number of ($\varepsilon_{major} + \varepsilon_{minor}$). It is noted that formability is the greatest under plane-strain stretching, during which the minor strain is zero. Therefore, a greater deformation of a sheet metal can be achieved in the ISF (Kim and Park 2002). The ISF is characterized by a local stretching deformation mechanics which determines a forming limit curve quite different from the

Fig. 8 A comparative plot of FLC in conventional forming and incremental forming

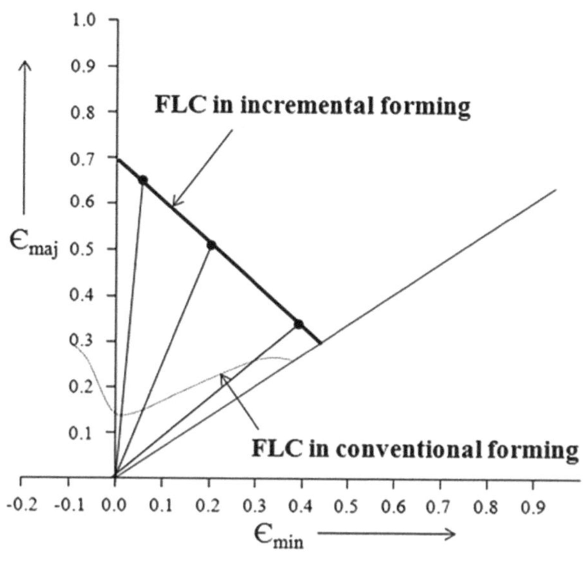

traditional one and such FLC has a linear shape with a negative slope in the positive ε_{minor} side of FLD (Filice et al. 2002).

The depth and diameter have no effect the likelihood of forming a part. Also, the material thickness, tool size, and the interaction between material thickness and tool size have a significant effect on maximum forming angle (Ham and Jeswiet 2006). The ball tool is more effective than the hemispherical head tool in terms of formability. The little friction at tool/sheet interface helps to improve the formability. They also found that formability increases with decreasing feed rate (Kim and Park 2002). The formability is enhanced while deforming the metal sheets in warm conditions, and the role of tool diameter is negligible as compared to the influence of temperature and tool depth (Ambrogio et al. 2008). The deformation mechanisms of both SPIF and TPIF are increasing stretching and shear in the radial–axial plane (perpendicular to the tool direction) and shear in the tool direction (Jackson and Allwood 2009). Hussain et al. (2007) found that the formability increases as the radius of curvature decreases. Ham and Jeswiet (2006) formalized the two designs of experiments for the forming parameters critical in SPIF and the degree to which they affect formability.

Ambrigio et al. investigate that in hot ISF, the quality of bottom surface is better as compared to the one in contact with the punch. This is due both the lower temperature reached on the bottom side that reduces oxidation phenomenon and the absence of mechanical actions on the sheet. Anyway because of low sheet thickness, the thermal gradient is very low and this reduces the difference between the two sheet sides (Ambrogio et al. 2012).

Attananio et al. optimized of the tool path in two-point sheet incremental forming, with a full die in a particular asymmetric sheet incremental forming configuration. They carried out the experimental evaluation of the tool path, which is able to reproduce an automotive component with the best dimensional accuracy, the best surface quality, and the lowest sheet thinning (Hussain et al. 2007).

The geometries produced by ISF represent some errors along the oblique walls. In particular, a sort of distortion is also obtained, generating a curvature on the expected straight sides. This phenomenon is due to elastic spring back whose effect is lower in correspondence of the edges, where the geometrical stiffness is higher than in other areas (Cerro et al. 2006; Filice et al. 2002; Ham and Jeswiet 2007). The results obtained in geometric accuracy measurements with the process model, as compared to experimental results obtained by testing in the CNC machine, are approximately same (Cerro et al. 2006). Azaouzi and Lebaal (2012) gave optimal solution provided an improvement of about 7 % regarding the sheet thickness distribution at the maximum forming depth.

Azauzi et al. (2012) found that the forming forces depend largely on the proper design of the tool path. The forming force is slightly lower than the experimental values, but results are very good (Cerro et al. 2006). Forming forces obtained by numerical simulation show good correlation with measured values. However, it was a slight underestimation of the axial forces during thinning. It is assumed the influence of grain size and softening on the material behavior (Thibaud et al. 2012).

Araghi et al. (2009) investigated that the thinning in ISF depends on the wall angle α and is given by the sine law $t_1 = t_0 \sin(90° - \alpha)$. Cerro et al. (2006) investigated that the results obtained in thickness measurements with the process model, in comparison with experimental results obtained by testing in the CNC machine, are approximately same. Increasing stretching and shear perpendicular to the tool direction account for differences between the sine law and measured wall thickness for SPIF and TPIF (Jackson and Allwood 2009). The prediction of thickness distribution is close to that obtained on the real part (Thibaud et al. 2012).

The literature review reveals that ISF method is well investigated, but there are still some issues not well understood in incremental forming like strain distribution on the sheet over the length of deformation, effect of step size, tool size, etc., in ISF.

2 Simulation of Single-Point Incremental Forming Process

In deforming sheets by incremental forming, methodology of deformation plays an important role. The current literature survey in the area of ISF has been kept into consideration in order to adopt a suitable forming methodology. To carry out a dedicated simulation study of the process, a blank holding arrangement for ISF has been prepared as shown in Fig. 9, using DEFORM 3D.

For the successful implementation of ISF, a proper forming methodology has to be followed as shown in Fig. 10. The first step in ISF is to identify need for the components. As soon as the need analysis is done, the geometrical dimensions of the product to be manufactured are decided. Based on geometrical dimensions, a 3D

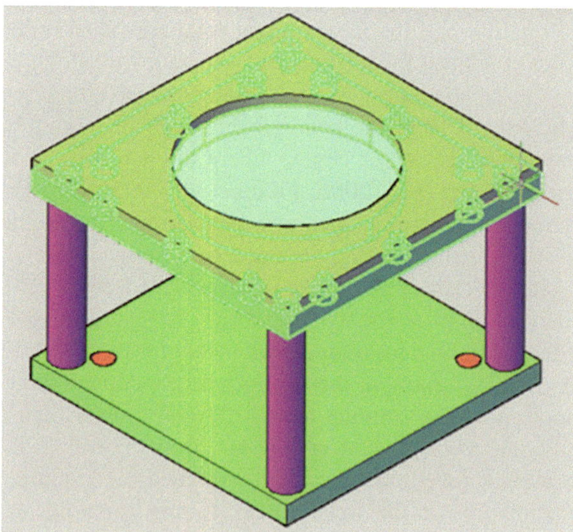

Fig. 9 Blank and blank holding arrangement

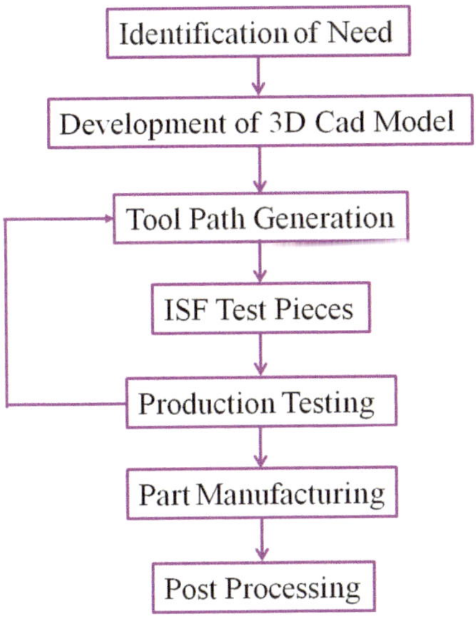

Fig. 10 Forming methodology of 3D components by ISF

CAD model of the product is developed using CAD/CAM package. Moving forward with the help of CAD/CAM package, the tool path, suitable for movement of the tool is generated. In the next step, the workpiece is held in the proper blank holding, and test pieces are produced using the generated tool path. The accuracy of test pieces is tested against the desired profile.

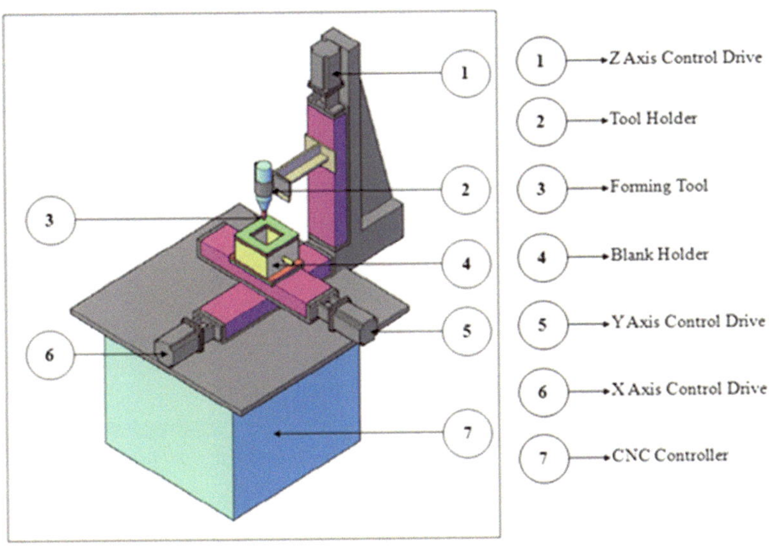

Fig. 11 CAD model of single-point ISF machine

The metals as well as alloys can be used for the implementation of ISF process. For the current investigations, the initial blank of material brass is selected for finding the capabilities of ISF process brass has good formability at room temperature. The sheet having dimensions 80 mm × 80 mm × 5 mm has been used for current case study. A CAD model of the single-point ISF can is represented in Fig. 11.

2.1 Simulation Results

Based on deforming methodology, a product as shown in Fig. 12 is produced using simulation run under ISF process. The stresses are found to be distributed in normal. The experimental results have also validated same. The effective stresses

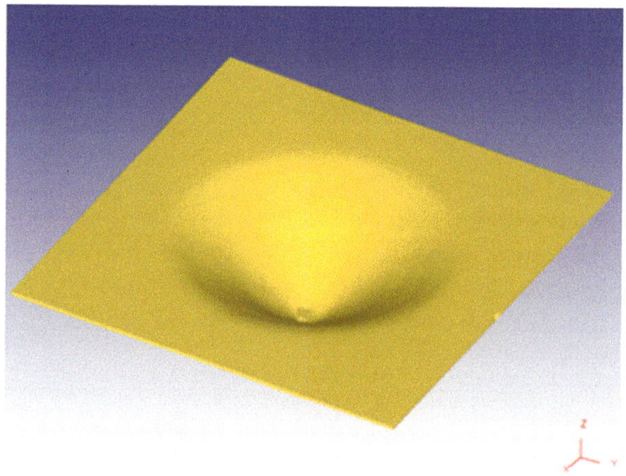

Fig. 12 Defect-free product (wall angle = 30°)

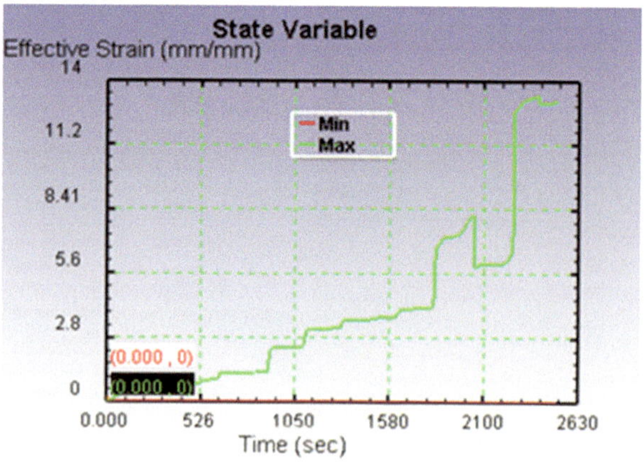

Fig. 13 Variation of effective strain with time

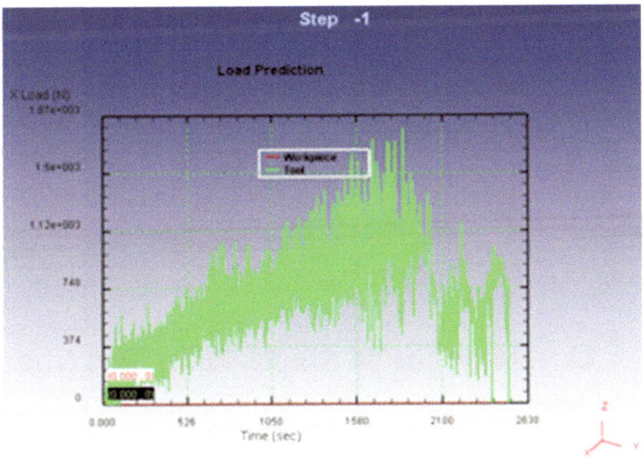

Fig. 14 Force prediction in *X*-direction

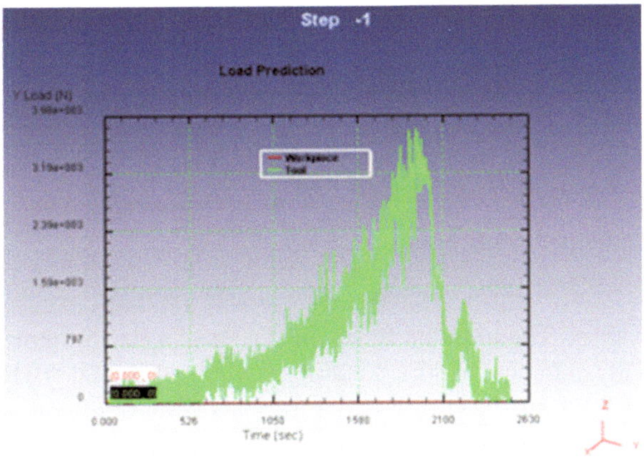

Fig. 15 Force prediction in *Y*-direction

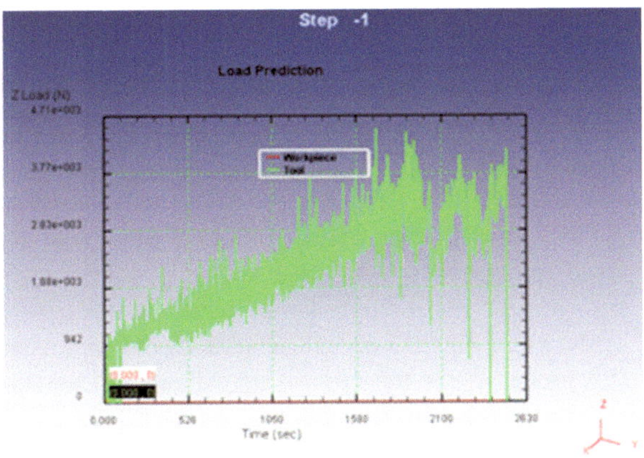

Fig. 16 Force prediction in *Z*-direction

are found increase as the DT path going to finish, i.e., the stresses increase with respect to time. This is obvious due to the strain hardening. Figure 13 shows the evolution of effective strain during the process, which is increasing with time.

The forming forces in X- and Y-direction are repetitive in nature. The forces in X- and Y-directions were also increased gradually. As the tool moves in Z-direction, the deforming force in Z-direction is found increasing. The force predicted in X-, Y-, and Z-directions are represented in Figs. 14, 15, and 16, respectively.

3 Development of Single-Point Incremental Forming Machine

The major elements of ISF are identified as follows: (i) a sheet metal blank, (ii) a blank holder, (iii) a single-point forming tool or DT, and (iv) CNC machine. The path of the DT is responsible for the shape, size and accuracy of the final product. Thus, the quality of the product depends on the proper tool path planning. The development of different parts of ISF machine has been discussed as below:

3.1 Sheet Metal Blank

The ISF machine is basically used for deforming product from metal sheets. The current research is mainly focused on the development of ISF process for light alloy. The material for metal sheets can be aluminum alloys, brass, titanium alloys, etc. For testing the machine developed at IIT (BHU), a brass sheet of blank size 100 mm × 100 mm × 0.5 mm has been used for current experimental study.

The shape, size, and accuracy of the finished product are compared with the simulation model. The shape and size of the processed model as compared to the simulation model has been found approximately same.

3.2 Blank Holder or Blank Holding Arrangement

Blank holder is the second most important element of the ISF machine. The proper blank holding arrangement is necessary for properly holding the blank sheet. A CAD model of the blank holding arrangement as shown in Fig. 11 is developed for simulation study. Also based on the simulation study, a modified blank holding arrangement and a single-point incremental forming machine have been developed as shown in Figs. 17 and 18. The metals as well as alloys can be deformed easily by ISF process. For the current investigations, the initial blank of material brass is selected for finding the capabilities of ISF process because brass is having good formability at room temperature.

Fig. 17 Blank and blank holding arrangement

Fig. 18 Single-point incremental forming machine

3.3 Single-Point Forming Tool or Deforming Tool

The single-point forming tool also known as DTis another major element or most important element of the ISF machine. The DT may be spherical or elliptical or conical in shape. For the current research, a conical tool, having hemispherical shape at tip, has been used as shown in Fig. 17.

3.4 CNC Machine

The CNC machine is needed in order to control the tool path. A GALIL-make motion card having capability of controlling 8 independent axes has been used. A 3-axis in-house CNC machine has been developed for controlling the tool path. The completed experimental setup of single-point incremental forming machine has been represented in Fig. 18.

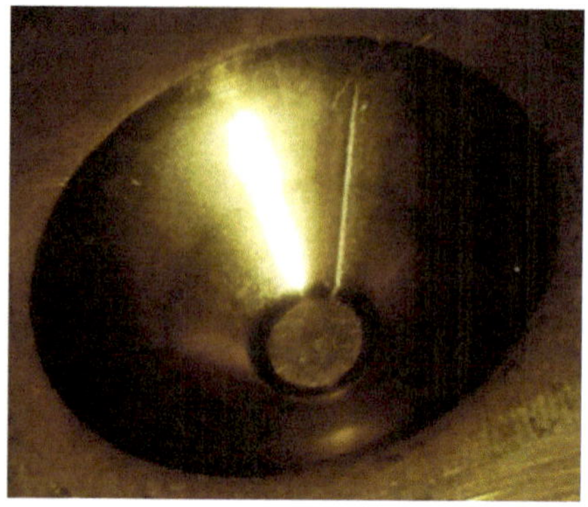

Fig. 19 Defect-free product (brass)

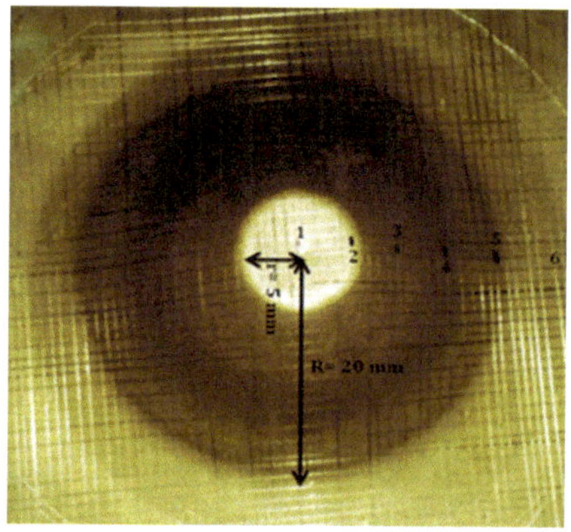

Fig. 20 Grid pattern on the bottom surface

4 Experimental Results

In order to investigate the strain distribution in the deformed product, the grid pattern was prepared on the bottom surface of sheet as shown in Fig. 20 having resolution of 1 mm. The metal sheet was divided into the 5 portions. Each portion is having 5 grids, i.e., 5 mm in length. The cone having top circle radius $R = 20$ mm and bottom circle radius = 5 mm has been successfully produced by ISF (Fig. 19).

For the analysis of strain distribution in the sheet deformed by ISF process, the following 5 regions were identified on the deformed sheet:

Region A: Starting from center point (Point 1) of bottom circle next 5 grids (Point 2) were identified as the region A.

Region B: Starting from Point 2 of next 5 grids, i.e., up to Point 3 were identified as the region B.

Region C: Starting from Point 3 of next 5 grids, i.e., up to Point 4 were identified as the region C.

Region D: Starting from Point 4 of next 5 grids, i.e., up to Point 5 were identified as the region D.

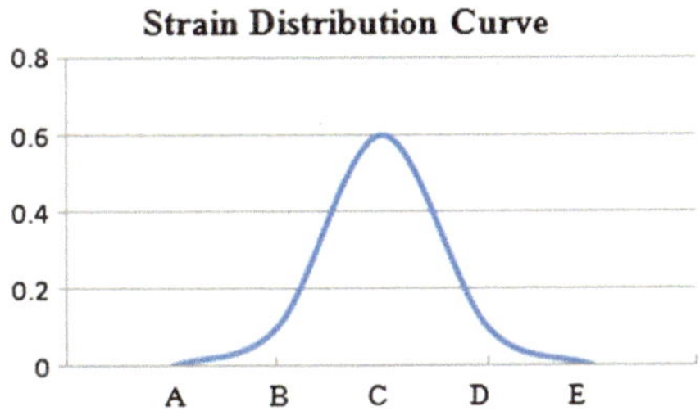

Fig. 21 Strain distribution

Fig. 22 Product formed by dieless forming machine (*top row* all aluminum and *bottom row* from *left* to *right* 1st brass, 2nd and 3rd aluminum, and 4th titanium)

Region E: The portion beyond Point 5 was identified as the region E.

The strain was found to be distributed normally as shown in Fig. 21.

Based on the simulation study, the experiments were carried out to get the defect-free products as shown in Fig. 22.

5 Conclusions

The ISF process is a flexible forming process, and it can be easily used for producing 3D complex shapes. The process can be used for larger forming angles with proper forming methodology. The laboratory setup of single-point ISF machine has been developed using motion card to control the 3 servo motors giving capability of individual 3-axis controlling. The defect-free products of brass, aluminum, and titanium have been produced using the ISF. The strain distribution over the length of deformation has been computed, and it has been found to be distributed over the length of the deformation.

Acknowledgments The authors acknowledge the help in running the simulation at IIT Kanpur.

References

Ambrogio G, Filice L, Manco GL (2008) Warm incremental forming of magnesium alloy AZ31. CIRP Ann Manuf Technol 57:257–260

Ambrogio G, Filice L, Gagliardi F (2012) Formability of lightweight alloys by hot incremental sheet forming. Mater Des 34:501–508

Araghi BT, Manco GL, Bambach M, Hirt G (2009) Investigation into a new hybrid forming process: incremental sheet forming combined with stretch forming. CIRP Ann Manuf Technol 58:225–228

Azaouzi M, Lebaal N (2012) Tool path optimization for single point incremental sheet forming using response surface method. Simul Model Pract Theor 24:49–58

Cerro I, Maidagan E, Arana J, Rivero A, Rodriguez PP (2006) Theoretical and experimental analysis of the dieless incremental sheet forming process. J Mater Process Technol 177:404–408

Filice L, Fratini L, Micari F (2002) Analysis of material formability in incremental forming. CIRP Ann Manuf Technol 51(1):199–202

Ham M, Jeswiet J (2006) Single point incremental forming and the forming criteria for AA3003. CIRP Ann Manuf Technol 55(1):241–244

Ham M, Jeswiet J (2007) Forming limit curves in single point incremental forming. CIRP Ann Manuf Technol 56(1):277–280

Hussain G, Gao L, Hayat N, Qijian L (2007) The effect of variation in the curvature of part on the formability in incremental forming: an experimental investigation. Int J Mach Tools Manuf 47:2177–2181

Jackson K, Allwood J (2009) The mechanics of incremental sheet forming. J Mater Process Technol 209:1158–1174

Juneja BL (2013) Fundamental of metal forming processes. New Age International (P) Limited, Delhi

Kim YH, Park JJ (2002) Effect of process parameters on formability in incremental forming of sheet metal. J Mat Process Technol 130:42–46

Kim TJ, Yang DY (2000) Improvement of formability for the incremental sheet metal forming process. Int J Mech Sci 42:1271–1286

Thibaud S, Hmida RB, Richard F, Malécot P (2012) A fully parametric toolbox for the simulation of single point incremental sheet forming process: numerical feasibility and experimental validation. Simul Model Pract Theor 29:32–43

Effect of Yield Strength, Pre-strain, and Curvature on Stiffness and Static Dent Resistance of Formed Panel

G. Manikandan, Rahul K. Verma, Marcel Lansbergen, Abhishek Raj and Deepak Deshpande

Abstract Increased importance on weight reduction is driving automotive industries to reduce thickness of the steel panels without compromising the vehicle safety and performance. High strength steels are looked at as a candidate for automotive applications. To overcome the limitation of less formability in high strength steel, steel makers introduced the bake hardening steel (BH) grades. This study compares the formability of extra deep draw (EDD) steel grades which are mainly used for body panels with that of bake hardening steel. The influence of material yield strength, pre-strain (ε_o), and curvature (R) of product on static dent resistance (SDR) and its stiffness is studied experimentally. It was found that higher yield strength provides higher dent resistance, whereas high panel thickness and smaller curvature resulted in higher stiffness. Higher dent resistance observed in bake-hardened steel is due to the increase in strength by bake hardening process. The use of bake-hardened steel in automotive applications represents a good opportunity for weight reduction, increased stiffness, and dent resistance. The experimental result was used to validate the SDR regression formulae developed by Tata Steel and then same formulae used to predict and compare the dent resistance value of back hardening and high strength interstitial free steel (HIF) grades.

Keywords High strength steel · Bake hardening · Formability · Dent resistance · Stiffness

---

G. Manikandan (✉) · R.K. Verma · A. Raj
Research and Development, Tata Steel Limited, Jamshedpur 831001, India
e-mail: manikandan.g@tatasteel.com

M. Lansbergen
Research, Development and Technology, Tata Steel Europe, Ijmuiden 1970, CA, Netherlands

D. Deshpande
Flat Product Technology Group, Tata Steel Limited, Jamshedpur 831001, India

R.G. Narayanan and U.S. Dixit (eds.), *Advances in Material Forming and Joining*,
Topics in Mining, Metallurgy and Materials Engineering,
DOI 10.1007/978-81-322-2355-9_3

1 Introduction

The practice of using thin sheets and new steel grades in the automotive industries is increasing in the last few years in order to improve safety and fuel efficiency. In the present competitive market, cost reduction with superior performance is the main objective in engineering design. Lightening the structures without decreasing its performances such as stiffness, noise vibration harshness (NVH), durability, and dent resistance is done by incorporation of high strength steels in design, if affordable (Cole and Sherman 1995).

In last few decades, steel makers continuously introduced new grades with high strength and good formability to meet the target of stiff and light automotive body. The poor formability of high strength steels is a major concern while considering its candidature for automotive panels (Hayashi and Nakagawa 1994). Material for outer body panels such as fenders, doors, hood, and deck lid requires good formability to make the design attractive. As the above components are external parts, chances of getting dented from stone impact and careless opening of an adjacently parked vehicle door is high. Hence, they require high static dent resistance (SDR) property, which in turn demands high strength. SDR is a measure of resistance to permanent deformation caused by static forces.

The bake hardening steel (BH) provides a good formability before stamping and enhanced its strength after baking. Hence, BH steel is widely used in vehicles, thus leading to a reduction in vehicle weight and improved safety (Alihosseini and Dehghani 2012). Bake hardening steels derive their increase in strength from a strain aging process that takes place on paint baking at 180 °C.

Previous studies regarding the dent resistance and stiffness of automotive panels reported some contradictory results. Dicello and George (1974) came to the conclusion that the lower stiffness of the panel, the better the dent resistance. Yutori et al. (1980) found that the higher the stiffness is, the higher the dent resistance.

This study focuses on the formability, SDR, and stiffness property of the Bake hardening steel and understanding the relation between them. The influence of material yield strength, pre-strain, and the effect of varying product curvature on the stiffness and static dent behavior of formed panels is also discussed in detail. SDR tests were conducted for three different pre-strain levels (1, 2, and 4 %) and two different curvatures (2000 and 4000 mm) and the results are discussed. Effect of change in mechanical properties due to bake hardening process was also experimentally studied.

2 Experimental Work

2.1 Mechanical Properties Characterization

Tensile tests were carried out using specimens machined as per ASTM standard specification. Three specimens were tested along three different directions with

Table 1 Mechanical properties

Grade	*T* (mm)	YS (MPa)	UTS (MPa)	EL (50 mm G.L.)	*n*	UEL (%)	*R*-bar
HIF	0.70	175	362	40	0.24	25	1.75
BH180	0.70	200	342	38	0.22	22	1.64
BH220	0.70	217	372	36.3	0.209	22.2	1.67

YS yield strength, *T* thickness, *UTS* tensile strength, *n* strain hardening exponent, *UEL* uniform elongation, *R-bar* plastic strain ratio, and *G.L* gauge length

respect to the rolling direction, with the tensile axis being parallel, diagonal, and transverse. The mechanical properties of the material are given in Table 1.

2.2 Forming Limit Diagram

Press forming is one of the most important processes used in automotive companies. To design a robust press forming process, it is important to have the formability data such as forming limit diagram (FLD) and tensile data along different directions. Moreover, most auto companies require such data during die design, so that they can use the data in finite element simulations.

The limit or failure strains in sheet-metal forming is represented best by a forming limit curve (FLC) which indicates the onset of localized necking over all possible combinations of strain in the plane of the sheet. To evaluate the strain ratios at necking, rectangular strips of different widths are stretched over a rigid hemispherical punch with the ends of the strip being held down by blank holder.

Test was conducted as per the International standard ISO 12004-2: Determination of forming limit curves in the laboratory (ISO 12004-2 2008). For evaluating the FLD experimentally, hemispherical punch with a diameter of 101.6 mm was used. The speed of the punch was 1.0 mm/s. Forming limit diagram was generated for high strength IF and bake hardening grade.

2.3 Static Dent Resistance Test

Dent resistance and stiffness are important property requirements for the vehicle body panel and hence these requirements need to be considered in designing or material selection stage itself. Before choosing a particular grade knowledge on its dent resistance is necessary. Despite these needs, the test methods for determining panel stiffness and dent resistance are not standardized. Automotive industries are using indigenous testing methodology (Hodgins 2001). Whichever system is used, the principle of force application and measuring the plastic deformation remains the same. Force and displacement measurements are generally incorporated into a data acquisition system. SDR test setup is shown in Fig. 1 and experimental procedures used in this study are applicable to Tata Steel.

Fig. 1 Static dent resistance test setup

The test product was representative of automotive exterior parts and the product geometry should preferably be pressed on stamps with a specified bottom radius. The thickness after forming the product was measured by ultrasonic thickness measurement technique.

The center of the product is loaded using a steel sphere to a predetermined force at a speed of 10 mm/min and then relieved with the same velocity. After a waiting time, height difference between the unloaded and loaded products is measured by means of three vertical displacement sensors. This process is repeated until a permanent deformation of 0.2 mm is obtained. Maximum dent power of the first load cycle is 10 N. Increase of indentation force per cycle was 10 N as shown in Fig. 2. Static dent resistance is calculated at the position where the indentation depth is 0.1 mm. The pre-strain levels were 1, 2, and 4 %, and curvatures were 2000 and 4000 mm. For each test, three samples were chosen to check repeatability.

Fig. 2 Load versus number of cycle

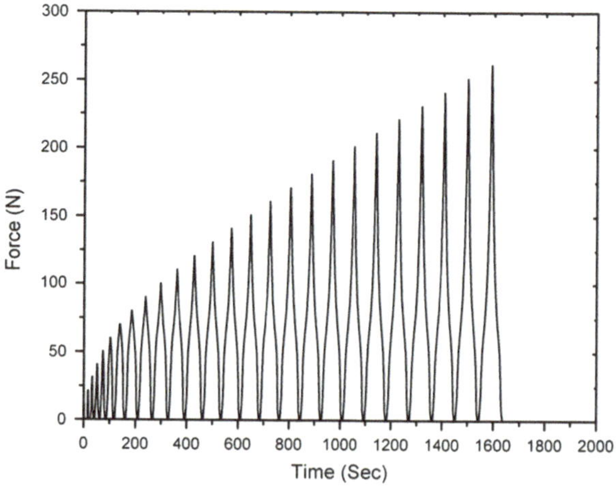

Fig. 3 A typical auto body panel stiffness plot

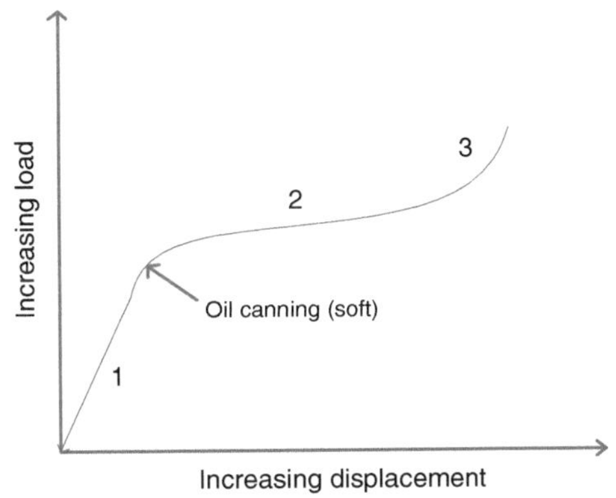

In this study, the static dent resistance of BH220 has been determined. Before the initiation of test, the material was stored in freezer to prevent the occurrence of aging process. Then, all the specimens have undergone a bake hardening treatment at 180 °C for 20 min before the SDR was determined.

Tensile samples (30 × 230 mm) were cut from the center of the product, which is parallel to the direction of rolling direction for determining the mechanical properties. During the test, the dent-shaped indentation is situated in the middle of the pull strip.

2.4 Panel Stiffness

The panel stiffness is defined as the resistance of a panel to elastic deformation. This is the ratio of the exerted force and deflection (i.e., initial slope of the force-deflection curve). Figure 3 is a typical stiffness plot showing three stiffness regions in a curved panel and the point at which the oil-canning response occurs (Thomas 2001).

3 Results and Discussion

3.1 Formability of BH Steel

Figure 4 shows the formability diagram of different high strength automotive steel grades. It can be seen that the FLC of BH steel is very close to that of high strength interstitial free steel (HIF). Formability of BH 180 is found to be a little more than BH 220 due to the fact that higher strength in BH220 reduces its formability.

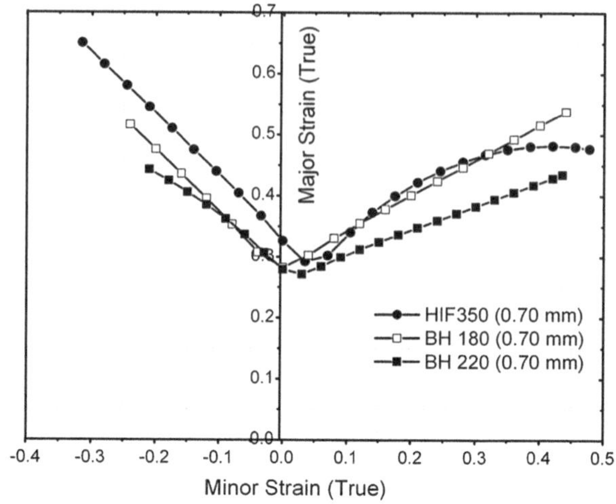

Fig. 4 Forming limit diagram

3.2 Effect of Pre-strain and Curvature on Dent Resistance

Figure 5 shows the force required to produce a permanent deformation of 0.1 mm on product with various pre-strains and curvatures. It can be observed that an increase in pre-strain level and curvature of the product increases the dent resistance level of the formed component. For a pre-strain level of 2 % when the product curvature is increased from 2000 to 4000 mm, the force required to make permanent deformation of 0.1 mm increased by 10 %. Similarly for a product curvature of 2000 mm, an increase in pre-strain level from 2 to 4 % resulted in an increase in dent resistance by 22 %. From this, it can be concluded that the effect of pre-strain on dent resistance is more than the product curvature. Plastic deformation during the forming process results in work hardening of the material. This work hardening

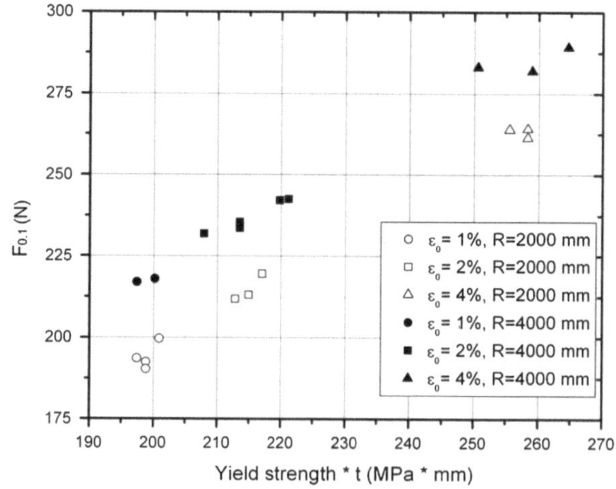

Fig. 5 The dent resistance force as a function of the panel yield strength multiplied by the panel thickness

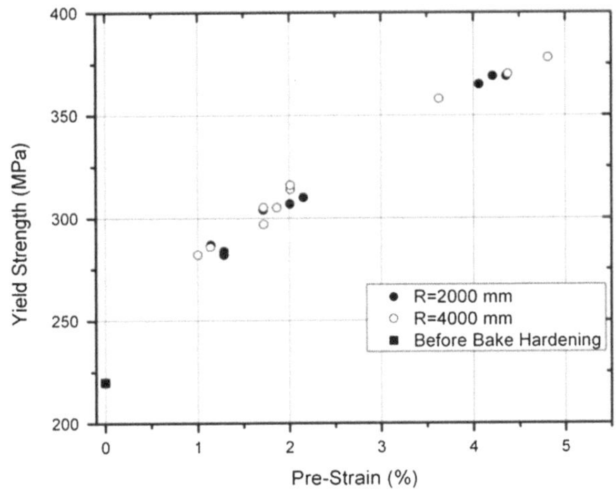

Fig. 6 Effect of pre-strain on yield strength after bake hardening

means that the yield stress is increased for subsequent loading. The increase in yield stress is beneficial to the dent resistance. This is supported by the study of Holmberg and Thilderkvist (2002).

3.3 Effect of Pre-strain on Increase in Yield Strength

Figure 6 shows a comparison of the yield strength of the pressed component with the initial strength level in the virgin material at corresponding strain levels. Initial yield strength of BH grade was 240 MPa. The bake hardening process increased the strength of the components by 54 %. Increase in strength also depends on the amount of pre-strain. The yield strengths observed for 1, 2, and 4 % pre-strains were 285, 310, and 370 MPa, respectively. The fact that strength increases due to strain hardening during stamping and bake hardening processes is reinforced.

Strength increment after baking process is attributed to diffusion of carbon atoms to the core of the dislocations formed during cold deformation that will result in pinning of dislocations, thereby preventing their movement. The effect of curvature on improvement of yield strength is much less pronounced, that for a given pre-strain even an increase of 2000 mm in curvature resulted in no significant improvement in yield strength. The effect of pre-stain and curvature on yield strength is shown in Table 2.

3.4 Effect of Pre-strain and Curvature on Panel Stiffness

Figure 7 shows the load and displacement cycle during the entire test for different levels of pre-strain and product curvature including the elastic displacement and oil-

Table 2 Static dent resistance test results

Specimen no.	Pre-strain (ε_0) (%)	Curvature (R) (mm)	Yield strength (MPa)	$F_{0.1}$ (N)
1	1	2000	282	194
2	1	2000	284	192
3	1	2000	284	190
4	2	2000	304	212
5	2	2000	307	213
6	2	2000	310	220
7	4	2000	369	261
8	4	2000	369	264
9	4	2000	365	264
10	1	4000	282	217
11	1	4000	286	218
12	2	4000	297	232
13	2	4000	305	234
14	2	4000	314	242
15	4	4000	378	289
16	4	4000	358	283
27	4	4000	370	282

$F_{0.1}$ Force required to make permanent deformation depth of 0.1 mm

Fig. 7 Panel stiffness

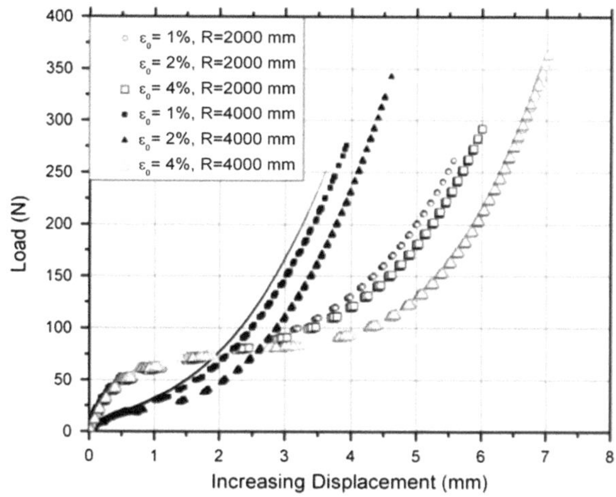

canning effect. Results show that panel stiffness is mainly dependent on product curvature and thickness. Panel stiffness increases with reduction of curvature from 4000 to 2000 mm. Over all, the panel stiffness increases with decrease in pre-strain from 4 to 1 %, the reason being variation in thickness according to pre-strain. Same phenomenon was observed in the study of Ekstranda and Asnafib (1998). But static dent energy absorption is good for 4 % pre-strained component owing to higher strength due to work hardening.

From this study, it is understandable that by changing the product curvature and pre-strain level, one can improve either panel dent resistance or stiffness, sacrificing the other in that process. This is supported by the study of Dicello and George (1974). A possible way of improving dent resistance without affecting stiffness is switching over to higher strength material. On the other hand, stiffness is not strength dependent and hence improving this is possible only by increasing the initial thickness which subsequently increases the final thickness of the panel.

3.5 *Effect of Bake Hardening on Material Property*

Bake hardening process increases the yield strength of the as received steel. Figure 8 shows the plot between true-stress and true-strain for as received and pre-strained, bake-hardened steel. While forming, some ductility of the material is utilized and subsequently strength is also improved to certain level by bake hardening treatment. Stretcher strain is observed in bake-hardened steel in the load–displacement curve.

From the previous section, we could understand that curvature does not play a major role on strength. A same phenomenon is observed in this study as well. Irrespective of the product curvature, subsequent yielding of all the products follows similar work hardening behavior for a given pre-strain and bake hardening process.

3.6 *Effect of Pre-strain and Curvature on Each Loading Cycle*

Load required to make permanent deformation on the product for every loading cycle is shown in Fig. 9. It shows that load required for the pre-strain level of 4 % is

Fig. 8 Effect of back hardening on mechanical properties

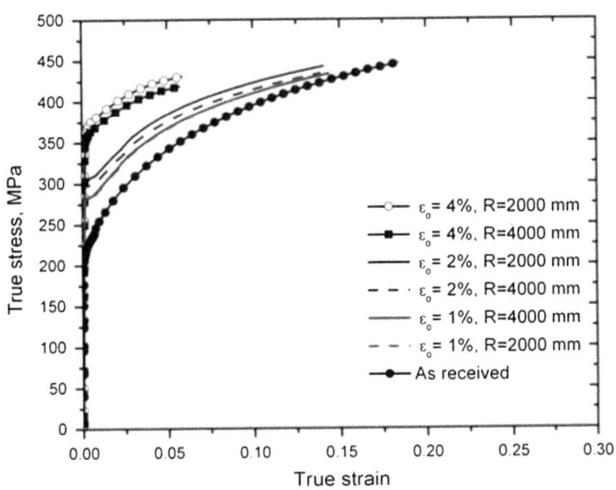

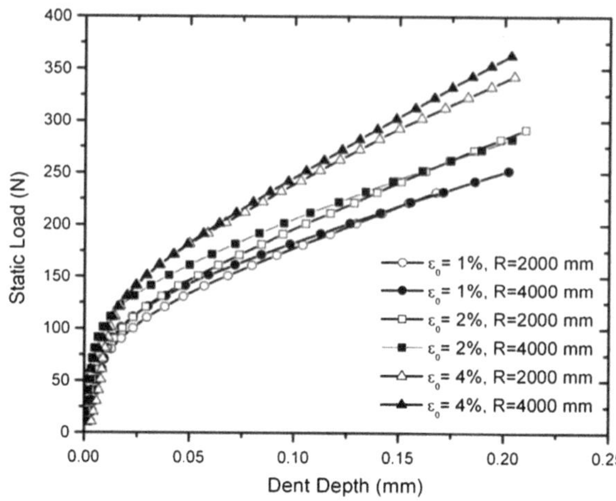

Fig. 9 Static load on each cycle

higher than the 1 and 2 % due to the higher amount of work hardening in case of 4 % pre-strain. Similarly, the load required for curvature of 4000 mm is higher than for 2000 mm, same trend has been observed in the SDR test results as well.

3.7 Validation of SDR Regression Formulae with Experimental Results

These test results validate the empirical model developed by TATA Steel for predicting the force required to initiate a dent (Tata Steel Internal report).

$$F_{0.1\,\mathrm{mm}} = b_{\mathrm{grade}}\left(\mathrm{YS}_{p0.2} \times t\right)^{\alpha} R^{\beta} \tag{1}$$

where

$F_{0.1\,\mathrm{mm}}$	local static dent resistance (N);
$\mathrm{YS}_{\mathrm{p0.2}}$	actual yield point;
$\mathrm{YS}_{\mathrm{p0.2}}$	after forming and baking (MPa);
t	actual thickness from tensile test (mm);
R	actual product curvature (mm);
b_{grade}	grade depending factor;
HIF Grades: $b_{\mathrm{IF\text{-}HS}}$	0.158;
BH Grades: b_{BH}	0.182;
α	1.18;
β	0.102.

It can be observed from Fig. 10 that both the results coincide very well. Experimentally finding static dent resistance force is time consuming and a costly

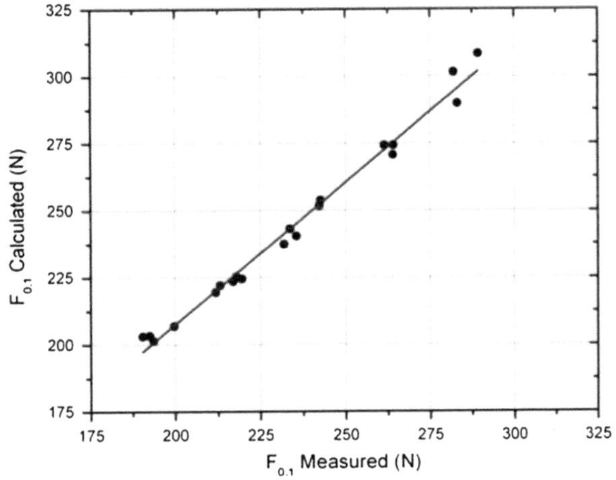

Fig. 10 Validation with SDR regression formulae

process. It has been proven that empirical SDR regression model could predict the dent resistance as well as experiments. So, for a comparison of two different grades, this theoretical method can be applied to save time and cost.

3.8 Comparison of Dent Resistance Between BH 220 and HIF350

Figure 11 shows the dent resistance force for both BH (0.7 mm) and HIF (0.7 mm) steel calculated by SDR regression formula. It is clear that the BH steel provides better dent resistance than the high strength IF steel, similar to experimental

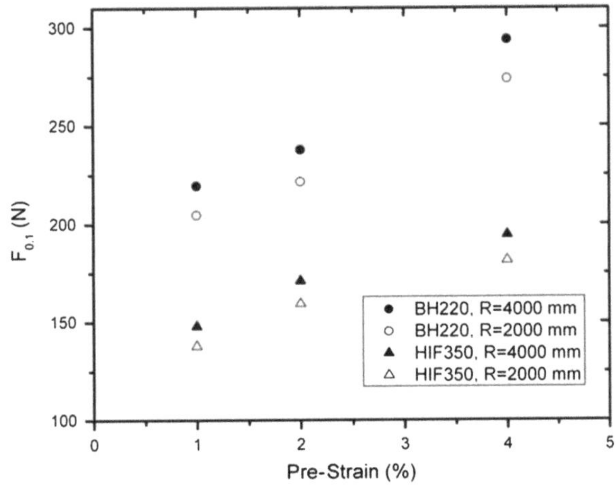

Fig. 11 Calculated dent resistance force with SDR regression formulae

findings. This model also shows that dent resistance of the component is directly proportional to the pre-strain and product curvature.

4 Conclusions

The experimental FLC was plotted for BH180/220 (0.7 mm) grades and compared with that for HIF. SDR test was conducted after baking at 180 °C for about 20 min in various pre-strain conditions and curvatures. From the experimental test results, following conclusions could be arrived at.

- Formability of BH steel grade is close to that of HIF.
- Dent resistance is directly related to yield strength. Yield strength of the product increases with pre-strain level. Stiffness on the other hand is directly proportional to final product thickness and inversely proportional to product curvature but independent of the strength of the component. High stiffness does not always entail high dent resistance. In short, dent resistance is mostly material strength dependent and stiffness is component design (curvature and final thickness) dependent.
- If a component requires both high dent resistance and high stiffness with reduced panel weight for automotive, an elegant solution is to select a bake hardening steel grade.
- Significant increase in strength level of the component was observed after bake hardening of BH220. When compared to HIF 350 steel, it was found that BH steel is a better option for the component that requires high dent resistance.

From this study, it is evident that weight saving and better dent resistance can be achieved by choosing Bake hardening(BH) steel grade, which will enable us to save energy and enhance safety with greater dent resistance of vehicle panels.

Acknowledgments The authors acknowledge the experimental support of Product Application Center, Netherlands, Tata Steel Europe. They are also thankful to Tata Steel for the permission to publish the results.

References

Alihosseini H, Dehghani K (2012) Bake hardening of ultra-fine grained low carbon steel produced by constrained groove pressing. Mater Sci Eng A 549:157–162

Cole GS, Sherman AM (1995) Lightweight materials for automotive applications. Elsevier Publication, Ford Motor Company, Dearborn, Ml 48121-2053

DiCello JA, George RA (1974) Design criteria for the dent resistance of auto body panels. SAE 740081

Ekstranda G, Asnafib N (1998) On testing of the stiffness and the dent resistance of auto body panels. Mater Des 19:145–156

Hayashi H, Nakagawa T (1994) Recent trends in sheet metals and their formability in manufacturing automotive panels. J Mater Process Technol 46:455–487

Hodgins B (2001) Numerical prediction of the dent resistance of medium scale aluminum structural panel assemblies. Dissertation, University of Waterloo

Holmberg S, Thilderkvist P (2002) Influence of material properties and stamping conditions on the stiffness and static dent resistance of automotive panels. Mater Des 23(8):681–691

Internal report no: IJTC/AUT-MPP/K/R/6504/02/Z Empirical relation for local static dent resistance, Tata steel, Ijmuiden, Netherlands

ISO 12004-2 (2008) Determination of forming-limit curves in the laboratory, 1st edn

Thomas D (2001) Numerical prediction of panel dent resistance incorporating panel forming strains. Dissertation, University of Waterloo

Yutori Y (1980) Studies on the static dent resistance. Mem Scientifiques Rev Metall 77:561–569

Formability Prediction and Springback Evaluation of Adhesive-Bonded Steel Sheets

V. Satheeshkumar, Avinash Kumar Yadav and R. Ganesh Narayanan

Abstract The mechanical properties of adhesive were evaluated by two different approaches to predict the formability of adhesive-bonded steel sheets by finite element (FE) simulations. The epoxy adhesive properties were varied by changing the hardener/resin (*H*/*R*) ratios. The true stress–strain behaviour of adhesive was evaluated from (i) direct tensile testing of adhesive samples (Approach-1) and (ii) rule of mixtures' concept applicable to adhesive-bonded sheets (Approach-2). The evaluated mechanical properties of adhesive through both the approaches were incorporated during simulation of tensile test and in-plane plane-strain formability test. The forming limit strains of adhesive-bonded sheets were evaluated during predictions by necking criteria, namely thickness gradient necking criterion (TGNC) and effective strain rate ratio based criterion (ESRC), and validated with the experimental limit strains. It is found that the limit strains predicted by both the approaches are moderately accurate with respect to experimental data. There is not much difference in limit strains of adhesive-bonded sheets obtained from Approach-1 and Approach-2. The TGNC shows better prediction as compared to ESRC. Further, the present study investigates the springback of adhesive-bonded sheets during V-bending with the influence of *H*/*R* ratio of adhesive, shims, and wire reinforcement. The results show that there is a considerable influence of *H*/*R* ratio on springback and about 8–20° reduction depending on the presence or absence of shim. But the wire reinforcement shows insignificant effect on V-bending of adhesive-bonded sheets.

Keywords Adhesive-bonded sheet · Steel sheets · Rule of mixtures · Prediction · Limit strain · Springback · V-bending · Formability

V. Satheeshkumar · A.K. Yadav · R.G. Narayanan (✉)
Department of Mechanical Engineering, IIT Guwahati, Guwahati 781039, India
e-mail: ganu@iitg.ernet.in

R.G. Narayanan and U.S. Dixit (eds.), *Advances in Material Forming and Joining*,
Topics in Mining, Metallurgy and Materials Engineering,
DOI 10.1007/978-81-322-2355-9_4

1 Introduction

Among different joining technologies available, the adhesive bonding technology, which is categorized under tailor-made concept, helps to accomplish uniform stress transfer, increased fatigue life, joining of dissimilar materials, and a reduction in structural weight that are required in aircraft and automotive sectors. The adhesives show their unique suitability for solving numerous joining problems that are commonly encountered in welding, riveting, and the use of other mechanical fasteners. In the modelling of adhesive-bonded blanks, Czamocki and Piekarski (1986) predicted the joint strength of adhesive-bonded joint by finite element (FE) method. Four different criteria including maximum normal stress criterion, modified distortion energy criterion and two criteria for polymers in biaxial state of stress were used in the prediction. The results showed that there is about 10 % difference in joint strength between experiment and predictions. Kim et al. (2011) introduced the superimposed FE method for adhesive-bonded joints. A cohesive zone model was used to model the interface between substrates and adhesive layer. The complicated behaviour of the interface was characterized into one cohesive layer. It was found that the simulation of adhesive joints with the superimposed FE method is efficient in view of the FE discretization and shows accurate results. May et al. (2014) introduced a rate-dependent constitutive law for cohesive interface elements for the adhesive considering both the rate dependency of the initiation stress and the rate dependency of the fracture toughness. The material model was validated by comparing FE predictions showing good agreement to the experimental data. It was demonstrated that the use of developed rate-dependent cohesive zone models is suitable for modelling the response of adhesively bonded metallic structures subjected to crash loading. Most of the studies on adhesive-bonded blanks are mainly aimed on predicting the joint strength and fracture mechanisms.

Aghchai et al. (2008) carried out a formability study on a two-layered metallic sheet (Al1100/St12) bonded by polyurethane adhesive. The two-layer sheet was assumed as an equivalent to a one-layer sheet by deriving equivalent mechanical parameters. In the other method, the mechanical properties of each layer were separately utilized. It was observed that the FLD of the two-layer sheet is better than the lower formable material. The two-layer sheet improves the formability of a sheet which has low formability. In this context, Satheeshkumar and Ganesh Narayanan (2014, 2015) investigated the influence of adhesive properties by varying hardener/resin ratio on the formability of adhesive-bonded steel sheets made of dissimilar steel sheets. The results showed that the elongation of adhesive-bonded sheets increases with increase in hardener/resin ratio. This is because of changeover of resin-rich formulation to hardener-rich formulation of adhesive systems, making the adhesive more ductile. It was also observed that the limit strains of sheets constituting adhesive-bonded blanks improve with increase in hardener/resin ratio of adhesive. Further, the forming behaviour of adhesive-bonded sheets was predicted with the effect of different hardener/resin ratios through FE method. There was a good agreement in the stress–strain behaviour of adhesive-bonded blanks between experiments and predictions. It was stated that the discrepancies in the limit strains of base materials constituting adhesive-bonded blanks between experiment and prediction are due to the absence of interfacial bonding between the adhesive layer and base materials (Satheeshkumar and Ganesh Narayanan 2014, 2015).

Abdullah et al. (2001) used the rule of mixtures' method to determine the weld metal properties from the tensile specimens in which the weld lies parallel to the axis of tension. It was found that the rule of mixtures is a useful method of weld metal property determination. Though the work of Abdullah et al. is not exactly relevant to the present work, the concept of rule of mixtures used for welded sheets has been applied for adhesive-bonded sheets in the present work. The work done by Perton et al. (2011) describes a new method involving the propagation of high-amplitude ultrasonic waves to evaluate the adhesive bond strength at high strain rate. The resulting tensile forces normal to the interface were used to estimate the delamination inside the laminates. The same was used to evaluate the bond strength. The effect of surface roughness on the bonding strength of adhesives such as epoxy and cyanoacrylate was studied by Uehara and Sakurai (2002). It was found that the optimum value of roughness is in the range of 3–6 μm with respect to tensile strength of the adhesion. There is no clear relationship observed between the peel strength and the surface roughness for different types of adhesives. Pereira et al. (2010) investigated the effect of surface treatment and adhesive thickness on the shear strength behaviour of aluminium alloy adhesive lap joints. The lap joints were fabricated by using a high-strength epoxy adhesive. The surface treatments using sodium dichromate–sulphuric acid etch and abrasive polishing showed improved joint shear strength as compared to caustic etch and Tucker's reagent etch. The shear strength of single-lap joints was found to be increased with decrease in surface roughness, increase in base sheets thickness and overlap length. Ravi and Ganesh Narayanan (2012) carried out a study on the influence of bonding width, adhesive thickness, and blank surface roughness on the tensile behaviour of adhesive-bonded sheets. It was found that the engineering stress increases with increase in surface roughness at same amount of strain during forming of bonded blanks. The extension of the bonded blank at the failure is less for higher surface roughness base materials. There is no considerable effect due to adhesive thickness on the elongation at failure. Takiguchi and Yoshida (2001) analysed the plastic bending of adhesive-bonded sheet metals through V-bending experiments. The specific increase of shear strain, as well as of the punch load which related directly to the change of the die–sheet contact boundary conditions, was analysed. The condition of deforming sheet at the initial stage supported by two edges of die was so-called air-bending condition. It was recommended to use air-bending operation for adhesive-bonded sheet metals for suppressing the shear deformation of the adhesive layer to within an acceptable limit. Further, the same authors proposed a new technique of plastic bending of adhesive-bonded sheet metals. The effect of forming speed on the deformation characteristics of adhesively bonded aluminium sheets was analysed through V-bending experiments and numerical simulations using a rate-dependent constitutive model of plasticity for the adhesive. It was found that the large shear deformation and the geometrical imperfection caused by large transverse shear deformation occurring in the adhesive layer are suppressed by high-speed forming since the deformation resistance becomes higher at high strain rate (Takiguchi and Yoshida 2003).

From the above literature study, it is highlighted that the bonded blanks are influenced significantly by the adhesive properties, bond strength, and surface treatment of base materials constituting bonded blanks during forming process. In the demonstration of FLD of the two-layer sheet which is better than the lower formable material by Aghchai et al. (2008), the influence of adhesive layer was not addressed. In their study, the FLD of two-layer sheet lies in between the FLD of monolayer higher formable sheet and lower formable sheet during biaxial stretching test. In the investigation carried out by Satheeshkumar and Ganesh Narayanan (2014, 2015), there is a significant improvement in the limit strains of the lower formable material (DDQ steel sheet) and also a moderate improvement in the limit strains of higher formable material (SS 316L sheet) due to the improved plasticity of adhesive by hardener-rich formulation. It is understood that there is no necessary condition of the final forming limit strain of the base materials constituting adhesive-bonded sheets to be lain between those of the lower formable material and the higher formable material. It was also emphasized that the discrepancies between experimental and predicted limit strains of base materials constituting adhesive-bonded blanks are due to the absence of interface bonding between adhesive and base materials (Satheeshkumar and Ganesh Narayanan 2014, 2015). Still appropriate modelling of adhesives with their adhesion virtues is a challengeable task.

In the present study, firstly, two different approaches are followed to evaluate the adhesive mechanical properties in the adhesive-bonded blanks to predict their forming behaviour. In the first method of evaluation (Approach-1), the mechanical properties of adhesive with different hardener/resin ratios are evaluated from simple tensile test and incorporated during the modelling of adhesive-bonded blanks to predict their formability. The interfacial bonding between adhesive and base materials is not taken into account in this approach. In this case, the adhesive-bonded blanks cannot be termed simply to the extreme as parallel strips during tensile testing, but the adhesive-bonded blanks with the defect in the adhesion properties. In the second method of evaluation (Approach-2), the tensile properties of adhesive is evaluated using rule of mixtures' method from the global tensile behaviour of adhesive-bonded blanks through experimentation. It is believed that the tensile ($\sigma - \varepsilon$) behaviour of adhesive derived from the tensile behaviour of adhesive-bonded blanks using Approach-2 is the constitutive tensile behaviour of adhesive and its interfacial bond strength between base materials, which may improve the accuracy of formability of adhesive-bonded blanks during prediction. In the predictions, the limit strains in adhesive-bonded blanks are evaluated by using thickness gradient based necking criterion (TGNC) and effective strain-rate based necking criterion (ESRC). These two necking criteria are used to check the applicability of predicting the limit strains in adhesive-bonded blanks, which are being used generally for predicting limit strains of un-bonded sheet metals. Finally, the predicted limit strains by TGNC and ESRC through Approach-1 and Approach-2 are validated with the experimental results.

Secondly, an important issue called 'Springback' during sheet-forming processes is considered in the present study, and an attempt has been made to minimize the springback of adhesive-bonded sheets by changing the adhesive properties. The adhesive property is varied by changing the hardener/resin ratio of adhesive with different levels and of the effect is investigated on the springback during V-bending of adhesive-bonded sheets made of same base material (DDQ steel). In addition, the effect of wire reinforcement and presence of shims between base materials are also presented. The springback angle and the failure pattern are monitored and analysed.

2 Methodology

2.1 Base Materials and Adhesive

Two dissimilar materials, namely deep drawing quality (DDQ) cold rolled steel and stainless steel (SS 316L) sheets, were used as base materials in the fabrication of adhesive-bonded blanks. The thickness of base materials is 0.6 mm each. Table 1 shows the nominal chemical composition of two base materials as obtained from the suppliers. The base materials were bonded by using commercially available Bisphenol-A-Epicholorhydrin-type epoxy resin (Araldite-AW106 → Part A) and Polyamidoamine-type hardener (Araldite-HV953IN → Part B). The stoichiometric ratio of hardener/resin ratio of epoxy adhesive system is 0.8:1. All the tensile tests of base materials were carried out as per ASTM E646-98 standard. The mechanical properties of DDQ steel and SS 316L sheet materials are listed in Table 2 along the zero degree rolling direction of the sheets. The plastic strain ratio of base materials as per ASTM E517 standard in different rolling directions is also tabulated.

2.2 Approach-1: Experimental Evaluation of Adhesive Mechanical Properties

The epoxy adhesive samples with different hardener/resin ratios (by weight) such as 0.6:1, 0.7:1, 0.8:1, 0.9:1, and 1:1 were prepared by using appropriate moulds and tested as per ASTM D 638-I standard. The adhesive specimens were allowed to cure for more than 24 h at room temperature before testing. The average thickness of adhesive layers was found to be 6 ± 0.5 mm. The tensile test of adhesive samples was carried out at a nominal cross-head speed of 1 mm/min, at room temperature of 25 °C in an INSTRON machine. The load–extension behaviour was recorded for all adhesive specimens with different hardener/resin ratios and converted into true stress–strain behaviour by standard procedure.

Table 1 Nominal chemical composition of DDQ steel and SS 316L sheets in weight %

Materials	C %	Si %	Mn %	S %	P %	Cr %	Ni %	Mo %	Cu %	Co %	Ti %	V %	Fe %
DDQ steel	0.100	0.12	0.600	0.035	0.040	–	–	–	–	–	–	–	Remaining
SS 316L	0.016	0.335	1.209	0.002	0.024	16.413	10.222	2.164	0.405	0.032	0.012	0.034	69.070

Table 2 Mechanical properties of DDQ steel and SS 316L sheets (0.6 mm thickness)

Materials	Yield strength MPa	Ultimate tensile strength MPa	Total elongation % in 50 mm gauge length	Strain hardening coefficient (n)	Strength coefficient K MPa	Plastic strain ratio		
						R_0	R_{45}	R_{90}
DDQ steel	189 ± 5	352 ± 5	39 ± 2	0.251 ± 0.005	557 ± 25	1.32	1.13	1.58
SS 316L	250 ± 5	610 ± 15	39 ± 2	0.466 ± 0.013	1352 ± 25	1.37	1.54	1.66

2.3 Approach-2: Rule of Mixtures for Evaluating Adhesive Mechanical Properties

The adhesive-bonded sheets were fabricated as per ASTM E646-07e1 standard. The base materials DDQ steel and SS 316L were bonded by epoxy adhesive with different hardener/resin ratios using appropriate moulds to maintain uniform thickness. The specimens were allowed to cure for more than 24 h. The average thickness of adhesive layers in the adhesive-bonded sheets was found to be 1 ± 0.015 mm. The tensile test was carried out at a nominal cross-head speed of 1 mm/min, at room temperature of 25 °C. The load–extension behaviour was recorded for all adhesive-bonded specimens and converted into true-strain behaviour. The $\sigma - \varepsilon$ data of adhesive-bonded sheets till adhesive failure were taken into account to evaluate the $\sigma - \varepsilon$ behaviour of adhesive layer by applying the rule of mixtures' concept (Fig. 1).

$$\text{Let, total load, } P_\mathrm{T} = P_\mathrm{d} + P_\mathrm{a} + P_\mathrm{s} \tag{1}$$

where

P_T total load during tensile test of adhesive-bonded blank from experiment,
P_d load on DDQ steel sheet in adhesive-bonded blanks,

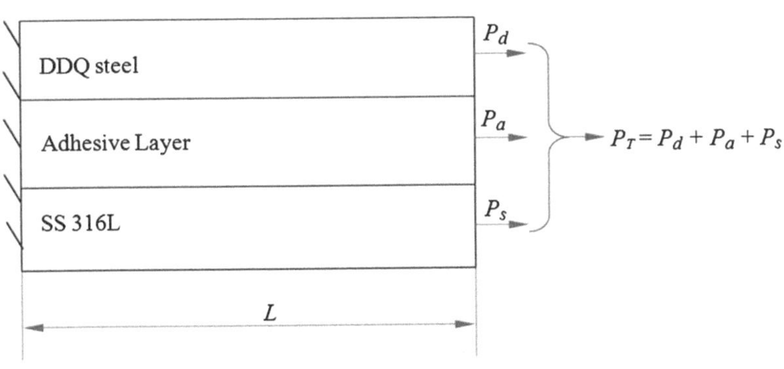

Fig. 1 Schematic of load sharing by adhesive-bonded sheets during tensile test

P_a load on adhesive layer in adhesive-bonded blanks, and
P_s load on SS 316L sheet in adhesive-bonded blanks.

The total load is written in terms of true stress as Eq. (2)

$$P_T = \sigma_a A_a + \sigma_d A_d + \sigma_s A_s \tag{2}$$

Based on the Hollomon's strain hardening law, Eq. (2) is modified as,

$$\sigma_a = \frac{P_T - (K_d \varepsilon_d^{n_d}) A_d - (K_s \varepsilon_s^{n_s}) A_s}{A_a} \tag{3}$$

Since adhesive and base materials are bonded along the longitudinal direction and assuming iso-strain condition ($\varepsilon_a = \varepsilon_d = \varepsilon_s$) till failure of adhesive, the true stress–strain relation of adhesive layer is given by Eq. (4)

$$\sigma_a = \frac{P_T - (K_d \varepsilon_a^{n_d}) A_d - (K_s \varepsilon_a^{n_s}) A_s}{A_a} \tag{4}$$

where

K_d and K_s	strength coefficients of DDQ steel and SS 316L, respectively.
n_d and n_s	strain hardening exponent of DDQ steel and SS 316L, respectively.
σ_a, σ_d, and σ_s	true stress of adhesive layer, DDQ steel and SS 316L sheets, respectively.
A_a, A_d and A_s	instantaneous cross-sectional area of adhesive layer, DDQ steel and SS 316L, respectively (Here $A_a = A_T - A_d - A_s$).
A_T	total instantaneous cross-sectional area of adhesive-bonded blanks.

The true stress–strain data obtained by Approach-1 and Approach-2 were incorporated during simulation of adhesive-bonded sheets to predict the forming limit strains during tensile test and in-plane plane-strain (IPPS) formability test.

The main focus of IPPS formability test is to analyse the critical forming limit of adhesive-bonded sheets prediction at the plane-strain condition. About 150 formability tests were simulated with different dimensions and geometries, and strain paths were predicted using a commercial available elasto-plastic, dynamic explicit FE code. The optimized geometry of the IPPS specimen which is close to plane-strain deformation mode is shown in Fig. 2, and the same has been used for fabricating adhesive-bonded samples. The same geometry was used for SS 316L sheet. The average thickness of adhesives in adhesive-bonded sheets was found to be 1 ± 0.013 mm. The IPPS formability test was carried out at a nominal cross-head speed of 1 mm/min in the INSTRON machine. The limit strains of adhesive-bonded sheets from both tensile and IPPS formability tests were evaluated by measuring

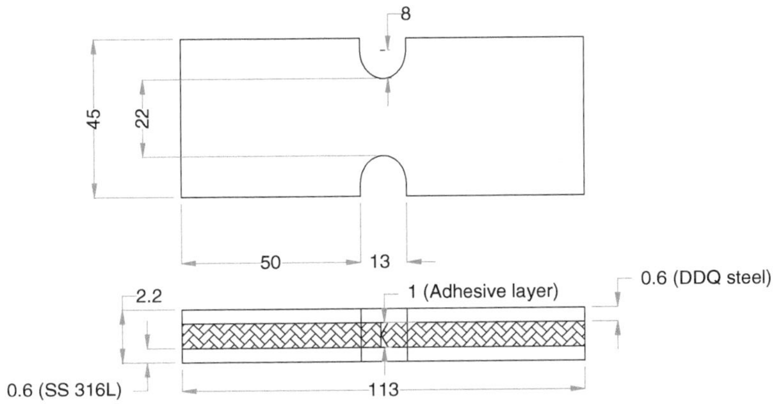

Fig. 2 Optimized geometry of IPPS sample used in the fabrication of adhesive-bonded sheets, and all dimensions are in mm and not to scale

major axis and minor axis of the circular grids printed on the sheet materials before and after deformation using digitized optical profile projector. The average initial diameter of circular grids was measured as 2.8 ± 0.05 mm.

2.4 Prediction of Limit Strains of Adhesive-Bonded Sheets

The CAD models of the tensile specimen were generated by Pro-E® software and imported as a surface model as per ASTM E646-07e1 into the commercially available elasto-plastic, dynamic explicit FE code for pre-processing, solving, and post-processing. The CAD model was meshed using the facility available in the code. The meshing was done with quadrilateral shell elements of the Belytschko–Tsay formulation. The average mesh size of about 2 mm was used throughout the tensile specimen and 1 mm was used for IPPS specimen. In order to generate adhesive-bonded blanks for FE simulations, three similar specimens were generated on the same plane and positioned one above other as shown in Fig. 3a, b. The top and bottom layers are base materials (DDQ steel and SS 316L), and the centre layer is adhesive. The base materials properties of DDQ steel and SS 316L given in Table 2 were incorporated during FE simulations. Hollomon's power law ($\sigma = K\varepsilon^n$; where K is the strength coefficient and n is the strain hardening exponent) was used to describe the strain hardening behaviour of base materials. Hill's 1948 isotropic hardening yield criterion (Banabic 2010) was used as the plasticity model for DDQ steel and SS 316L materials. The adhesive layer positioned between base materials was considered as an isotropic material. One end of the specimen having three layers was fixed stationary and the other end was displaced with a velocity of 1 mm/min (Fig. 3a, b). The bonding between adhesive and base materials was not modelled in the present work. The forming limit strains were predicted from FE

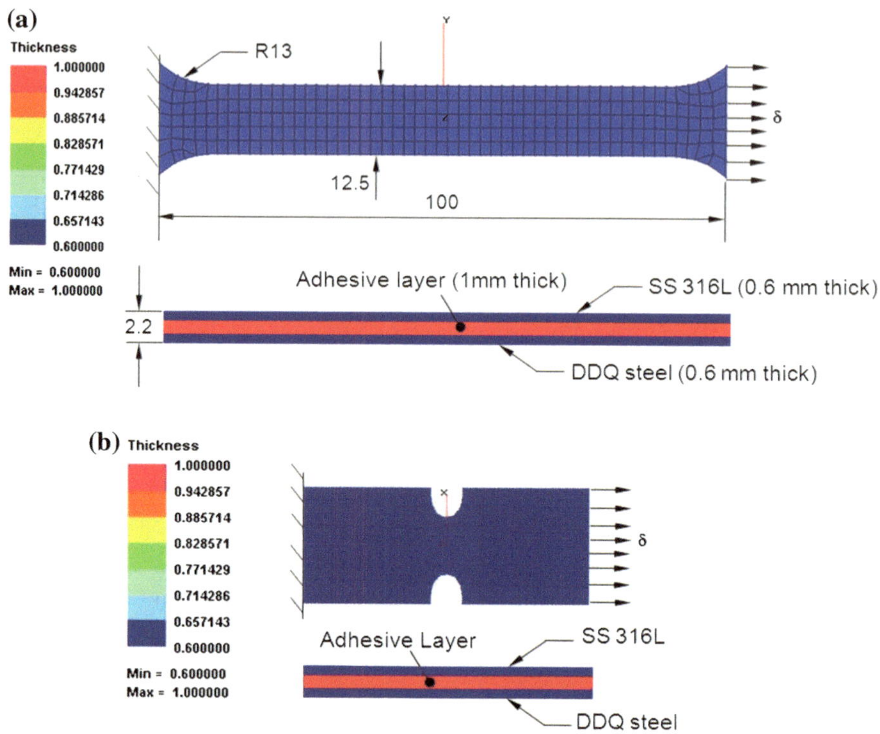

Fig. 3 Boundary conditions of adhesive-bonded blank in the simulation: **a** tensile specimen and **b** IPPS specimen. All dimensions are in mm and not to scale

simulations of tensile and IPPS forming tests. The limit strains are predicted using effective strain rate necking criterion (ESRC) and TGNC and validated with experimental results.

Two different necking criteria are used to predict the forming limit strain of base materials constituting adhesive-bonded sheets. They are described as below.

ESRC is the ratio of effective strain rate in the neck to that in the safe region (or bulk) (Narasimhan and Wagoner 1991) and is defined as:

$$\text{ESRC} = \frac{\text{Effective strain rate in the neck region}}{\text{Effective strain rate in the safe region}} \geq 4 \Rightarrow \text{Material failure/necking} \quad (5)$$

The development of strain in the bulk (or safe) region saturates (or tends to saturate) earlier during forming when compared with straining in the notch region (or neck) is based on the necking criteria. The ratio 4 is a 'lower bound value' below which limit strains are not reached, and the strain rate ratio increases unstably once the criterion is reached, which is the indication for the occurrence of necking.

The corresponding major strain and minor strain in the bulk element indicate the limit strain in that strain path.

Generally, necking is understood as a localized thinning phenomenon and TGNC basically determines localized region within the deforming sheet where a thickness gradient develops. It is found from the literature that necking occurs in the sheet metal when the thickness gradient falls below 0.92 and is also applied to predict the FLC (Sujit et al. 1994). A local critical thickness gradient must exist at the onset of a local visible neck. This criterion holds ease of applicability in sheet-forming processes to identify the onset of necking. In the present simulations, element pairs where the thickness ratio equals or falls below 0.92 are considered as necked elements. The major strains in all the thicker elements are noted. The largest major strain and the corresponding minor strain of such elements are treated as the forming limit strain. The thickness gradient necking criterion is given as,

$$\text{TGNC} = \frac{\text{Thickness}_{\text{Necking element}}}{\text{Thickness}_{\text{Neighbouring element}}} = \frac{t_n}{t_{n-1}} \text{ or } \frac{t_n}{t_{n+1}} \leq 0.92 \Rightarrow \text{Material failure/necking} \quad (6)$$

The ESRC and TGNC criteria were applied for un-bonded metal sheets earlier and also for welded sheets in which the sheets are located on the same plane. In the present work, the applicability of ESRC and TGNC for adhesive-bonded sheets will be checked and validated.

2.5 Experimental Evaluation of SpringBack

The DDQ steel sheet of 0.4 mm thickness as base material and epoxy adhesive were used in the fabrication of bonded sheets. During wire reinforcement of adhesive-bonded sheets, commercially available aluminium wire with Ø0.3 mm was used. The dimensions of die and punch used during V-bending experiment are shown in Fig. 4. The mechanical properties of DDQ steel were evaluated in different rolling

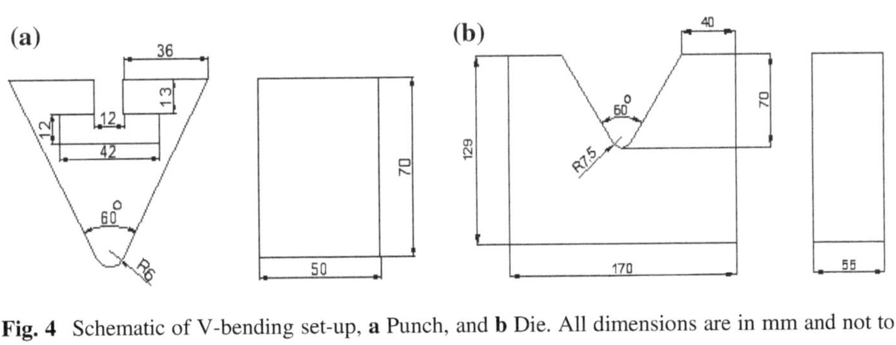

Fig. 4 Schematic of V-bending set-up, **a** Punch, and **b** Die. All dimensions are in mm and not to scale

Table 3 Material properties of DDQ steel with 0.4 mm thickness

Rolling direction (°)	Yield strength in MPa	Ultimate tensile strength in MPa	Uniform elongation %	Elongation % at failure	Strain hardening exponent (*n*)	Strength coefficient *K* in MPa	Plastic strain ratio (*R*)
0	224	372	28	34	0.23	647	1.12
45	216	360	26	34	0.23	633	1.37
90	201	344	25	32	0.24	608	1.47

directions such as 0, 45, and 90° through tensile test (as per ASTM E646-07e1 standard) in the INSTRON machine (Max. load capacity = 100 kN) at 1 mm/min cross-head speed. Three samples were tested in each case, and the load–extension data were recorded in order to evaluate engineering stress–strain and true stress–strain data. The plastic strain ratio (*R*) was found in accordance with ASTM E517 standard. The base material properties thus evaluated are shown in Table 3. The aluminium wires were initially tested as per ASTM A931–08 (2013) in the INSTRON machine (Max. load capacity = 5 kN) at 1 mm/min cross-head speed in order to determine the ductility and load-carrying capacity of the wire. Three wires were tested with gauge length of 125 mm, and it was found that the average total extension of the wire was about 40 mm, and the average maximum load attained was about 10 N.

The DDQ steel sheets of size 170 mm × 50 mm were cut along 0° rolling direction for fabricating adhesive-bonded sheets. The sheets were cleaned well, and the homogeneous mixture of hardener and resin parts of adhesive with different ratios was cast on the sheets. During fabrication of bonded sheets, appropriate moulds were used to maintain uniform thickness of the adhesive. The adhesive-bonded blanks were allowed to cure for more than 24 h at 25 °C. Three sets of samples were fabricated for springback tests: (i) samples made by using shims with size of about 50 mm × 30 mm at the extreme ends located between the top and bottom sheets with the help of commercially available thermoplastic glue which could impart quick fixability. The main aim of using shim is to maintain uniform thickness of adhesive layer during bonding; (ii) samples made without using shims, just to determine the effect of only epoxy adhesive. These two samples were made with different *H*/*R* ratios; and (iii) bonded samples with wire reinforced adhesive. In this case of wire reinforced adhesive, the length of single wire strand was maintained as 160 mm, and the strands were reinforced at distance of 2 mm each other. The number of wires was increased from 5 to 25 numbers with different levels. Three trials were performed in each case for checking the repeatability, and fourth trial was performed in the case of repeatability is not good.

The thickness of adhesive-bonded specimens was measured at different locations using digitized micrometre, and the average thickness was taken into account as a final thickness of the samples. The V-bending experiment was carried out in the universal testing machine (UTM). The un-bonded and adhesive-bonded sheets were tested during V-bending in order to compare the springback results. The cross-head

Table 4 Parameters and levels used in the experiments

Experimental parameters	Level				
	1	2	3	4	5
H/*R* ratio with shim	0.6:1	0.7:1	0.8:1	0.9:1	1:1
H/*R* ratio without shim	0.6:1	0.7:1	0.8:1	0.9:1	1:1
No. of wires	5	10	15	20	25

speed was maintained as 1 mm/min, and the testing was stopped at the maximum punch stroke of about 68 mm. At the completion of V-bending of bonded blanks, the springback angle was measured using digitized optical profile projector. The springback angle was measured at three different locations, and the average was considered as the final springback angle. During bending of adhesive-bonded sheets, delamination of sheet was observed and hence, springback of upper sheet and bottom sheets was measured separately. All springback measurements were performed within 24 h, after completion of experiment. The various parameters involved in the experimental study are given in Table 4.

3 Results and Discussion

In this section, the limit strain results of base materials constituting adhesive-bonded sheets predicted by Approach-1 and Approach-2 are compared. Further, the results of springback of adhesive-bonded sheets are discussed.

3.1 Comparison of Limit Strains of Base Materials Constituting Adhesive-Bonded Sheets Through Approach-1 and Approach-2

Figure 5a, b shows the comparison of true stress–strain behaviour of cured epoxy adhesive obtained by Approach-1 and Approach-2 with different hardener/resin ratios. It is observed that the true strain increases with increase in hardener/resin ratio during testing of adhesive samples. The same trend is seen in the tensile behaviour of adhesive layer derived from the adhesive-bonded blanks through Approach-2. There is about 2 MPa variation in true stress and about 0.5 % variation in true strain at failure between Approach-1 and Approach-2.

Figure 6a–d shows the comparison of limit strains in base materials constituting adhesive-bonded blanks with different hardener/resin ratios between experiments and predictions during tensile test. Similarly, the limit strains during IPPS formability test are shown in Fig. 7a–d. During simulations of tensile and IPPS forming

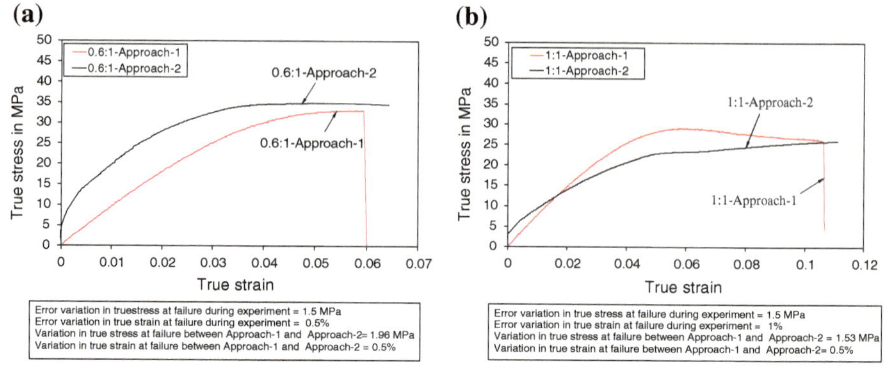

Fig. 5 Comparison of true stress–strain behaviour of epoxy adhesive obtained by Approach-1 and Approach-2 with different hardener/resin ratios: **a** H/R = 0.6:1, and **b** H/R = 1:1

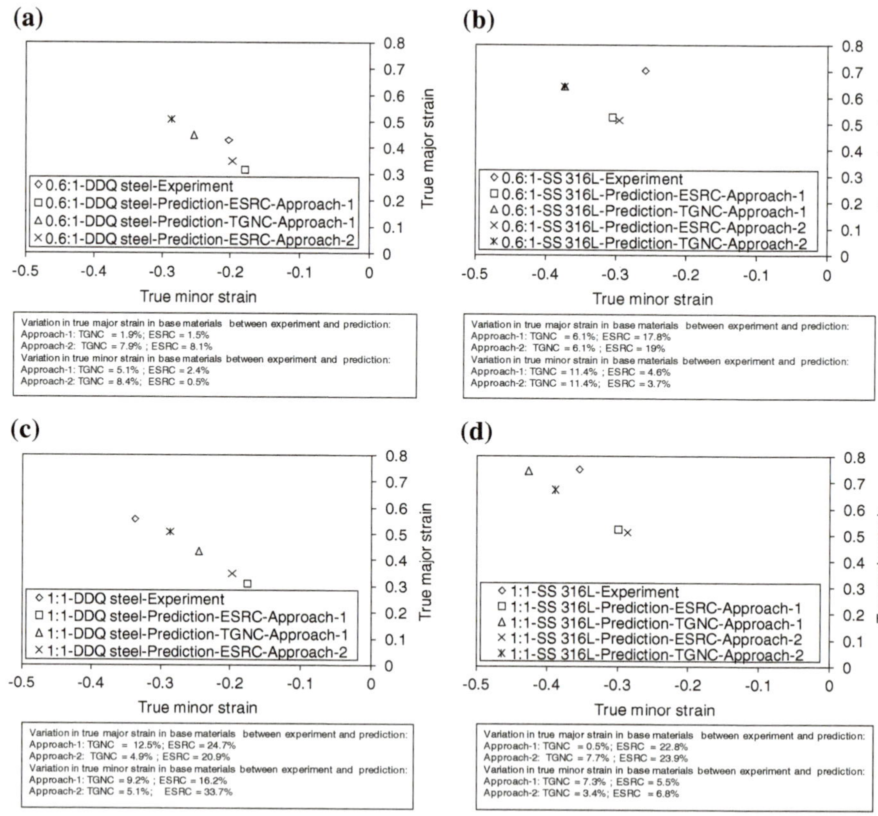

Fig. 6 Comparison of limit strain results of adhesive-bonded sheets with different H/R ratios between experiments and predictions during tensile test: **a** H/R = 0.6:1-DDQ steel, **b** H/R = 0.6:1-SS 316L, **c** H/R = 1:1-DDQ steel, and **d** H/R = 1:1-SS 316L

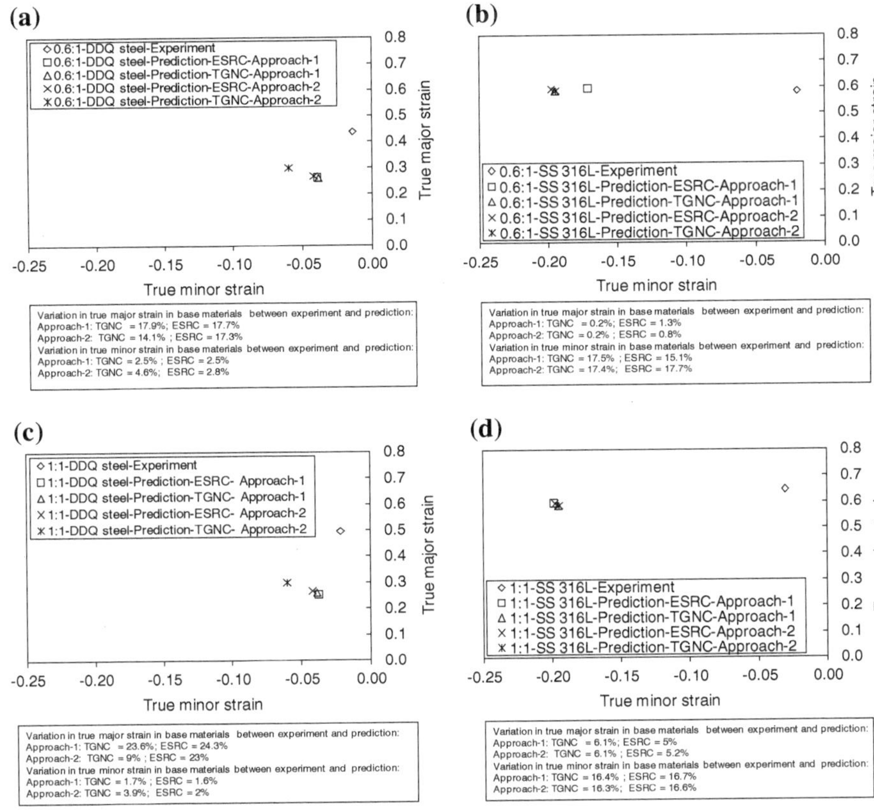

Fig. 7 Comparison of limit strain results of adhesive-bonded blanks with different *H*/*R* ratios between experiments and predictions during IPPS formability test: **a** *H*/*R* = 0.6:1-DDQ steel, **b** *H*/*R* = 0.6:1-SS 316L, **c** *H*/*R* = 1:1-DDQ steel, and **d** *H*/*R* = 1:1-SS 316L

tests, the limit strains evaluated by the necking criteria TGNC and ESRC through Approach-1 and Approach-2 are compared. It is observed that the limit strains evaluated through Approach-1 and Approach-2 during predictions exhibit almost same results. These results reveal that either Approach-1 or Approach-2 could be followed to evaluate adhesive mechanical properties so that limit strains of base materials can be predicted. It is noted that the limit strains evaluated by TGNC predictions in both the approaches show less difference with respect to experimental limit strains. It is also noted that the limit strains predicted by TGNC through both the approaches are close to each other in most of the cases. Similarly, in the case of ESRC predictions, the limit strains from Approach-1 and Approach-2 are very close to each other. Particularly, in the case of SS 316L during IPPS forming test, the limit strains from both necking criteria are almost same and there is a slight change in the strain paths.

3.2 Influence of Adhesive on SpringBack of Adhesive-Bonded Sheets

In this section, the influence of adhesive on the springback of adhesive-bonded blanks is discussed with the influence of different *H*/*R* ratios, Shims, which are introduced in between two base sheets to maintain uniform thickness of adhesive layer, and the metallic wire reinforcement.

3.2.1 Influence of Hardener/Resin Ratios of Epoxy Adhesive on SpringBack

Figure 8a, b shows the influence of *H*/*R* ratio of epoxy adhesive on the springback angle of adhesive-bonded sheets in the presence of shims that were initially glued with the bottom sheet and upper sheet to maintain uniform thickness of adhesive layer during fabrication of bonded blanks. Here, the base sheets constituting adhesive-bonded blanks are termed as upper sheet and bottom sheet with respect to the punch stroke towards die. In the case of un-bonded base material (DDQ steel sheet), the springback angle was found to be about 74°. During V-bending of bonded blanks, there was delamination of base materials at the location of shim (at the extreme ends) in response to the base sheet with which the shims were initially glued. The delamination occurs between the upper sheet and the shim during V-bending, which reveals that the joint strength between thermoplastic adhesive used for gluing shim and bottom sheet is more than the joint strength between epoxy adhesive and upper sheet. Because of the delamination, the upper sheet bends separately as compared to bottom sheet and adhesive. Hence, after load removal, the springback of the upper sheet is very different from that of bottom sheet and adhesive. In fact, it is expected that the upper sheet will bend like base material, but it is seen from Fig. 8a that the springback of upper sheet (i) is much lesser than that of bottom sheet and adhesive and (ii) is almost same as that of die angle of 60° indicating no springback. Even then, the springback increases with the increase in *H*/*R* ratio for both the upper and bottom sheets. There is cohesive (adhesive) failure at the point of application of load, mainly because of the limited ductility of adhesive at 0.6:1 *H*/*R* ratio of epoxy adhesive, while no adhesive failure is witnessed with increase in *H*/*R* ratio, specifically at 1:1 *H*/*R* ratio of epoxy adhesive.

In the case when the shim is initially glued with upper sheet during bending, the delamination occurred at the bottom side between the adhesive and the bottom sheet. The upper sheet and adhesive deformed together, while the bottom sheet deformed separately. As a result, two different springback angles were expected. But the springback of detached bottom sheet was restricted by the die and punch, and the upper sheet, resulting in almost the same springback behaviour as that of upper sheet/adhesive. Even in this case, the springback increases with increase in *H*/*R* ratio of epoxy adhesive (Fig. 8b). Moreover, in the first case (Fig. 8a), the springback of DDQ un-bonded base material is in between that of bottom sheet

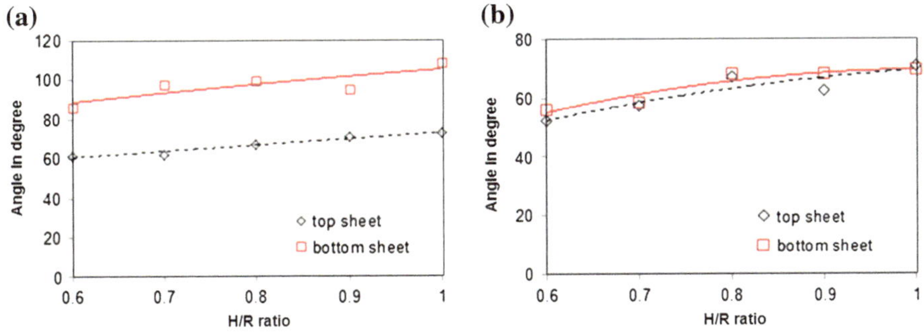

Fig. 8 Influence of *H*/*R* ratio of epoxy adhesive on the springback of bonded sheets when shim is initially glued with **a** the *bottom sheet* (average deviation: 4°) and **b** the *top sheet* (average deviation: 1.5°)

and upper sheet constituting adhesive-bonded sheets. While in the second case (Fig. 8b), the adhesive-bonded sheets exhibit a reduced springback (55–69°) as compared to DDQ un-bonded sheet, which could be due to the combined effect of increase in overall thickness and decrease in strength of the adhesive sample as compared to base material. Since there is not much difference in the springback angle when the shims are initially glued with the top sheets, the present study has been restricted without investigating further probable cases like gluing each shim in both sheets at alternative extremes. Figure 9 shows the influence of *H*/*R* ratio on the springback of adhesive-bonded sheets without shim. Due to presence of shims in the previous cases, there was a large variation in springback of upper sheet and bottom sheet. In order to determine the effect of only epoxy adhesive, the shims were removed and V-bending of bonded blanks was performed with different *H*/*R* ratios. It was observed that the delamination occurred between the bottom sheet and adhesive due to transverse shear, and there was no delamination between upper sheet and adhesive during testing. There was adhesive breakage at the point where the punch load was applied, which is due to the reduction in ductility of adhesive as compared to the base metal. The springback of both the upper sheet and lower sheet has slightly decreased (about 7°) with increase in *H*/*R* ratio of epoxy adhesive.

In this case, the springback of adhesive-bonded sheet is governed by change in Young's modulus, tensile strength, and ductility of adhesive, when *H*/*R* ratio is increased. It is observed from our earlier attempts (Satheeshkumar and Narayanan 2014, 2015) that the Young's modulus of adhesive decreases, tensile strength decreases, and ductility increases, with increase in *H*/*R* ratio of adhesive. The modifications in such mechanical properties influence the overall mechanical properties of the bonded sheets directly. It is seen from the available literature that with (i) decrease in elastic modulus, the springback increases; (ii) decrease in tensile strength, springback decreases, and (iii) increase in adhesive-bonded sheet ductility, springback increases because of more elastic recovery. The combined effect of all this is to reduce the springback of adhesive-bonded sheet slightly (Fig. 9), with increase in *H*/*R* ratio of epoxy adhesive.

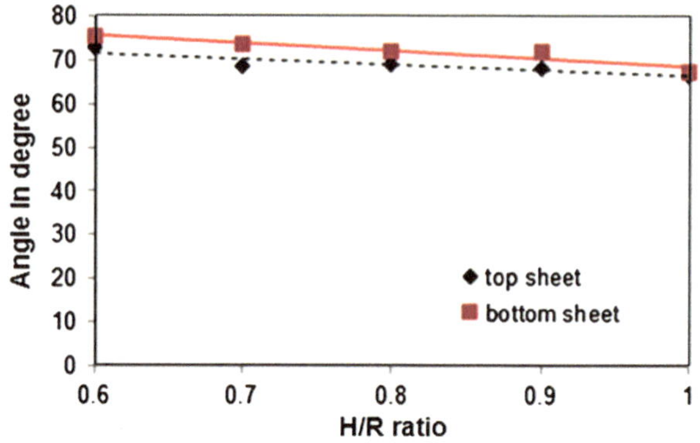

Fig. 9 Influence of *H*/*R* ratio on the springback of adhesive-bonded sheets without shim (average deviation: 2.3°)

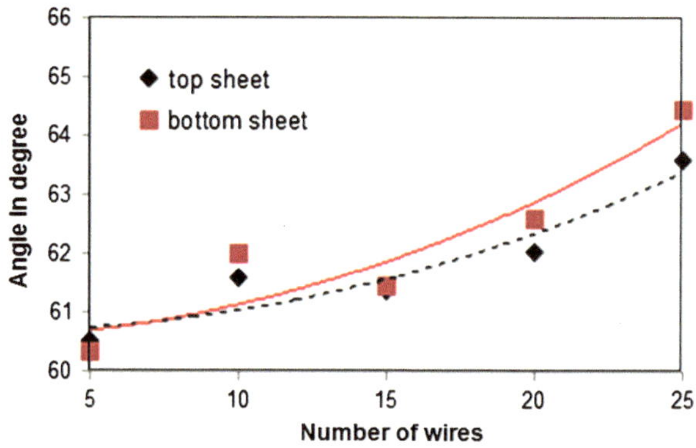

Fig. 10 Influence of number of wires on the springback of adhesive-bonded sheets at *H*/*R* = 1:1 (average deviation: 2.4°)

3.2.2 Influence of Wire Reinforcement on the SpringBack

Figure 10 shows the comparison of different level wire reinforced adhesive on the springback of bonded sheets. It is noticed that the springback increases (about 3°) with the increase in number of wire reinforcement in adhesive. This variation is negligible as compared to the springback experimental deviation among different trials. Initially, delamination of sheet occurs due to transverse shear deformation, but no adhesive failure was observed because of hardener-rich formulation of *H*/*R* ratio of 1:1, which has more ductility as compared to other lower hardener/resin ratios of adhesive. It is also observed that the maximum springback angle is about 64° in the case of wire reinforced adhesive-bonded sheets, which is slightly less than the unreinforced adhesive-bonded sheets that is about 67°.

4 Conclusions

From the present work, the following conclusions are drawn.

- It is suggested that the $\sigma - \varepsilon$ behaviour of adhesive obtained either from rule of mixtures approach or from experiments could be followed in the forming limit strain predictions of adhesive-bonded sheets. Limit strains predicted through these approaches are almost same.
- The limit strains evaluated by TGNC show less difference with respect to experimental limit strains, while following both the approaches. This suggests that prediction of forming limit strains of adhesive-bonded blanks by the necking criterion, TGNC, is more suitable than the necking criterion, ESRC. It is believed that the considerable difference is due to the absence of interface bonding between adhesive layer and base materials during simulation.
- The stress–strain behaviour evaluated from the rule of mixtures approach includes interfacial bonding, while it is not from experiments. Hence, if interfacial bonding is modelled during simulations, then the rule of mixtures approach would provide accurate formability results as compared to the experimentation approach, i.e. adhesive mechanical properties evaluated through experiments.
- It has been demonstrated from the present work that the springback of steel sheets can be controlled significantly with the help of adhesive, by changing the adhesive mechanical properties. Just by changing the hardener/resin ratio of adhesive, and by wire reinforcement of adhesive, the springback of bonded sheets can be drastically modified as compared to un-bonded DDQ steel sheet.
- The use of shim to maintain uniform adhesive thickness plays a vital role in deciding the springback of adhesive-bonded sheets, by changing the delamination location, i.e. between the upper sheet and adhesive, and between adhesive and lower sheet.
- The effect of *H*/*R* ratio controls the springback in the absence of shims also.
- The springback of bonded sheets decreases with number of wires as reinforcement of adhesive, but not significantly.
- It is suggested from the present study to use adhesives between two metal sheets so that the springback could be reduced and later the adhesive could be removed for using the remaining single sheet components for further applications.

References

Abdullah K, Wild PM, Jesweit JJ, Ghasempoor A (2001) Tensile testing for weld deformation properties in similar gage tailor welded blanks using the rule of mixtures. J Mater Process Technol 112:91–97

Aghchai AJ, Shakeri M, Dariani M (2008) Theoretical and experimental formability study of two-layer metallic sheet (Al1100/St12). Proc Inst Mech Eng Part B J Eng Manuf 222:1131–1138

Banabic D (2010) Sheet metal forming processes constitutive modelling and numerical simulation. Springer, Heidelberg

Czamocki P, Piekarski K (1986) Fracture strength of an adhesive-bonded joint. Int J Adhes Adhes 6:93–95

Kim YT, Lee MJ, Lee BC (2011) Simulation of adhesive joints using the superimposed finite element method and a cohesive zone model. Int J Adhes Adhes 31:357–1362

May M, Vob H, Hiermaier S (2014) Predictive modelling of damage and failure in adhesively bonded metallic joints using cohesive interface elements. Int J Adhes Adhes 49:7–17

Narasimhan K, Wagoner RH (1991) Finite element modelling simulation of in-plane forming limit diagrams of sheets containing finite defects. Metall Trans A 22:2655–2665

Pereira AM, Ferreira JM, Antunes FV, Bártolo PJ (2010) Analysis of manufacturing parameters on the shear strength of aluminium adhesive single-lap joints. J Mater Process Technol 210:610–617

Perton M, Blouin A, Monchalin JP (2011) Adhesive bond testing of carbon–epoxy composites by laser shockwave. J Phy D Appl Phys. doi:10.1088/0022-3727/44/3/034012

Ravi K, Ganesh Narayanan R (2012) In-plane testing behavior of adhesive-bonded steel sheets: influence of sheet surface roughness, adhesive thickness, and bonding width. J Test Eval 40 (2):97–106

Satheeshkumar V, Ganesh Narayanan R (2014) Investigation on the influence of adhesive properties on the formability of adhesive bonded steel sheets. Proc Inst Mech Eng Part C J Mech Eng Sci 228(3):405–425

Satheeshkumar V, Ganesh Narayanan R (2015) Prediction of formability of adhesive bonded steel sheets and experimental validation. Arch Civ Mech Eng 15:30–41

Sujit K, Date PP, Narasimhan K (1994) A new criterion to predict necking failure under biaxial stretching. J Mater Process Technol 45:583–588

Takiguchi M, Yoshida F (2001) Plastic bending adhesive bonding of sheet metals. J Mater Process Technol 113:743–748

Takiguchi M, Yoshida F (2003) Analysis of plastic bending of adhesive-bonded sheet metals taking account of viscoplasticity of adhesive. J Mater Process Technol 140:441–446

Uehara K, Sakurai M (2002) Bonding strength of adhesives and surface roughness of joined parts. J Mater Process Technol 127:178–181

Feasibility Study of Thermal Autofrettage Process

S.M. Kamal and Uday Shanker Dixit

Abstract Usually, autofrettage is achieved either by applying hydraulic pressure to the bore of the thick-walled cylinder or by pushing an oversized mandrel through the bore to deform the cylinder plastically. This produces the compressive residual stresses on the inner side of the cylinder, enabling the cylinder to withstand higher working pressure. There are certain difficulties in these methods of autofrettage. This work proposes a novel method for achieving autofrettage that involves creating temperature gradient in the wall of the cylinder. The proposed process is analyzed theoretically considering strain hardening. The simulation results and preliminary experiments indicate an encouraging trend for the feasibility of the proposed process.

Keywords Autofrettage · Thick-walled cylinder · Thermal residual stresses · Strain hardening · Elastic–plastic interface

1 Introduction

Autofrettage is a metal fabrication technique which applies to components subjected to very high pressure such as gun barrels, rocket shells, high pressure piping, and high pressure containers. The process is employed to induce the compressive residual stresses at and around the inner wall of the thick-walled cylindrical vessel, thereby increasing the pressure-carrying capacity of the vessel. There are various ways of achieving this. The most common ways of achieving autofrettage are by applying ultra-high hydraulic internal pressure known as the hydraulic autofrettage and by mechanical swaging of the cylinder known as the swage autofrettage.

S.M. Kamal (✉) · U.S. Dixit
Department of Mechanical Engineering, Indian Institute of Technology Guwahati, Guwahati 781039, India
e-mail: k.seikh@iitg.ac.in

U.S. Dixit
e-mail: uday@iitg.ac.in

R.G. Narayanan and U.S. Dixit (eds.), *Advances in Material Forming and Joining*, Topics in Mining, Metallurgy and Materials Engineering,
DOI 10.1007/978-81-322-2355-9_5

In hydraulic autofrettage, a thick-walled metallic cylinder is subjected to an ultra-high internal hydraulic pressure at the bore. In the process, the material at and around the inner radius of the cylinder is stretched beyond its elastic limit causing plastic deformation and the material at the outer portion is subjected to elastic deformation. Later, when the internal pressure is released, the outer portion of the material which has undergone elastic deformation will attempt to return to its original state while the material at and around the inner radius which has been deformed plastically will attempt to remain in the deformed state. This results in a residual compressive stress at and around the inner radius of the cylinder, and thus when the cylinder is repressurized, the magnitude of the maximum stress reduces significantly enhancing its pressure-carrying capacity. As the residual stresses are self-equilibrating, the portion of the vessel from outer surface to some intermediate radius of the cylinders will have tensile residual stresses, but this does not pose any problem as the working pressure produces relative smaller hoop stresses in the outer portion of the vessel. The process was first suggested by Jacob, a French artillery officer, for pre stressing monobloc gun barrels in 1907 (Malik and Khushnood 2003).

Davidson et al. (1962) developed a new approach for achieving autofrettage which is called swage autofrettage, in which an oversized mandrel is forced through the inside diameter of a cylinder to plastically deform the inner wall and some portion beneath it. This process of achieving autofrettage is less complex than hydraulic autofrettage. In hydraulic autofrettage, much care has to be taken in applying ultra-high hydraulic pressure as it is to be controlled accurately to achieve the desired deformation. The process is slow, expensive, and at times dangerous. In swage autofrettage, the hydraulic pressure required to push the mandrel is much less than the conventional hydraulic autofrettage; however, the preparation of mandrel and swaging assembly makes the process complex. Another disadvantage associated with the swage autofrettage is that the axial force required for driving the ram increases with the increase in size of the mandrel (Iremonger and Kalsi 2003). The autofrettage can also be achieved by detonating an explosive charge inside the vessel, which is called explosive autofrettage (Mote et al. 1971; Ezra 1973). The modeling of this process involves dynamic elasto-plasticity and control of the process requires skill due to the involvement of explosive.

Residual stresses can also be produced due to thermal gradient in the wall of the cylinder. However, there is hardly any reference in the literature that analyzes this effect. Hussain et al. (1980) observed that stress distribution in a partially autofrettaged tube can be analyzed by a simulated thermal loading, such that thermo-elastic stresses produced due to simulated loading will be same as the stress distribution of partially autofrettaged tube. They did not study the thermo-elastic–plastic stresses, which might have indicated the feasibility of thermal autofrettage. Thus, there is no experimental or theoretical study on thermal autofrettage in open literature. Thermal autofrettage is a potential process offering many advantages compared to hydraulic and swage autofrettage, such as absence of high pressure and moving parts. This chapter analyzes thermal autofrettage for exploring the feasibility of its industrial exploitation.

2 Proposed Thermal Autofrettage Process

The proposed method of autofrettage is based on the stress field generated due to the temperature gradient developed between outer and inner walls of the cylinder or hollow disk. The schematic diagram of the method is shown in Fig. 1. In order to understand the mechanics of thermal autofrettage, plane stress assumption is employed in Sect. 3. Plane stress assumption is appropriate for a thin circular hollow disk. It may provide reasonable solution for short cylinders as well.

An open-ended thin hollow disk or short cylinder is considered having inner radius a and outer radius b. The disk is heated externally and flow of cold fluid is continuously regulated such that outer wall of the disk is subjected to a temperature T_b and inner wall is subjected to a temperature T_a. At a sufficiently large temperature difference between the outer and inner walls of the disk, the inner portion of the disk always yields first and it becomes plastic up to some intermediate radius, while the outer portion of the disk remains in the elastic state if the temperature difference is not large enough to cause the yielding of outer wall of the disk. When the temperature difference is increased beyond certain limit, the outer wall of the disk begins to yield and the plastic deformation of both the inner and outer portions of the disk undergoes simultaneously leaving the intermediate portion of the disk in the elastic state. When the entire cylinder is cooled to room temperature, compressive residual stresses are produced in the cylinder.

The proposed method is very simple and easy to handle compared to the existing methods of autofrettage. It requires no pressure and is expected to be inexpensive due to elimination of costly hydraulic power pack and/or explosives. Moreover, due to elimination of moving parts, maintenance load is also reduced. In the following section, thermal stress analysis of the process is carried out.

3 Thermal Stress Analysis in the Disk

Consider an open-ended thin circular hollow disk ($\sigma_z = 0$) of homogeneous material with inner radius a and outer radius b as shown in Fig. 2. The disk is subjected to Dirichlet temperature boundary conditions. The outer wall of the disk is maintained

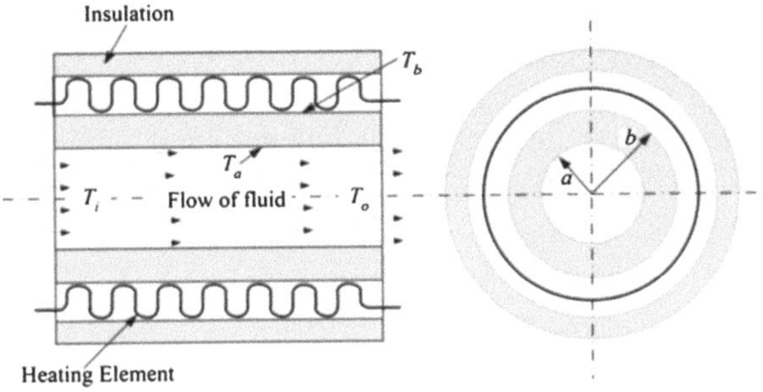

Fig. 1 A schematic diagram of thermal autofrettage process

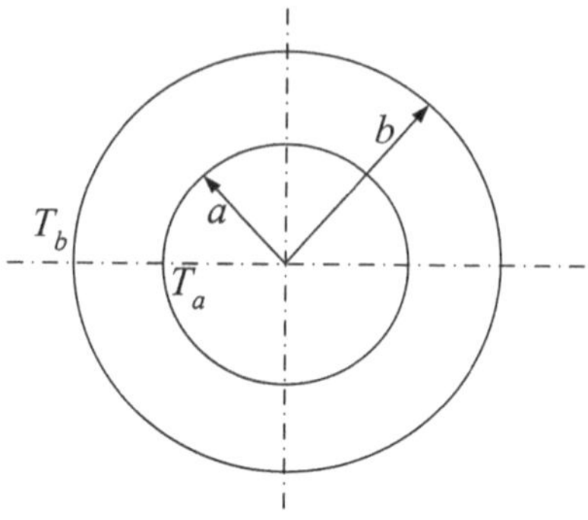

Fig. 2 Cross section of the disk with Dirichlet temperature boundary conditions

at temperature T_b and the inner wall is maintained at temperature T_a such that $T_b > T_a$. The steady-state temperature distribution in the disk is given by Chakrabarty (2006)

$$T = T_b + (T_a - T_b)\frac{\ln\left(\frac{b}{r}\right)}{\ln\left(\frac{b}{a}\right)}. \tag{1}$$

For a low temperature gradient ($T_b - T_a$), which is not sufficient to cause the yielding of the wall of the disk, the entire wall thickness ($b - a$) is in the elastic state. When ($T_b - T_a$) is increased further, at a certain threshold temperature gradient, the inner wall of the disk yields first and upon crossing this temperature gradient the wall deforms plastically to some intermediate radial position. When the temperature gradient ($T_b - T_a$) reaches a second threshold, the outer wall begins to yield simultaneously with the inner wall. Upon crossing the second threshold, the cylinder wall gets divided into an inner plastic, intermediate elastic, and outer plastic zone. In this work, it is assumed that the material yields as per Tresca criterion. The yielding behavior of the most metals is close to von Mises criterion. However, it is not uncommon to correct the residual stresses obtained using Tresca criterion by multiplying them by a factor of 1.15 to correspond to von Mises case (Parker 1980). In the following sections, solution for thermal stresses for elastic and different phases of elasto-plastic deformation are discussed.

3.1 Thermo-elastic Solution

With zero temperature as a reference and using generalized Hook's law, the radial and hoop strains under plane stress condition for a purely elastic body at temperature T are given as (Noda et al. 2003)

$$\varepsilon_r = \frac{1}{E}(\sigma_r - v\sigma_\theta) + \alpha T, \tag{2}$$

$$\varepsilon_\theta = \frac{1}{E}(\sigma_\theta - v\sigma_r) + \alpha T, \tag{3}$$

where σ_r is the radial stress, σ_θ is the hoop stress, v is the Poisson's ratio, E is the Young's modulus of elasticity, and α is the coefficient of thermal expansion. For radial displacement u as a function of r, the radial and hoop strains are given by Chakrabarty (2006)

$$\varepsilon_r = \frac{\mathrm{d}u}{\mathrm{d}r}, \quad \varepsilon_\theta = \frac{u}{r}. \tag{4}$$

Hence, the strain compatibility in cylindrical polar coordinate system can be written as

$$\varepsilon_r - \varepsilon_\theta = r\frac{\mathrm{d}\varepsilon_\theta}{\mathrm{d}r}. \tag{5}$$

The stresses are governed by the equilibrium equation given by Chakrabarty (2006)

$$\sigma_\theta - \sigma_r = r\frac{\mathrm{d}\sigma_r}{\mathrm{d}r}. \tag{6}$$

Using Eqs. (2) and (3) in Eq. (5) and solving it along with Eq. (6) provides the following differential equation:

$$\frac{\mathrm{d}}{\mathrm{d}r}\left\{\frac{1}{r}\frac{\mathrm{d}}{\mathrm{d}r}\left(r^2\sigma_r\right)\right\} + E\alpha\frac{\mathrm{d}T}{\mathrm{d}r} = 0, \tag{7}$$

where T is given by Eq. (1). The solution of Eq. (7) gives

$$\sigma_r = A - \frac{B}{r^2} + \frac{E\alpha(T_b - T_a)}{2\ln\left(\frac{b}{a}\right)}\left\{\frac{1}{2} - \ln\left(\frac{r}{a}\right)\right\}, \tag{8}$$

where A and B are constants. Using Eq. (8) in Eq. (6), the solution for hoop stress is given by

$$\sigma_\theta = A + \frac{B}{r^2} - \frac{E\alpha(T_b - T_a)}{2\ln\left(\frac{b}{a}\right)}\left\{\frac{1}{2} + \ln\left(\frac{r}{a}\right)\right\}. \tag{9}$$

Using the boundary condition of vanishing radial stress at inner and outer radii, i.e., at $r = a$, and $r = b$, $\sigma_r = 0$, in Eq. (8) yields

$$A = -\frac{1}{2}\frac{E\alpha(T_b - T_a)}{2\ln\left(\frac{b}{a}\right)} - \left(\frac{b^2}{a^2 - b^2}\right)\frac{E\alpha(T_b - T_a)}{2}, \tag{10}$$

$$B = -\left(\frac{a^2b^2}{a^2 - b^2}\right)\frac{E\alpha(T_b - T_a)}{2}. \tag{11}$$

Inserting Eqs. (10) and (11) in Eqs. (8) and (9), the elastic radial and hoop stresses are obtained as (Noda et al. 2003)

$$\sigma_r = \frac{E\alpha}{2}(T_b - T_a)\left\{-\frac{\ln\left(\frac{r}{a}\right)}{\ln\left(\frac{b}{a}\right)} + \left(1 - \frac{a^2}{r^2}\right)\frac{b^2}{b^2 - a^2}\right\}, \tag{12}$$

$$\sigma_\theta = \frac{E\alpha}{2}(T_b - T_a)\left\{-\frac{1 + \ln\left(\frac{r}{a}\right)}{\ln\left(\frac{b}{a}\right)} + \left(1 + \frac{a^2}{r^2}\right)\frac{b^2}{b^2 - a^2}\right\}. \tag{13}$$

3.2 The Thermo-elasto-plastic Behavior

A typical distribution of elastic stresses as a function of radial position given by Eqs. (12) and (13) is shown in Fig. 3 for an aluminum disk with a = 10 mm, b = 20 mm, E = 69 GPa, $(T_b - T_a)$ = 40 °C, and $\alpha = 22.2 \times 10^{-6}$/°C. Figure 3 suggests that the thermal stresses are distributed according to the inequality $\sigma_\theta > \sigma_r > \sigma_z$ at and around inner radius of the disk. At and around the outer radius, the stresses are distributed such that $\sigma_r > \sigma_z > \sigma_\theta$. The difference $|\sigma_\theta - \sigma_z|$ at inner radius is always greater than the difference $|\sigma_r - \sigma_\theta|$ at outer radius. Hence, for

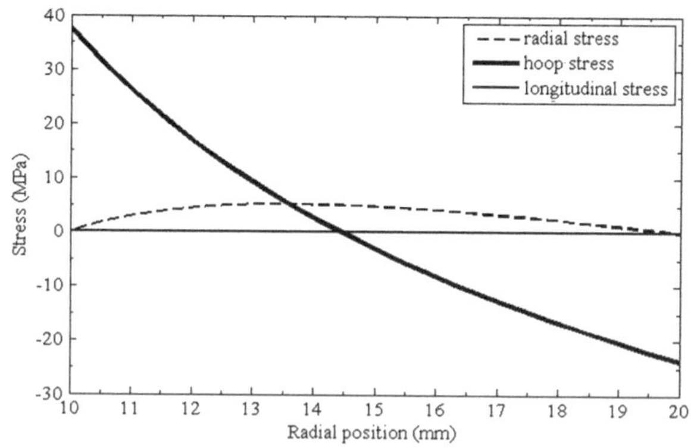

Fig. 3 Elastic thermal stress distributions in an aluminum hollow disk

increasing thermal gradient, the yielding of the disk begins first at the inner radius of the disk as per Tresca yield criterion. For initiation of yielding at the inner radius, $r = a$ Tresca yield criterion provides

$$(\sigma_\theta - \sigma_z)\big|_{r=a} = k_1 \sigma_Y, \tag{14}$$

where σ_Y is the yield stress and k_1 is a sign factor that can be +1 or −1 depending on whether $T_b > T_a$ or $T_b < T_a$. As $\sigma_z = 0$, evaluating σ_θ at $r = a$ from Eq. (13) and substituting it into Eq. (14), the temperature difference required for initial yielding at inner radius is given by

$$(T_b - T_a)_{Y_i} = \frac{2k_1\sigma_Y}{E\alpha\left\{\frac{\ln\left(\frac{b}{a}\right)-1}{\ln\left(\frac{b}{a}\right)} + \left(\frac{a^2}{b^2-a^2}\right)\left(\frac{b^2}{a^2}+1\right)\right\}}. \tag{15}$$

Equation (15) shows that for materials having large E and α and low yield stress, a small temperature difference can cause yielding at the inner radius.

If the temperature difference $(T_b - T_a)$ exceeds the temperature difference required for initial yielding at the inner radius given by Eq. (15), but is lesser than the temperature difference required for yielding the outer wall, the first phase of elasto-plastic deformation takes place. During the first phase of elasto-plastic deformation, a portion of the inner wall of the disk becomes plastic up to a radius c and beyond c the outer wall remains in the elastic state. Thus, an inner plastic zone $a < r < c$ and an outer elastic zone $c < r < b$ are formed within the disk. When the temperature difference keeps on increasing further, at some temperature difference, the yielding commences at the outer radius of the disk. Upon exceeding this temperature difference, the wall of the disk undergoes second phase of elasto-plastic deformation forming an inner plastic zone $a < r < c$ and an outer plastic zone $d < r < b$ separated by an intermediate elastic zone $c < r < d$. The second phase of elasto-plastic deformation avails in the wall of the disk as long as the temperature difference is not large enough to cause the plastic deformation of the whole wall. The different zones developed within the wall of the disk as a result of elasto-plastic deformation in general are shown schematically in Fig. 4. In the following subsections, the analysis of different zones during the first and second phase of elasto-plastic deformation is carried out.

3.2.1 Analysis of the First Phase of Elasto-plastic Deformation

During the first phase of elasto-plastic deformation, an inner plastic zone propagates outwards to a radius c. Thus, the inner plastic zone $a < r < c$ is surrounded by an outer elastic zone $c < r < b$ within the wall of the disk. The radius c is known as the radius of elastic–plastic interface. The inner plastic and outer elastic zones are analyzed as follows:

Elastic Zone: $c < r < b$

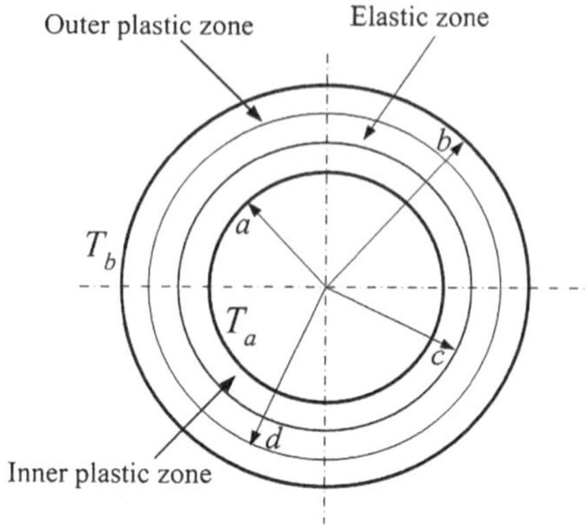

Fig. 4 The elastic and plastic zones in the hollow disk

For the outer elastic zone $c < r < b$, the solution for thermal stresses is still given by Eqs. (8) and (9). Using boundary condition of vanishing radial stress at the outer radius and $(\sigma_\theta)_{r=c} = k_1\sigma_Y$ in Eqs. (8) and (9), the constants A and B are evaluated and the expressions for radial and hoop stresses are given by

$$\sigma_r = \frac{E\alpha(T_b - T_a)}{2\ln\left(\frac{b}{a}\right)}\left[\ln\left(\frac{b}{r}\right) + \left(\frac{c^2}{b^2+c^2}\right)\left(1 - \frac{b^2}{r^2}\right)\left\{1 - \ln\left(\frac{b}{c}\right)\right\}\right] + \frac{c^2 k_1 \sigma_Y}{b^2+c^2}\left(1 - \frac{b^2}{r^2}\right), \tag{16}$$

$$\sigma_\theta = \frac{E\alpha(T_b - T_a)}{2\ln\left(\frac{b}{a}\right)}\left[\ln\left(\frac{b}{r}\right) - 1 + \left(\frac{c^2}{b^2+c^2}\right)\left(1 + \frac{b^2}{r^2}\right)\left\{1 - \ln\left(\frac{b}{c}\right)\right\}\right] + \frac{c^2 k_1 \sigma_Y}{b^2+c^2}\left(1 + \frac{b^2}{r^2}\right). \tag{17}$$

The radial displacement component, u, in the outer elastic zone is obtained using Eqs. (16) and (17) in Eq. (3) and second relation of Eq. (4) as

$$u = \frac{\alpha(T_b - T_a)}{2\ln\left(\frac{b}{a}\right)}\left[\begin{array}{l}(1-\nu)r\,\ln\left(\frac{b}{r}\right) - r + (1-\nu)r\left(\frac{c^2}{b^2+c^2}\right)\left\{1 - \ln\left(\frac{b}{c}\right)\right\} \\ +\left(\frac{c^2}{b^2+c^2}\right)\left\{1 - \ln\left(\frac{b}{c}\right)\right\}\left\{(1-\nu)r + (1+\nu)\frac{b^2}{r}\right\}\end{array}\right] + \frac{c^2 k_1 \sigma_Y}{E(b^2+c^2)}\left\{(1-\nu)r + (1+\nu)\frac{b^2}{r}\right\} + r\alpha T. \tag{18}$$

Plastic Zone: $a < r < c$

For the plastic zone ($a < r < c$), the effect of strain hardening of the material needs to be considered. When the material strain hardens, Tresca yield criterion in the inner plastic zone becomes

$$\sigma_\theta = k_1 \sigma_{\mathrm{eq}}, \tag{19}$$

where σ_{eq} ($>\sigma_Y$) is the equivalent stress in uniaxial tension or compression. It is assumed that the hardening of the disk material follows the Ludwik's hardening law (Dixit and Dixit 2008) given by

$$\sigma_{\mathrm{eq}} = \sigma_Y + K\left(\varepsilon_{\mathrm{eq}}^p\right)^n, \tag{20}$$

where K is the hardening coefficient and n is the strain hardening exponent. Using Eq. (19) in Eq. (6)

$$\frac{\mathrm{d}}{\mathrm{d}r}(r\sigma_r) = k_1\left\{\sigma_Y + K\left(\varepsilon_{\mathrm{eq}}^p\right)^n\right\}. \tag{21}$$

Integrating Eq. (21) in the plastic zone from r to $r = c$ and using the boundary condition, σ_r at $r = c$ from elastic zone (Eq. 16), the solution for radial thermal stress in the plastic zone is obtained as

$$\begin{aligned}\sigma_r = {} & k_1\sigma_Y\left\{1 - \left(\frac{2b^2}{b^2+c^2}\right)\left(\frac{c}{r}\right)\right\} \\ & + \left(\frac{c}{r}\right)\frac{E\alpha(T_b - T_a)}{2\ln\left(\frac{b}{a}\right)}\left[\ln\left(\frac{b}{c}\right) + \left(\frac{c^2}{b^2+c^2}\right)\left(1 - \frac{b^2}{c^2}\right)\left\{1 - \ln\left(\frac{b}{c}\right)\right\}\right] \\ & - k_1\frac{K}{r}\int_r^c \left(\varepsilon_{\mathrm{eq}}^p\right)^n \mathrm{d}r_1,\end{aligned} \tag{22}$$

where r_1 is a dummy variable. The thermal hoop stress in the plastic zone is given by Eq. (19).

In the plastic zone, the total strain is composed of elastic and plastic strains. Hence,

$$\varepsilon_r = \frac{\mathrm{d}u}{\mathrm{d}r} = \varepsilon_r^e + \varepsilon_r^p, \quad \varepsilon_\theta = \frac{u}{r} = \varepsilon_\theta^e + \varepsilon_\theta^p. \tag{23}$$

The elastic components ε_θ^e and ε_r^e are given by Eqs. (2) and (3). In view of Tresca's associated flow rule (Chakrabarty 2006), $\mathrm{d}\varepsilon_r^p = 0$, $\mathrm{d}\varepsilon_\theta^p = -\mathrm{d}\varepsilon_z^p$. Hence, ε_r is entirely elastic. Substituting Eq. (23) in Eqs. (2) and (3),

$$\sigma_r = \frac{E}{1-\nu^2}\left\{\frac{\mathrm{d}u}{\mathrm{d}r} + \nu\left(\frac{u}{r} - \varepsilon_\theta^p\right) - \alpha T(1+\nu)\right\}, \tag{24}$$

$$\sigma_\theta = \frac{E}{1-\nu^2}\left\{\left(\frac{u}{r} - \varepsilon_\theta^p\right) + \nu\frac{\mathrm{d}u}{\mathrm{d}r} - \alpha T(1+\nu)\right\}. \tag{25}$$

Inserting Eqs. (24) and (25) in Eq. (6), the following differential equation is obtained:

$$\frac{\mathrm{d}}{\mathrm{d}r}\left\{\frac{1}{r}\frac{\mathrm{d}}{\mathrm{d}r}(ur)\right\} = \nu\frac{\mathrm{d}\varepsilon_\theta^p}{\mathrm{d}r} + \alpha(1+\nu)\frac{\mathrm{d}T}{\mathrm{d}r} - \left(\frac{1-\nu}{r}\right)\varepsilon_\theta^p. \tag{26}$$

For the plastic zone, the total strain compatibility condition (Eq. 5) may be written as

$$r\frac{\mathrm{d}}{\mathrm{d}r}\left(\varepsilon_\theta^e + \varepsilon_\theta^p\right) = \left(\varepsilon_r^e - \varepsilon_\theta^e\right) - \varepsilon_\theta^p. \tag{27}$$

Putting ε_r^e and ε_θ^e from Eqs. (2) and (3) the following differential equation is obtained:

$$\frac{1}{E}\frac{\mathrm{d}\sigma_\theta}{\mathrm{d}r} + \frac{1}{E}\frac{\mathrm{d}\sigma_r}{\mathrm{d}r} + \alpha\frac{\mathrm{d}T}{\mathrm{d}r} + \frac{\mathrm{d}\varepsilon_\theta^p}{\mathrm{d}r} = -\frac{\varepsilon_\theta^p}{r}. \tag{28}$$

Substituting ε_θ^p/r from Eq. (32) in Eq. (30) and then integrating:

$$\frac{1}{r}\frac{\mathrm{d}}{\mathrm{d}r}(ur) = \frac{1-\nu}{E}(\sigma_\theta + \sigma_r) + 2\alpha\left\{T_a + (T_b - T_a)\frac{\ln\left(\frac{r}{a}\right)}{\ln\left(\frac{b}{a}\right)}\right\} + \varepsilon_\theta^p + C, \tag{29}$$

where C is integration constant. To determine the constant C in Eq. (29), the right- and left-hand side expressions are evaluated at $r = c$ by taking the corresponding values from the elastic zone. Equations (16) and (17) provide σ_r and σ_θ at $r = c$ and Eq. (18) provides

$$\begin{aligned}\left.\frac{\mathrm{d}}{\mathrm{d}r}(ur)\right|_{r=c} &= \frac{\alpha(T_b - T_a)}{2\ln\left(\frac{b}{a}\right)}\left[\begin{array}{l}2(1-\nu)c\ln\left(\frac{b}{c}\right) - c(1-\nu)\\ +2(1-\nu)\left(\frac{c^3}{b^2+c^2}\right)\left\{1-\ln\left(\frac{b}{c}\right)\right\} + 4c\ln\left(\frac{c}{a}\right)\end{array}\right]\\ &\quad + \frac{2(1-\nu)}{E}\left(\frac{c^3 k_1 \sigma_Y}{b^2+c^2}\right) + 2cT_a\alpha.\end{aligned} \tag{30}$$

Also,

$$\varepsilon_\theta^p\big|_{r=c} = 0. \tag{31}$$

Insertion of all expressions in Eq. (29) provides $C = 0$. Substituting C in Eq. (29) and integrating it in the plastic domain from r to $r = c$ gives

$$u|_{r=c}c - ur = \left(\frac{1-\nu}{E}\right)\left(c^2\sigma_r|_{r=c} - r^2\sigma_r\right) + \alpha T_a\left(c^2 - r^2\right) + \frac{\alpha(T_b - T_a)}{\ln\left(\frac{b}{a}\right)}\left[\left\{c^2\ln\left(\frac{c}{a}\right) - r^2\ln\left(\frac{r}{a}\right)\right\} - \frac{1}{2}\left(c^2 - r^2\right)\right] + \int_r^c r_1\varepsilon_\theta^p \mathrm{d}r_1 \tag{32}$$

Evaluating $u|_{r=c}$ from Eq. (18) and $\sigma_r|_{r=c}$ from Eq. (16) and then inserting these into Eq. (32) provide total hoop strain component as

$$\frac{u}{r} = \frac{\alpha(T_b - T_a)}{2\ln\left(\frac{b}{a}\right)}\left[\left\{\frac{2c^2b^2}{r^2(b^2+c^2)}\right\}\left\{1 - \ln\left(\frac{b}{c}\right)\right\} + 2\ln\left(\frac{r}{a}\right) - 1\right] + \frac{2}{E}\frac{k_1c^2b^2\sigma_Y}{r^2(b^2+c^2)} + \frac{1-\nu}{E}\sigma_r + \alpha T_a - \frac{1}{r^2}\int_r^c r_1\varepsilon_\theta^p \mathrm{d}r_1. \tag{33}$$

The plastic hoop strain field ε_θ^p is given by

$$\varepsilon_\theta^p = \frac{u}{r} - \varepsilon_\theta^e. \tag{34}$$

Inserting u/r, from Eq. (33), ε_θ^e from Eq. (3) and using Eqs. (19) and (22), the plastic hoop strain component is obtained as

$$\varepsilon_\theta^p = \frac{\alpha(T_b - T_a)}{2\ln\left(\frac{b}{a}\right)}\left[\begin{array}{l}\left\{\frac{2c^2b^2}{r^2(b^2+c^2)}\right\}\left\{1 - \ln\left(\frac{b}{c}\right)\right\} - 1 + \frac{c}{r}\ln\left(\frac{b}{c}\right) \\ + \left\{\frac{c^3}{r(b^2+c^2)}\right\}\left(1 - \frac{b^2}{c^2}\right)\left\{1 - \ln\left(\frac{b}{c}\right)\right\}\end{array}\right] + \frac{2}{E}\frac{k_1c^2b^2\sigma_Y}{r^2(b^2+c^2)} + \frac{k_1c^3\sigma_Y}{Er(b^2+c^2)}\left(1 - \frac{b^2}{c^2}\right) - \frac{k_1\sigma_Y}{E}\left(\frac{c}{r}\right) - k_1\frac{K}{Er}\int_r^c \left(\varepsilon_{\mathrm{eq}}^p\right)^n \mathrm{d}r_1 - \frac{1}{r^2}\int_r^c r_1\varepsilon_\theta^p \mathrm{d}r_1 - k_1\frac{K}{E}\left(\varepsilon_{\mathrm{eq}}^p\right)^n. \tag{35}$$

The equivalent plastic strain field $\varepsilon_{\mathrm{eq}}^{p}$ for the present case is given as

$$\varepsilon_{\mathrm{eq}}^{p}=\frac{2}{\sqrt{3}}\varepsilon_{\theta}^{p}. \tag{36}$$

Thus, Eq. (35) becomes

$$\begin{aligned}\varepsilon_{\theta}^{p}=&\frac{\alpha(T_b-T_a)}{2\ln\left(\frac{b}{a}\right)}\left[\begin{array}{l}\left\{\frac{2c^2b^2}{r^2(b^2+c^2)}\right\}\left\{1-\ln\left(\frac{b}{c}\right)\right\}-1+\frac{c}{r}\ln\left(\frac{b}{c}\right)\\+\left\{\frac{c^3}{r(b^2+c^2)}\right\}\left(1-\frac{b^2}{c^2}\right)\left\{1-\ln\left(\frac{b}{c}\right)\right\}\end{array}\right]+\frac{2}{E}\frac{k_1c^2b^2\sigma_Y}{r^2(b^2+c^2)}\\&+\frac{k_1c^3\sigma_Y}{Er(b^2+c^2)}\left(1-\frac{b^2}{c^2}\right)-\frac{k_1\sigma_Y}{E}\left(\frac{c}{r}\right)-k_1\left(\frac{2}{\sqrt{3}}\right)^n\frac{K}{Er}\int_r^c\left(\varepsilon_{\theta}^{p}\right)^n\mathrm{d}r_1-\frac{1}{r^2}\int_r^c r_1\varepsilon_{\theta}^{p}\mathrm{d}r_1\\&-k_1\left(\frac{2}{\sqrt{3}}\right)^n\frac{K}{E}\left(\varepsilon_{\theta}^{p}\right)^n.\end{aligned} \tag{37}$$

Equation (37) can be solved for ε_{θ}^{p} at any radial positions in the plastic domain. An iterative approach may be employed for this.

3.2.2 Analysis of the Second Phase of Elasto-plastic Deformation

During the second phase of elasto-plastic deformation, the wall of the disk consists of three zones: an inner plastic zone $a<r<c$, an outer plastic zone $d<r<b$, and an intermediate elastic zone $c<r<d$. The radii, c and d, are the radii of elastic–plastic interface. The elasto-plastic zones during the second phase of elasto-plastic deformation are analyzed as follows.

Intermediate Elastic Zone: $c<r<d$

Using the boundary conditions $(\sigma_\theta)_{r=c}=k_1\sigma_Y$ and $(\sigma_\theta-\sigma_r)_{r=d}=-k_1\sigma_Y$, in Eq. (9), the constants A and B are evaluated. Then substituting A and B in Eqs. (8) and (9), the stresses in the elastic zone $c<r<d$ are obtained as

$$\sigma_r=\frac{E\alpha(T_b-T_a)}{2\ln\left(\frac{b}{a}\right)}\left\{1+\ln\left(\frac{c}{r}\right)-\frac{d^2}{2c^2}-\frac{d^2}{2r^2}\right\}+k_1\sigma_Y\left(1+\frac{d^2}{2c^2}+\frac{d^2}{2r^2}\right), \tag{38}$$

$$\sigma_\theta=\frac{E\alpha(T_b-T_a)}{2\ln\left(\frac{b}{a}\right)}\left\{\ln\left(\frac{c}{r}\right)-\frac{d^2}{2c^2}+\frac{d^2}{2r^2}\right\}+k_1\sigma_Y\left(1+\frac{d^2}{2c^2}-\frac{d^2}{2r^2}\right). \tag{39}$$

The displacement component for this elastic zone is obtained using Eqs. (38) and (39) in Eq. (3) and second relation of Eq. (4) as

$$u = \frac{\alpha(T_b - T_a)}{2\ln\left(\frac{b}{a}\right)}\left\{(1+\nu)\frac{d^2}{2r} - (1-\nu)\frac{d^2 r}{2c^2} + (1-\nu)r\ln\left(\frac{c}{r}\right) - \nu r\right\}$$
$$+ \frac{k_1\sigma_Y}{E}\left\{(1-\nu)r + (1-\nu)\frac{d^2 r}{2c^2} - (1+\nu)\frac{d^2}{2r}\right\}$$
$$+ \alpha r\left\{T_a + (T_b - T_a)\frac{\ln\left(\frac{r}{a}\right)}{\ln\left(\frac{b}{a}\right)}\right\}. \quad (40)$$

Inner Plastic Zone: $a < r < c$

To find out the solution for radial thermal stress in the inner plastic zone ($a < r < c$), the differential Eq. (21) is used. Integrating Eq. (21) and using the boundary condition, σ_r at $r = c$ from elastic zone $c < r < d$ (Eq. 38), the expression for radial thermal stress is obtained as

$$\sigma_r = \left(\frac{c}{r}\right)\frac{E\alpha(T_b - T_a)}{2\ln\left(\frac{b}{a}\right)}\left(1 - \frac{d^2}{c^2}\right) + k_1\sigma_Y\left(1 + \frac{d^2}{cr}\right) - k_1\frac{K}{r}\int_r^c \left(\varepsilon_{\mathrm{eq}}^p\right)^n \mathrm{d}r_1. \quad (41)$$

The hoop stress component is given by Eq. (19).

For evaluating the total hoop strain component, Eq. (29) is utilized. In Eq. (29), the constant C is obtained by evaluating the right- and left-hand side expressions at $r = c$. For this, the values of σ_r and σ_θ at $r = c$ are evaluated from Eqs. (38) and (39) and using Eq. (40),

$$\left.\left\{\frac{1}{r}\frac{\mathrm{d}}{\mathrm{d}r}(ur)\right\}\right|_{r=c} = \frac{\alpha(T_b - T_a)}{2\ln\left(\frac{b}{a}\right)}\left\{(1-\nu) - (1-\nu)\frac{d^2}{c^2} + 4\ln\left(\frac{c}{a}\right)\right\}$$
$$+ \frac{k_1\sigma_Y}{E}\left\{2(1-\nu) + (1-\nu)\frac{d^2}{c^2}\right\} + 2\alpha T_a. \quad (42)$$

Substitution of all these expressions along with Eq. (31) in Eq. (29) provides $C = 0$. Inserting C in Eq. (29) and integrating it in the plastic domain from r to $r = c$ provides Eq. (32). Evaluating $u|_{r=c}$ from Eq. (40) and $\sigma_r|_{r=c}$ from Eq. (38) and then substituting these in Eq. (32) gives

$$\varepsilon_\theta = \frac{u}{r} = \frac{\alpha(T_b - T_a)}{2\ln\left(\frac{b}{a}\right)}\left\{\frac{d^2}{r^2} - 1 + 2\ln\left(\frac{r}{a}\right)\right\} - \frac{k_1\sigma_Y}{E}\frac{d^2}{r^2} + \left(\frac{1-\nu}{E}\right)\sigma_r$$
$$+ \alpha T_a - \frac{1}{r^2}\int_r^c r_1\varepsilon_\theta^p \mathrm{d}r_1. \quad (43)$$

Inserting u/r from Eq. (43), ε_θ^e from Eq. (3) in Eq. (34) and using Eqs. (19), (41), and (36) the plastic part of hoop strain is obtained as

$$\varepsilon_\theta^p = \frac{\alpha(T_b - T_a)}{2\ln\left(\frac{b}{a}\right)}\left\{\frac{d^2}{r^2} - 1 + \left(\frac{c}{r}\right)\left(1 - \frac{d^2}{c^2}\right)\right\} + \frac{k_1\sigma_Y}{E}\left(\frac{d^2}{cr} - \frac{d^2}{r^2}\right)$$
$$- \frac{1}{r^2}\int_r^c r_1\varepsilon_\theta^p \mathrm{d}r_1 - k_1\left(\frac{2}{\sqrt{3}}\right)^n \frac{K}{rE}\int_r^c \left(\varepsilon_\theta^p\right)^n \mathrm{d}r_1 - k_1\left(\frac{2}{\sqrt{3}}\right)^n \frac{K}{E}\left(\varepsilon_\theta^p\right)^n. \quad (44)$$

Outer Plastic Zone: $d < r < b$

For the plastic zone ($d < r < b$), considering the effect of strain hardening of the material, Tresca yield criterion is given by

$$\sigma_\theta - \sigma_r = -k_1\sigma_{\mathrm{eq}} \quad (45)$$

where σ_{eq} is given by Eq. (20). Using Eq. (45) and Eq. (6),

$$\frac{\mathrm{d}\sigma_r}{\mathrm{d}r} = -k_1\left\{\frac{\sigma_Y}{r} + \frac{K}{r}\left(\varepsilon_{\mathrm{eq}}^p\right)^n\right\}. \quad (46)$$

Integrating Eq. (46) in the plastic domain from $r = d$ to r and using the boundary condition, σ_r at $r = d$ from elastic zone $c < r < d$ (Eq. 38), the solution for radial thermal stress is given by

$$\sigma_r = \frac{E\alpha(T_b - T_a)}{2\ln\left(\frac{b}{a}\right)}\left\{\frac{1}{2} + \ln\left(\frac{c}{d}\right) - \frac{d^2}{2c^2}\right\} + k_1\sigma_Y\left\{\frac{3}{2} + \frac{d^2}{2c^2} - \ln\left(\frac{r}{d}\right)\right\}$$
$$+ k_1K\int_r^d \frac{1}{r_1}\left(\varepsilon_{\mathrm{eq}}^p\right)^n \mathrm{d}r_1. \quad (47)$$

Inserting Eq. (47) in Eq. (45), the solution for hoop stress is given by

$$\sigma_\theta = \frac{E\alpha(T_b - T_a)}{2\ln\left(\frac{b}{a}\right)}\left\{\frac{1}{2} + \ln\left(\frac{c}{d}\right) - \frac{d^2}{2c^2}\right\} + k_1\sigma_Y\left\{\frac{1}{2} + \frac{d^2}{2c^2} - \ln\left(\frac{r}{d}\right)\right\}$$
$$+ k_1K\int_r^d \frac{1}{r_1}\left(\varepsilon_{\mathrm{eq}}^p\right)^n \mathrm{d}r_1 - k_1K\left(\varepsilon_{\mathrm{eq}}^p\right)^n. \quad (48)$$

In the outer plastic zone, the stresses are such that $\sigma_\theta < \sigma_z < \sigma_r$. Thus in view of Tresca's associated flow rule, $\mathrm{d}\varepsilon_z^p = 0,\ \mathrm{d}\varepsilon_\theta^p = -\mathrm{d}\varepsilon_r^p$. Hence, plastic incompressibility gives $\varepsilon_r^p + \varepsilon_\theta^p = 0$. The total strain in the plastic zone is composed of both elastic and plastic strain components and thus

$$\varepsilon_r + \varepsilon_\theta = \varepsilon_r^e + \varepsilon_r^p + \varepsilon_\theta^e + \varepsilon_\theta^p = \varepsilon_r^e + \varepsilon_\theta^e \tag{49}$$

The elastic strain components in Eq. (49) are given by Eqs. (2) and (3). Using Eq. (4) and inserting Eqs. (2) and (3) in Eq. (49), the following differential equation is obtained:

$$\frac{\mathrm{d}}{\mathrm{d}r}(ur) = 2\alpha\left\{T_a + (T_b - T_a)\frac{\ln\left(\frac{r}{a}\right)}{\ln\left(\frac{b}{a}\right)}\right\}r + \frac{1}{E}(1-v)\frac{\mathrm{d}}{\mathrm{d}r}\left(r^2\sigma_r\right) \tag{50}$$

Integration of Eq. (50) provides,

$$ur = \alpha T_a r^2 + 2\alpha\frac{(T_b - T_a)}{\ln\left(\frac{b}{a}\right)}\left\{\ln\left(\frac{r}{a}\right)\frac{r^2}{2} - \frac{r^2}{4}\right\} + \frac{1}{E}(1-v)r^2\sigma_r + D \tag{51}$$

where D is an integration constant. To find out the constant D, the left- and right-hand sides of Eq. (51) are evaluated at $r = c$. For this, σ_r and u at $r = c$ are evaluated using Eqs. (38) and (40) and are inserted into Eq. (51). This yields

$$D = \frac{\alpha(T_b - T_a)}{2\ln\left(\frac{b}{a}\right)}d^2 - \frac{k_1\sigma_Y}{E}d^2. \tag{52}$$

Substituting the value of D from Eq. (52) in Eq. (51), the total hoop strain component in the plastic zone $d < r < b$ is obtained as

$$\frac{u}{r} = \alpha T_a + \frac{\alpha(T_b - T_a)}{2\ln\left(\frac{b}{a}\right)}\left\{2\ln\left(\frac{r}{a}\right) - 1 + \frac{d^2}{r^2}\right\} + \frac{1}{E}(1-v)\sigma_r - \frac{k_1\sigma_Y}{E}\frac{d^2}{r^2}. \tag{53}$$

Inserting u/r from Eq. (53) and ε_θ^e from Eq. (3) in Eq. (34) and using Eqs. (47), (48) and (36), the plastic part of hoop strain is obtained as

$$\varepsilon_\theta^p = \frac{\alpha(T_b - T_a)}{2\ln\left(\frac{b}{a}\right)}\left\{\frac{d^2}{r^2} - 1\right\} + \frac{k_1\sigma_Y}{E}\left(1 - \frac{d^2}{r^2}\right) + \frac{k_1 K}{E}\left(\frac{2}{\sqrt{3}}\right)^n\left(\varepsilon_\theta^p\right)^n. \tag{54}$$

4 Residual Thermal Stresses

When the temperature difference induced in the hollow disk due to heating vanishes after cooling, residual thermal stresses are generated in the disk. It is assumed that the unloading process is completely elastic and linear devoid of Bauschinger effect. In the following subsections, residual thermal stresses generated in different zones within the wall of the disk during the first and second phase of elasto-plastic deformation are given.

4.1 Residual Stresses in the Zones During the First Phase of Elasto-plastic Deformation

As the unloading process is entirely elastic, the expressions for the residual radial and hoop stresses in the outer elastic zone ($c < r < b$) are obtained by subtracting Eqs. (12) and (13) from Eqs. (16) and (17) as

$$(\sigma_r)_{\text{res}} = \frac{E\alpha(T_b - T_a)}{2\ln\left(\frac{b}{a}\right)}\left[\left(\frac{c^2}{b^2+c^2}\right)\left(1-\frac{b^2}{r^2}\right)\left\{1-\ln\left(\frac{b}{c}\right)\right\}+\left(\frac{a^2}{b^2-a^2}\right)\ln\left(\frac{b}{a}\right)\left(\frac{b^2}{r^2}-1\right)\right] + \frac{c^2k_1\sigma_Y}{b^2+c^2}\left(1-\frac{b^2}{r^2}\right), \tag{55}$$

$$(\sigma_\theta)_{\text{res}} = \frac{E\alpha(T_b - T_a)}{2\ln\left(\frac{b}{a}\right)}\left[\left(\frac{c^2}{b^2+c^2}\right)\left(1+\frac{b^2}{r^2}\right)\left\{1-\ln\left(\frac{b}{c}\right)\right\}-\left(\frac{a^2}{b^2-a^2}\right)\ln\left(\frac{b}{a}\right)\left(1+\frac{b^2}{r^2}\right)\right] + \frac{c^2k_1\sigma_Y}{b^2+c^2}\left(1+\frac{b^2}{r^2}\right). \tag{56}$$

Similarly, the residual stresses in the inner plastic zone ($a < r < c$) are obtained by subtracting Eqs. (12) and (13) from Eqs. (22) and (19) and are given by

$$(\sigma_r)_{\text{res}} = k_1\sigma_Y\left\{1-\left(\frac{2b^2}{b^2+c^2}\right)\left(\frac{c}{r}\right)\right\} - k_1\frac{K}{r}\int_r^c\left(\varepsilon_{\text{eq}}^p\right)^n \mathrm{d}r + \frac{E\alpha(T_b-T_a)}{2\ln\left(\frac{b}{a}\right)}\begin{bmatrix}\left(\frac{c}{r}\right)\ln\left(\frac{b}{c}\right)+\left(\frac{c^2}{b^2+c^2}\right)\left(\frac{c}{r}\right)\left(1-\frac{b^2}{r^2}\right)\left\{1-\ln\left(\frac{b}{c}\right)\right\}\\ -\ln\left(\frac{b}{r}\right)-\left(\frac{a^2}{b^2-a^2}\right)\ln\left(\frac{b}{a}\right)\left(\frac{b^2}{r^2}-1\right)\end{bmatrix}, \tag{57}$$

$$(\sigma_\theta)_{\text{res}} = k_1\left\{\sigma_Y + K\left(\varepsilon_{\text{eq}}^p\right)^n\right\} + \frac{E\alpha(T_b-T_a)}{2\ln\left(\frac{b}{a}\right)}\left\{1+\ln\left(\frac{r}{a}\right)-\left(1+\frac{a^2}{r^2}\right)\ln\left(\frac{b}{a}\right)\left(\frac{b^2}{b^2-a^2}\right)\right\}. \tag{58}$$

4.2 Residual Stresses in the Zones During the Second Phase of Elasto-plastic Deformation

The expressions for the residual stresses for elastic zone ($c < r < d$) are obtained by subtracting Eqs. (12) and (13) from Eqs. (38) and (39) as

$$(\sigma_r)_{\text{res}} = \frac{E\alpha(T_b - T_a)}{2\ln\left(\frac{b}{a}\right)}\left\{1 + \ln\left(\frac{c}{a}\right) - \frac{d^2}{2c^2} - \frac{d^2}{2r^2} - \left(1 - \frac{a^2}{r^2}\right)\ln\left(\frac{b}{a}\right)\left(\frac{b^2}{b^2 - a^2}\right)\right\}$$
$$+ k_1\sigma_Y\left\{1 + \frac{d^2}{2c^2} + \frac{d^2}{2r^2}\right\}, \tag{59}$$

$$(\sigma_\theta)_{\text{res}} = \frac{E\alpha(T_b - T_a)}{2\ln\left(\frac{b}{a}\right)}\left\{1 + \ln\left(\frac{c}{a}\right) - \frac{d^2}{2c^2} + \frac{d^2}{2r^2} - \left(1 + \frac{a^2}{r^2}\right)\ln\left(\frac{b}{a}\right)\left(\frac{b^2}{b^2 - a^2}\right)\right\}$$
$$+ k_1\sigma_Y\left\{1 + \frac{d^2}{2c^2} - \frac{d^2}{2r^2}\right\}. \tag{60}$$

The residual radial thermal stress in the inner plastic zone ($a < r < c$) is obtained by subtracting Eq. (12) from Eq. (41) as

$$(\sigma_r)_{\text{res}} = \left(\frac{c}{r}\right)\frac{E\alpha(T_b - T_a)}{2\ln\left(\frac{b}{a}\right)}\left\{\frac{c}{r} - \frac{d^2}{cr} + \ln\left(\frac{r}{a}\right) - \left(1 - \frac{a^2}{r^2}\right)\ln\left(\frac{b}{a}\right)\left(\frac{b^2}{b^2 - a^2}\right)\right\}$$
$$+ k_1\sigma_Y\left(1 + \frac{d^2}{cr}\right) - k_1\frac{K}{r}\int_r^c \left(\varepsilon_{\text{eq}}^p\right)^n \mathrm{d}r, \tag{61}$$

The residual thermal hoop stress in this zone is given by Eq. (58).

In the outer plastic zone ($d < r < b$), the residual thermal stresses are obtained by subtracting Eqs. (12) and (13) from Eqs. (47) and (48) as

$$(\sigma_r)_{\text{res}} = \frac{E\alpha(T_b - T_a)}{2\ln\left(\frac{b}{a}\right)}\left\{\frac{1}{2} + \ln\left(\frac{c}{d}\right) - \frac{d^2}{2c^2} + \ln\left(\frac{r}{a}\right) - \left(1 - \frac{a^2}{r^2}\right)\ln\left(\frac{b}{a}\right)\left(\frac{b^2}{b^2 - a^2}\right)\right\}$$
$$+ k_1\sigma_Y\left\{\frac{3}{2} + \frac{d^2}{2c^2} - \ln\left(\frac{r}{d}\right)\right\} + k_1K\int_r^d \frac{1}{r_1}\left(\varepsilon_{\text{eq}}^p\right)^n \mathrm{d}r_1, \tag{62}$$

$$(\sigma_\theta)_{\text{res}} = \frac{E\alpha(T_b - T_a)}{2\ln\left(\frac{b}{a}\right)}\left\{\frac{3}{2} + \ln\left(\frac{c}{d}\right) - \frac{d^2}{2c^2} + \ln\left(\frac{r}{a}\right) - \left(1 + \frac{a^2}{r^2}\right)\ln\left(\frac{b}{a}\right)\left(\frac{b^2}{b^2 - a^2}\right)\right\}$$
$$+ k_1\sigma_Y\left\{\frac{1}{2} + \frac{d^2}{2c^2} - \ln\left(\frac{r}{d}\right)\right\} + k_1K\int_r^d \frac{1}{r_1}\left(\varepsilon_{\text{eq}}^p\right)^n \mathrm{d}r_1 - k_1K\left(\varepsilon_{\text{eq}}^p\right)^n. \tag{63}$$

5 Solution Methodology

During the first phase of elasto-plastic deformation, the radius of elastic–plastic interface, c is unknown. The elastic–plastic interface radii, c and d, are the unknowns during the second phase of elasto-plastic deformation. The boundary condition of vanishing radial stress at the inner radius during the first phase of elasto-plastic deformation provides

$$k_1\sigma_Y\left\{1-\left(\frac{2b^2}{b^2+c^2}\right)\left(\frac{c}{a}\right)\right\}+\left(\frac{c}{a}\right)\frac{E\alpha(T_b-T_a)}{2\ln\left(\frac{b}{a}\right)}\left[\ln\left(\frac{b}{c}\right)+\left(\frac{c^2}{b^2+c^2}\right)\left(1-\frac{b^2}{c^2}\right)\left\{1-\ln\left(\frac{b}{c}\right)\right\}\right]$$
$$-k_1\frac{K}{a}\int_a^c\left(\varepsilon_{\mathrm{eq}}^p\right)^n\mathrm{d}r_1=0. \tag{64}$$

Solution of Eq. (64) will give the value of unknown radius c. The boundary conditions of vanishing radial stress both at inner and outer radii during the second phase of elasto-plastic deformation gives

$$\left(\frac{c}{a}\right)\frac{E\alpha(T_b-T_a)}{2\ln\left(\frac{b}{a}\right)}\left(1-\frac{d^2}{c^2}\right)+k_1\sigma_Y\left(1+\frac{d^2}{ca}\right)-k_1\frac{K}{a}\int_a^c\left(\varepsilon_{\mathrm{eq}}^p\right)^n\mathrm{d}r_1=0, \tag{65}$$

$$\frac{E\alpha(T_b-T_a)}{2\ln\left(\frac{b}{a}\right)}\left\{\frac{1}{2}+\ln\left(\frac{c}{d}\right)-\frac{d^2}{2c^2}\right\}+k_1\sigma_Y\left\{\frac{3}{2}+\frac{d^2}{2c^2}-\ln\left(\frac{b}{d}\right)\right\}$$
$$+k_1K\int_b^d\frac{1}{r_1}\left(\varepsilon_{\mathrm{eq}}^p\right)^n\mathrm{d}r_1=0. \tag{66}$$

To find out the values of unknown radii c and d, Eqs. (65) and (66) need to be solved simultaneously. The Eqs. (64, 65, and 66) are nonlinear in nature. Therefore, to evaluate c or c and d, from Eq. (64) or Eqs. (65) and (66), a numerical solution methodology as described in the following is adopted.

Step 1: For the first phase of elasto-plastic deformation, the initial estimate for radius of elastic–plastic interface, c is obtained by making $K = 0$ in Eq. (64) and solving it using fzero function (based on bisection method) in MATLAB. However, during the second phase of elasto-plastic deformation, the initial estimates for c and d are obtained by making $K = 0$ in Eqs. (65) and (66) and solving them for c and d using fsolve function (the least square method of solving simultaneous nonlinear equations) in MATLAB. The initial guess value of ε_θ^p is taken as zero.

Step 2: The updated plastic strain field ε_θ^p at different radial positions is found out from Eq. (37) during the first phase of elasto-plastic deformation and from Eqs. (44) and (54) during the second phase of elasto-plastic deformation. For the fixed c, in Eq. (37)

and for the fixed values of c and d, in Eqs. (44) and (54), values of ε_θ^p are updated further and this process is repeated till convergence in ε_θ^p is achieved. This basically amounts to solving for ε_θ^p using fixed point iteration method (Gerald and Wheatley 1994). The integrations involved in Eqs. (37, 44 and 54) are evaluated numerically using trapezoidal rule. Using Eq. (36), the updated estimate of $\varepsilon_{\text{eq}}^p$ is obtained.

Step 3: Now using the values of ε_θ^p at different radial positions in the corresponding plastic zones, the integral terms in Eqs. (64, 65 and 66) are evaluated and the expressions are solved for new estimated values of c or c and d. If the new estimated values of c or c and d are same as the previous estimated values, go to Step 4. Otherwise, go to Step 2 and repeat the process till the convergence for c or c and d is achieved.

Step 4: Using the latest updated values of ε_θ^p, c or c and d, the radial and hoop stresses in the plastic zones for the corresponding cases are calculated from Eqs. (22), (19), (41), (47) and (48). The thermal residual stresses are estimated using the relevant equations of Sect. 4.

6 Numerical Simulation

The proposed autofrettage process is simulated numerically using the equations developed in Sects. 3.2.1 and 3.2.2. For this purpose, two cases are considered. The corresponding equations are solved following the methodology discussed in Sect. 5. The simulated results are discussed in the following sections.

6.1 Case of Aluminum Disk

An aluminum disk is considered with inner radius, a = 10 mm, and outer radius, b = 20 mm. The material properties of aluminum are as follows: Young's modulus of elasticity, E = 69 GPa; yield stress, σ_Y = 50.3 MPa; and coefficient of thermal expansion, $\alpha = 22.2 \times 10^{-6}$/°C. The temperature difference required for initial yielding at inner wall of the disk is calculated using Eq. (15) and is obtained as 53.66 °C. For achieving autofrettage in the aluminum disk, two temperature differences are considered—a temperature difference of 80 °C to cause the first phase of elasto-plastic deformation and a temperature difference of 130 °C to cause the second phase of elasto-plastic deformation in the wall of the disk. The radius of elastic–plastic interface, c, is obtained as 11.2089 mm when the temperature difference is 80 °C. The estimates of c and d are obtained as 12.4754 mm and 17.4997 mm when the temperature difference is 130 °C. The hardening coefficient K and strain hardening exponent n for the disk material are taken as 58.18 MPa and

0.482 in Ludwik's hardening law given by Eq. (20). The various stresses generated in the disk as a result of autofrettage due to temperature difference are simulated and are shown as a function of radial position.

6.1.1 Elasto-plastic Thermal Stress Pattern

The elasto-plastic thermal stresses for 80 °C temperature difference are obtained using Eqs. (16), (17), (19) and (22). For the temperature difference of 130 °C, the elastic–plastic stresses are estimated using Eqs. (38), (39), (41), (19), (47) and (48). The resulting elasto-plastic stress distribution for temperature difference of 80 and 130 °C is shown in Fig. 5. It is observed that the magnitude of radial stresses is quite small compared to hoop stresses. The radial stresses are always tensile, while the hoop stresses change from tensile to compressive along the positive radial direction. This trend gets reversed when the temperature difference is removed. It is also observed that strain hardening effect is not very significant in the present case.

6.1.2 Residual Stress Pattern

On removal of the temperature difference, residual thermal stresses are generated in the disk. The residual stress variation with the radial position is shown in Fig. 6 for the temperature differences of 80 and 130 °C. It is observed that the residual compressive hoop stresses are generated at and around the inner wall of the disk, while residual tensile stresses are generated at the outer region of the disk. For the temperature difference of 130 °C, the magnitudes of compressive residual stresses are higher compared to the case of temperature difference of 80 °C. The maximum magnitude of residual stress for the temperature difference of 80 °C is about 23 MPa (compressive), generated at the inner radius of the cylinder. For the case of 130 °C temperature difference, this magnitude is about 70 MPa (compressive). Hence, when the disk is subjected to loading by inducing internal pressure, then the

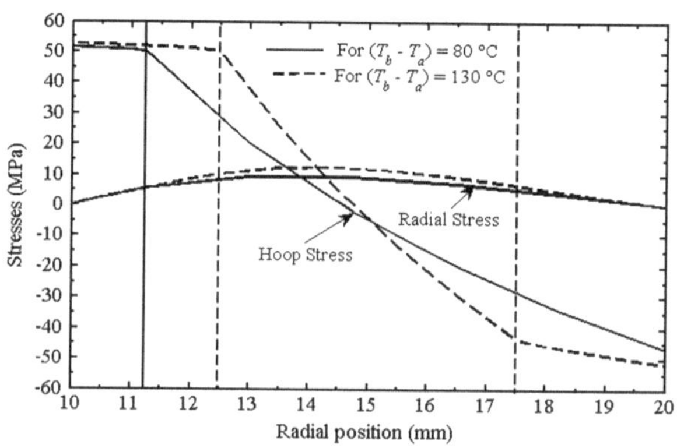

Fig. 5 Elasto-plastic stress distribution in aluminum disk

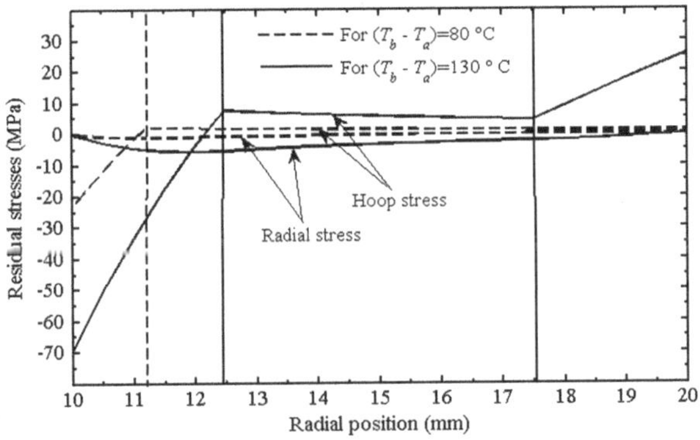

Fig. 6 Residual stress distribution in aluminum disk

resulting residual compressive stress leads to decrease in the maximum value of stress occurring in the disk.

The magnitudes of residual tensile stresses at the outer portion of the disk are observed to be small compared to residual compressive stresses for the case of 80 °C. For the 130 °C temperature difference, the magnitudes of tensile residual stresses show an increasing trend at the outer plastic zone. For the temperature difference of 80 °C, the maximum tensile residual stress is generated at the elastic–plastic interface (at $r = c$). For the temperature difference of 130 °C, the maximum tensile residual stress occurs at the outer radius. Moreover, its magnitude is larger in comparison with 80 °C temperature difference case. In corrosive environment, the larger tensile residual stress at the outer radius may cause stress corrosion cracking. Therefore, autofrettage by 130 °C temperature difference is not desirable in corrosive environment.

6.1.3 Overall Stresses with and Without Autofrettage

Assume that the aluminum disk which is autofrettaged by inducing temperature difference between inner and outer walls is subjected to an internal working pressure of 18 MPa for both the temperature difference case. The overall stresses are then obtained by adding stresses due to internal working pressure to the residual thermal stresses. The stresses that occur in the disk due to internal working pressure are given by well-known Lame's equations (Chakrabarty 2006).

$$\sigma_r = \frac{p_i a^2}{b^2 - a^2}\left(1 - \frac{b^2}{r^2}\right), \tag{67}$$

$$\sigma_\theta = \frac{p_i a^2}{b^2 - a^2}\left(1 + \frac{b^2}{r^2}\right), \tag{68}$$

where p_i is the internal working pressure. If the same disk is considered only under the condition of internal working pressure without autofrettage, then the elastic stresses generated in the disk are given by Eqs. (67) and (68) only. The resulting stresses with and without autofrettage for the case of 80 °C temperature difference is shown in Fig. 7a. For 130 °C temperature difference, they are shown in Fig. 7b. The equivalent Tresca stress ($|\sigma_\theta - \sigma_r|/2$, in the present cases) is observed to be maximum at the elastic–plastic interface ($r = c$), in the autofrettaged disk during the pressurization. Therefore, for the autofrettaged disk, yielding will first take place at the elastic–plastic interface ($r = c$).

When the temperature difference is 80 °C, the autofrettaged disk can withstand a maximum pressure of 22 MPa without yielding. For the temperature difference of 130 °C, the maximum pressure-carrying capacity of the autofrettaged disk is 21 MPa. On the other hand, the non-autofrettaged disk can withstand a maximum pressure of 18 MPa. It is observed that the maximum equivalent Tresca stress is reduced by 13.91 % in the disk when it is autofrettaged by inducing temperature difference of 80 °C as compared to the disk without autofrettage. For the temperature difference of 130 °C, this reduction is observed to be 8.52 %. Ali et al. (2010) provided a typical example for the hydraulic autofrettage of aluminum cylinder in

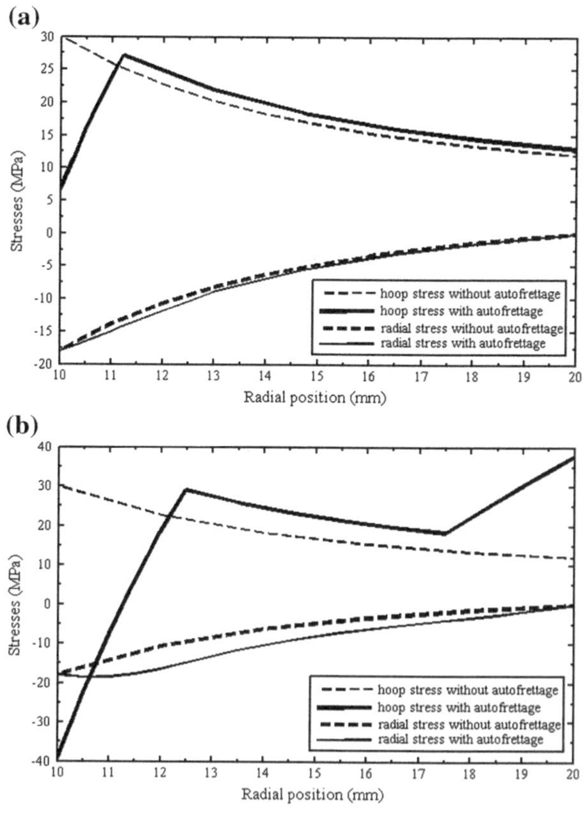

Fig. 7 Comparison of stresses with and without autofrettage for **a** $(T_b - T_a) = 80$ °C, **b** $(T_b - T_a) = 130$ °C in aluminum disk

which the reduction in the maximum equivalent stress is of similar order. Underwood and Miller (1988) have also advocated partial autofrettage even in the case of hydraulic autofrettage to circumvent the problem of stress corrosion cracking. Nevertheless, compared to hydraulic autofrettage, the level of thermal autofrettage is limited, because the maximum temperature needs to be kept well below the recrystallization temperature of the material.

6.2 *Case of Mild Steel Disk*

A mild steel disk with inner radius, a = 10 mm, and outer radius, b = 30 mm, is simulated when it undergoes autofrettage process due to a temperature difference between the outer and inner surfaces. Material properties considered for steel are as follows: Young's modulus of elasticity, $E = 200$ GPa; yield stress, $\sigma_Y = 324$ MPa; and coefficient of thermal expansion, $\alpha = 13 \times 10^{-6}$/°C. The temperature difference required for initial yielding at the inner wall of the disk for this case is 186 °C which is much lower than the recrystallization temperature of mild steel. However, a large temperature difference is required to cause the yielding of the outer wall of the disk. The large temperature difference may greatly change the material properties. Therefore, it is not advisable to induce such a large temperature difference that will cause the yielding of both the inner and outer walls of the disk for achieving autofrettage. Hence, the disk is simulated for a temperature difference of 270 °C for achieving autofrettage. At 270 °C temperature difference, the wall of the disk undergoes first phase of elasto-plastic deformation. The radius of elastic–plastic interface, c, is obtained as 11.709 mm. The hardening coefficient K and strain hardening exponent n for the disk are taken as 226.98 MPa and 0.43. The stresses generated in the disk as a result of autofrettage are shown as a function of radial position.

6.2.1 Elastic–plastic Thermal Stress Pattern

The elastic–plastic thermal stresses for the disk are obtained using Eqs. (16), (17), (19) and (22). The similar trend, as obtained in the case of aluminum disk with 80 °C temperature difference, is observed. The results are shown in Fig. 8. Here also, the strain hardening is not significant.

6.2.2 Residual Stress Pattern

The residual stress pattern for the autofrettaged steel disk is shown in Fig. 9. It is observed that a significant amount of compressive residual hoop stresses are generated in and around the inner radius of the disk, and hence, it reduces the magnitude of resultant stress when it is subjected to internal working pressure. The magnitudes of tensile residual stresses generated at the outer region of the disk are very small.

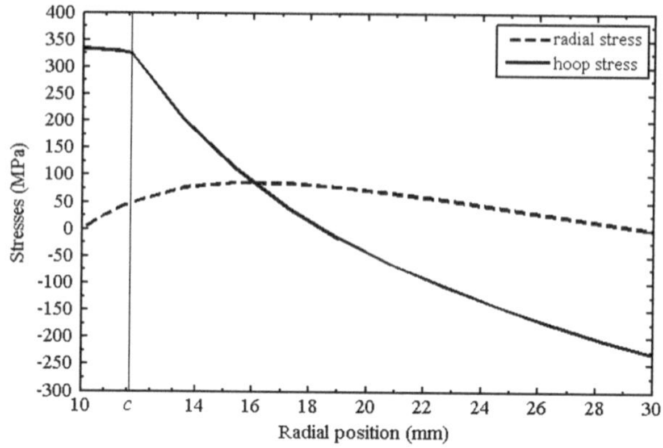

Fig. 8 Elasto-plastic stresses in mild steel disk

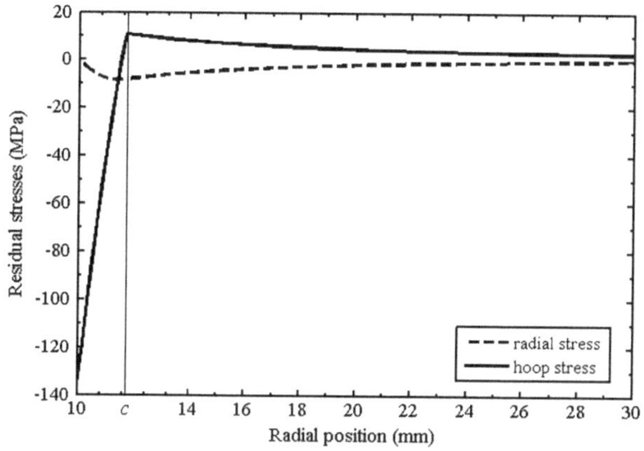

Fig. 9 Residual stresses in mild steel disk

6.2.3 Overall Stresses with and Without Autofrettage

The autofrettaged steel disk is subjected to an internal working pressure of 100 MPa and the resultant stress pattern is shown in Fig. 10. The corresponding stresses without autofrettage are also shown using Eqs. (67) and (68). In case of the autofrettaged disk, the magnitude of the maximum equivalent Tresca stress is found to be 18.82 % less than that in case of the disk without autofrettage during the pressurization. The maximum equivalent Tresca stress exists at the radius of elastic–plastic interface for the autofrettaged case. Therefore, yielding of the disk during the pressurization will first take place at the elastic–plastic interface. The maximum pressure-carrying capacity of the autofrettaged disk in this case is 186 MPa. The non-autofrettaged disk can withstand a maximum pressure of 144 MPa only.

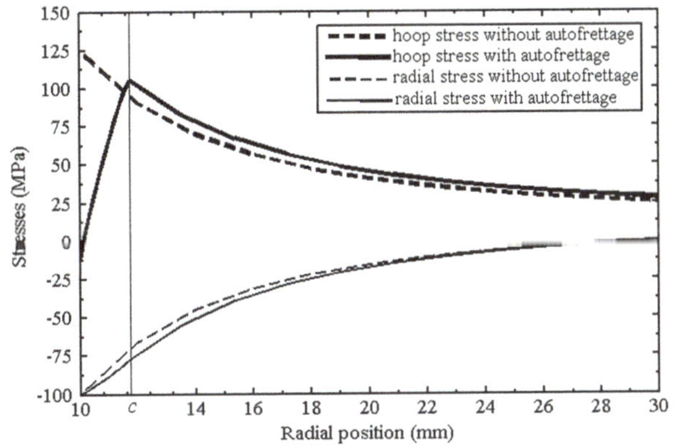

Fig. 10 Comparison of stresses with and without autofrettage in mild steel disk

The following points are observed in general from the simulation of stresses for both aluminum and steel disk:

1. Due to residual compressive hoop stress at the inside wall of the disk, the resultant hoop stress becomes significantly lower in case of autofrettaged disk than the original hoop stresses set up in the same disk due to internal pressure.
2. On the application of internal working pressure, the distribution of radial stresses in the autofrettaged disk is very close to the radial stresses developed in the non-autofrettaged disk.

7 Preliminary Experiments

In order to assess the feasibility of thermal autofrettage, an experimental setup has been developed at IIT Guwahati and is shown in Fig. 11. For creating temperature gradient in the cylinder, the cylinder is heated in the heating chamber by means of electrical heating element and the inner surface is made cooler by regulating continuous flow of cold water. In the present experiment, the length of the cylinder

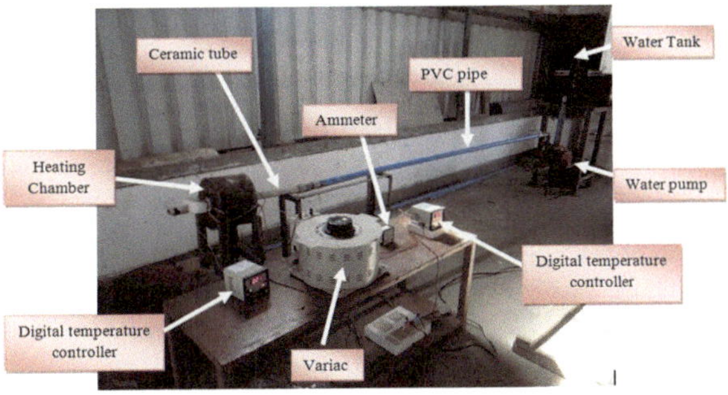

Fig. 11 The experimental setup

considered is not large enough and no significant temperature variation of cold water flow from inlet to outlet is observed. Thus, there will be uniform yielding along the length of the cylinder. For the experiment, a mild steel disk (short cylinder!) of inner radius 5 mm and outer radius 20 mm is considered. The length of the cylinder is 50 mm. A temperature difference of 268 °C was generated between the outer and inner walls of the disk, which is sufficient for achieving thermal autofrettage. After that, the cylinder was allowed to cool to the room temperature. The micro-hardness tests were conducted on the outer and inner surfaces of the autofrettaged disk using Vickers micro-hardness tester at a load of 500 gf. The micro-hardness test of the non-autofrettaged disk material was also carried out. The average micro-hardness of the non-autofrettaged disk was 170.64 HV with standard deviation 5.25 HV. The hardness of the outer surface of the autofrettaged disk was 157.28 HV with standard deviation 7.32, whereas the hardness of the inner surface was 219.52 HV with standard deviation 12.84. It is envisaged that tensile residual stresses on the outer surface cause a decrease in the micro-hardness. Similarly, the compressive residual stresses on the inner surface cause an increase in the micro-hardness. Thus, it appears that the partial thermal autofrettage has been achieved.

8 Conclusion

In this chapter, a novel autofrettage process by inducing temperature difference in the hollow circular disk (or short cylinder) is proposed. The process is analyzed theoretically considering the strain hardening of the disk during the plastic deformation. Simulations are carried out for thermal autofrettage of aluminum and mild steel disks. The overall simulation results show trends similar to those observed by other researchers in case of hydraulic autofrettage. Some preliminary experiments are also carried out to see the feasibility of the proposed process. The results of the preliminary experiment are encouraging. Compared to hydraulic and swage autofrettage, the achievable level of autofrettage is limited by the maximum allowable temperature at the outer surface of the disk. However, the simplicity of the process has potential to make it a competitive process.

References

Ali ARM, Ghosh NC, Alam TE (2010) Optimum design of pressure vessel subjected to autofrettage process. World Acad Sci Eng Technol 46:667–672

Chakrabarty J (2006) Theory of plasticity, 3rd edn. Butterworth-Heinemann, Burlington

Davidson TE, Barton CS, Reiner AN, Kendall DP (1962) New approach to the autofrettage of high-strength cylinders. Exp Mech 2:33–40

Dixit PM, Dixit US (2008) Modeling of metal forming and machining processes: by finite element and soft computing methods. Springer, London

Ezra AA (1973) Method and apparatus for explosive autofrettage. U.S. Patent No. 3,751,954

Gerald CF, Wheatley PO (1994) Applied numerical analysis, 5th edn. Addison-Wesley, England

Hussain MA, Pu SL, Vasilakis JD, O'Hara P (1980) Simulation of partial autofrettage by thermal loads. ASME J Press Vessel Technol 102:314–318

Iremonger MJ, Kalsi GS (2003) A numerical study of swage autofrettage. J Press Vessel Technol 125:347–351

Malik MA, Khushnood S (2003) A review of swage-autofrettage process. In: Proceedings of 11th international conference on nuclear engineering, Tokyo, pp 20–23

Mote JD, Ching LKW, Knight RE, Fay RJ, Kaplan MA (1971) Explosive autofrettage of cannon barrels, AMMRC CR 70-25. Army Materials and Mechanics Research Center, Watertown

Noda N, Hetnarski RB, Tanigawa Y (2003) Thermal stresses, 2nd edn. Taylor and Francis, New York

Parker AP (1980) Autofrettage of open-end tubes-pressures, stresses, strains, and code comparisons. ASME J Press Vessel Technol 123:271–281

Underwood JH, Miller JJ (1988) Stress corrosion cracking of A723 steel pressure vessels: two case studies. Technical report ARCCB-TR-88002, US Army ARDEC, Benet Laboratories, Watervliet

Numerical Analysis of Warm Deep Drawing for Ti–6Al–4V Alloy

Nitin Kotkunde, Aditya D. Deole, Amit Kumar Gupta and Swadesh Kumar Singh

Abstract The objective of this study was to focus on the current state of finite element methods with respect to reliability in modeling of sheet metal forming processes. The trustworthiness of the FE simulations largely depends on the input material models used and correctness of the input material data. Formability parameters required for finite element analysis have been determined for Ti–6Al–4V alloy at temperature ranging from room temperature to 400 °C at intervals of 50 °C. Based on these formability parameters, various anisotropic yield criteria, namely, von-Mises, Hill 1948, Barlat 1989, Barlat 1996 and Cazacu–Barlat, have been implemented for Ti–6Al–4V alloy. In addition to that, circular deep drawing experiments have been performed in order to study the formability of Ti–6Al–4V alloy sheet at warm condition. It has been observed that formability of the material is very poor at room temperature. At temperatures above 150 °C till 400 °C, the limiting drawing ratio is found to be 1.8, which is substantially lesser than other structural alloys. Additionally, in the properly drawn cups, thickness distribution and earing tendency have been studied over the range of temperature. In order to validate the experimental results, finite element analysis has been done using commercially available software DYNAFORM version 5.6.1 with LSDYNA solver version 971. Also, failure regions during the experimentation have been identified using FE analysis. Comparison of yield criteria based on thickness distribution, earing profile, complexity in material parameter identification, and computational time has shown Cazacu–Barlat to be well suited for deep drawing of Ti–6Al–4V alloy.

Keywords Ti–6Al–4V alloy · Warm deep drawing · Yield criteria · Finite element analysis

N. Kotkunde (✉) · A.D. Deole · A.K. Gupta
Department of Mechanical Engineering, BITS Pilani, Hyderabad Campus, Hyderabad, Telangana 500078, India
e-mail: nitinkotkunde@gmail.com

S.K. Singh
Department of Mechanical Engineering, GRIET, Hyderabad, Hyderabad, Telangana 500072, India

R.G. Narayanan and U.S. Dixit (eds.), *Advances in Material Forming and Joining*, Topics in Mining, Metallurgy and Materials Engineering,
DOI 10.1007/978-81-322-2355-9_6

1 Introduction

Metal forming is a general term, for a large group, that includes a wide variety of manufacturing processes (Altan et al. 1983). Metal forming processes are characteristic in that the metal being processed is plastically deformed in order to shape it into a desired geometry (Marciniak 2002). Metal forming processes are commonly used in the industry because it eradicates costly operations such as welding, machining, and manufactures parts with reduced weight and good mechanical properties with high production rates (Kotkunde et al. 2014a). Therefore, metal forming is the backbone of modern manufacturing industry (Altan et al. 1983; Marciniak 2002; Kotkunde et al. 2014a). Moreover, many sheet metal components are produced using a deep drawing process. It is one of the most important and popular processes in assessment of formability of sheet metal (Beal et al. 2006).

In the manufacturing industry, low carbon steels, austenitic stainless steels, have been popularly used for a long time due to their exceptional formability at room temperature, strength, good surface finish, and low cost (Singh et al. 2010; Hussaini et al. 2014). However, despite the excellent properties, lightweight alloys such as aluminum, titanium, and magnesium were ranked far behind than steels because of cost and formability issues at room temperature (Toros et al. 2008; Bolt et al. 2001; Cole and Sherman 1995; Naka and Yoshida 1999). Nowadays, because of superior properties, demand of lightweight alloys in automotive and aerospace industries has increased significantly (Bolt et al. 2001). But, formability of these alloys at room temperature is a big challenge (Cole and Sherman 1995). Formability of these light alloys can be greatly improved by warm forming (Kotkunde et al. 2014a; Kurukuri et al. 2009; Odenberger et al. 2013). Since elevated temperature results in decreased flow stress and increased ductility in the sheet, it allows deeper drawing and more stretching to form products (Kotkunde et al. 2014a).

In sheet metal industries, during the last two decades, finite element (FE) simulations have gained special attention. Finite element analysis in sheet metal forming is beneficial to reduce inaccurate and expensive tryouts (Takuda and Mori et al. 2003). Nowadays, more attentions are given for finite element analysis at warm conditions (Bong et al. 2013). The trustworthiness of the FE simulations largely depends on the input material models used and correctness of the input material data (Kotkunde et al. 2014b). Particularly, selection of an appropriate yield criterion is essential because it provides an accurate prediction of the observed initial and subsequent yield behaviors of a material (Naka and Yoshida 1999; Kotkunde et al. 2014b). In addition to that, selection of appropriate hardening model is crucial for accurate FE simulations in metal forming (Lin and Chen 2011; Kotkunde et al. 2014c, d; Gupta et al. 2013).

Since the selection of an appropriate material model is essential in finite element simulations, a considerable effort has been made for experimental observations of yield loci on various types of metals (Banabic 2010; Hill 1950). In the last few years, several efforts have been made for the development of anisotropic yield criteria which consider plastic anisotropy. Therefore, FE simulation for warm forming involves highly nonlinear yielding and flow behavior, thermo-mechanical coupled characteristics which make the FE simulation complicated (Kurukuri et al. 2009).

Hill's 1948 model, Barlat's 1989 model, or the von-Mises isotropic yield functions were used in many applications because of the simplicity in determining the material constants (Banabic 2010; Hill 1950). Barlat yield function with power law work-hardening relation was used for finite element simulations of circular deep drawing of AA 5042-H32 alloy (Keum et al. 2001). Boogaard et al. proposed the flow stress behavior of AA5754-O with the modified power law model and also with the Bergstrom model. The yield surface locus was assumed constant with variation of temperature (Bergström 1983). Van den Boogaard and Huétink (2006) proposed Vegter yield criterion with Bergström models for warm forming simulations of cylindrical cup deep drawing. Fully coupled thermo-mechanical FE simulation for aluminum sheet was proposed by Abedrabbo et al. In this FE analysis, they proposed Barlat yield function (YLD2000-2d) and a phenomenological Holloman work-hardening model (Abedrabbo et al. 2007). Very recently, Vegter yield criterion with physical-based work-hardening Nes model was used for analysis of warm deep drawing of Al–Mg sheet. They used an in-house implicit finite element code DiekA for the analysis (Kurukuri et al. 2009). Therefore, very few efforts were reported about the implementation of various anisotropic yield criteria in FE analysis for warm forming applications.

The objective of the present work is to study applicability of different anisotropic yield criteria in finite element simulation of deep drawing of Ti–6Al–4V alloy at warm condition.

2 Experimental Determination of Formability Parameters

Determination of mechanical properties accurately is one of the key issues in analyzing the deformation behavior of the materials (Kotkunde et al. 2014b). Numerous techniques are available to obtain the required flow stress curves such as tensile, compression, torsion, shear, biaxial tensile, and hydraulic bulge test (Lange 1985). The flow stress curves determined using different tests and test conditions do not replicate each other due to effects of stress state, yield criterion assumption, anisotropy effect, Bauschinger effect, experimental inaccuracies, temperature, and general weakness of the modeling (Koç et al. 2011). Hence, none of the test methods can be named as the best or optimal (Lange 1985; Koç et al. 2011). Each has its specific field of application due to definite straining paths. Therefore, the selection of required test is one of the very important and crucial steps for finite element analysis (Güner et al. 2009).

Among these test systems, the most widely used one is uniaxial tensile test that is standardized by ASTM E8-04 and DIN EN 10 002 (Wessel 2004). The important formability parameters required for FE analysis are strain-hardening exponent (n), strain rate sensitivity (m), Lankford parameters (r), yield strength (σ_y), ultimate tensile strength, and ductility (Hannon and Tiernan 2008). The procedure to determine the important formability parameters is given below.

2.1 Strain-hardening exponent (n):

Strain-hardening coefficient (n) is determined by the dependence of the flow (yield) stress on the level of strain. In materials with a high n value, the flow stress increases rapidly with strain. This tends to distribute further strain to regions of lower strain and flow stress. A high n value leads to a large difference between yield strength and ultimate tensile strength which is an indication of good formability. The strain-hardening characteristics of a material are usually dependent on strain, strain rate, and temperature (Mishra et al. 1989). If strain rate and temperatures are assumed constant, the plastic-state equation can be approximated by the constitutive Eq. (1) proposed by Hollomon (1945).

$$\sigma = k\varepsilon^n \tag{1}$$

This constitutive equation provides good approximation for most steels, but is not very accurate for dual-phase steels and some aluminum alloys. For these materials, two or three n values may be needed to be calculated for the low-, intermediate-, and high-strain regions (Davis 2004). However, for specific cases, few other constitutive equations such as Ludwik (1909), Swift (1952), and Voce (1948) are also used for determination of the n value. The detail test procedure for determining n value is as per ASTM E 646–07 standard.

2.2 Strain rate sensitivity (m):

The material flow stress curve is dependent on strain rate and temperature (Hosford and Canddell 2014). The strain rate sensitivity is, however, calculated in most publications based on the strain rate dependency of flow curves. The logarithmic strain rate sensitivity m is originally defined from the extended Hollomon Eq. (2).

$$\sigma = k \cdot \varepsilon^n \cdot \dot{\varepsilon}^m \tag{2}$$

There are two commonly used methods of determining m value. One is to obtain continuous stress–strain curves at several different strain rates and compare the levels of stress at a fixed strain using Eq. (2). The other is to make abrupt changes of strain rate during a tension test and use corresponding level of $\Delta\sigma$. This method is popularly known as rate change test or jump test (Davis 2004). Generally, use of continuous stress–strain curves yields larger value of m than jump test. The jump test has an advantage that several strain rate changes can be made on a single specimen, whereas continuous stress–strain curves require a specimen for each strain rate (Hosford and Canddell 2014). For the present study, continuous stress–strain curves at several different strain rates approach have been considered.

2.3 Lankford Coefficient (r)

The variation of the plastic deformation behavior with direction is assessed by a quantity called Lankford parameter or anisotropy coefficient (Banabic 2010). Sheet metals generally exhibit a considerable anisotropy due to their crystallographic texture. This anisotropy has a great influence on the formability and shape of the specimen during and after the deformation (Marciniak et al. 1973). Sheet metal deformation processes are challenged due to the difficulties such as wrinkling, earing, and failure that are tightly connected to the anisotropy and hardening of the material (Pearce 1968). Lankford parameter is generally determined by uniaxial tensile tests. The anisotropy coefficient r is defined by Eq. (3).

$$r = \frac{\varepsilon_{\text{width}}}{\varepsilon_{\text{Thickness}}} \tag{3}$$

The above equation can be rewritten as Eq. (4)

$$r = \frac{\ln \frac{w}{w_0}}{\ln \frac{t}{t_0}} \tag{4}$$

where w and t are the final width and thickness, respectively, and w_0 and t_0 are the initial width and thickness of the specimen, respectively. As the thickness of the specimen is very small compared to its width, the relative errors of measurement of the two strains will be quite different. Therefore, the above relationships are replaced by considering the length of the specimen. Equation (4) is rearranged as Eq. (5).

$$r = \frac{\ln \frac{w}{w_0}}{\ln \frac{l_0 w_0}{l w}} \tag{5}$$

Generally, anisotropic is expressed in two ways. Normal anisotropy (r_n) is nothing but the average of r values obtained for different directions in the plane of the sheet. It is mathematically expressed as Eq. (6). It indicates the drawability of sheet.

$$r_n = \frac{r_0 + 2r_{45} + r_{90}}{4} \tag{6}$$

Table 1 Formability parameters of Ti–6Al–4V alloy with variation of temperature

Temp. (°C)	σ_0 (MPa)	σ_{45} (MPa)	σ_{90} (MPa)	r_0	r_{45}	r_{90}	n	m	σ_{ut} (MPa)
Room temp.	879	824	875	0.4698	1.4325	0.2939	0.0335	0.001	945
50	876	806	882	0.4832	1.4235	0.3025	0.0337	0.003	941
100	845	780	842	0.5278	1.3925	0.3489	0.0415	0.006	913
150	795	740	784	0.5625	1.3711	0.3694	0.0463	0.009	842
200	728	642	757	0.6036	1.6144	0.5605	0.0575	0.010	811
250	707	617	727	0.7964	1.1854	0.7093	0.0594	0.010	761
300	687	598	702	0.7594	1.1523	0.4896	0.0636	0.011	751
350	683	595	693	0.7139	1.0727	0.5329	0.0677	0.012	742
400	681	591	691	0.6010	1.2632	0.5124	0.0679	0.012	737

where r_0, r_{45}, and r_{90} are the Lankford parameters in 0°, 45°, and 90°, respectively, orientation with respect to the rolling direction of the sheet.

However, planer anisotropy (Δr) can be expressed by difference strain ratio values in various directions of the sheet plane. It is responsible for the formation of ears in the drawn cups as well uneven thinning (Davis 2004). The mathematically, it is expressed as Eq. (7).

$$\Delta r = \frac{r_0 - 2r_{45} + r_{90}}{2} \tag{7}$$

In addition to that, yield strength in different directions, ultimate tensile strength, and ductility are considerable parameters for forming FE analysis. Table 1 indicates formability parameters calculated for Ti–6Al–4V alloy.

3 Yield Criteria for Finite Element Analysis

The transition from the elastic to plastic state occurs when a material reaches its yield point. For uniaxial tensile-state case, it is very easy to understand and predict the yield point. However, for multiaxial loading case, it is more difficult to define and predict the yielding behavior of a material (Banabic 2010; Hill 1950). Therefore, mathematical relationships are essential to define the yield function. Generally, it is a relation between the principal stresses in multiaxial loading case and yield stress in uniaxial case. Its relation is expressed as Eq. (8)

$$F(\sigma_1, \sigma_2, \sigma_3, \sigma_Y) = 0 \tag{8}$$

where σ_1, σ_2, and σ_3 are the principal stresses and σ_Y is the yield stress obtained from a simple tension/compression/shear test (Banabic 2010).

Equation (8) signifies the mathematical description of a surface in the three-dimensional space of the principal stresses generally called the yield locus. The most commonly used yield criteria for isotropic materials have been proposed by Huber–von-Mises (Banabic 2010). The Following are a few important yield criteria which are available in popular FE analysis software for sheet metal forming.

3.1 von-Mises Yield Function

In the ideal case of isotropic materials, von-Mises equation for 3D case is given by Eq. (9)

$$(\sigma_1 - \sigma_2)^2 + (\sigma_2 - \sigma_3)^2 + (\sigma_3 - \sigma_1)^2 = 2\sigma_Y^2 \tag{9}$$

For plane stress condition, the above equation reduces to

$$\sigma_1^2 - \sigma_1 \cdot \sigma_2 + \sigma_2^2 = \sigma_Y^2 \tag{10}$$

The representative yield loci for von-Mises yield criterion for Ti–6Al–4V alloy are shown in Fig. 1.

3.2 Hill 1948 Yield Criterion

Hill 1948 (Hill 1950) is one of the most popular anisotropic yield criteria for finite element simulation of sheet metal forming processes. The plane stress yield function is given by Eq. (11)

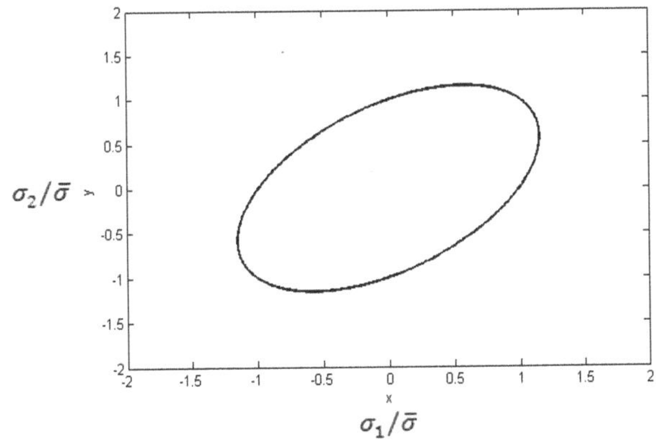

Fig. 1 Representative von-Mises yield criterion for Ti–6Al–4V alloy

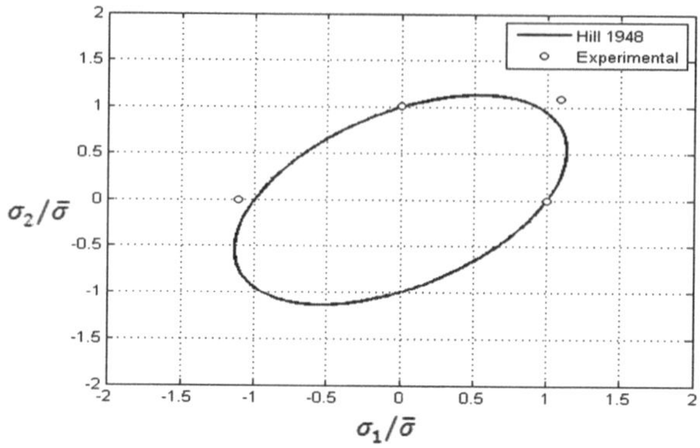

Fig. 2 Hill 1948 yield locus for Ti–6Al–4V alloy at 400 °C

$$F(\sigma) = \left(\sigma_{11}^2 + \sigma_{22}^2 - \frac{2R}{R+1}\sigma_{11}\sigma_{22} + 2\frac{2R+1}{R+1}\sigma_{12}\right)^{1/2} \tag{11}$$

where R is normal anisotropy coefficient and σ_{11}, σ_{22} are the principal stresses. The material response after yielding is governed by an elastic–plastic constitutive relation with plastic hardening modulus and yield stress of the material as an input. The representative yield locus for Ti–6Al–4V alloy at 400 °C is shown in Fig. 2.

3.3 Barlat 1989 Yield Criterion

Anisotropic yield function developed by Barlat and Lian (1989) for modeling sheets under plane stress condition is expressed in Eq. (12). The material response after yielding is governed by exponential hardening rule; m is the material parameter which depends on the crystal structure of a material. For Ti–6Al–4V alloy m value is taken as 8 for HCP structure materials.

$$\Phi = a|K_1 + K_2|^m + a|K_1 - K_2|^m + c|2K_2|^m = 2\sigma_y^m \tag{12}$$

where $K_1 = \frac{\sigma_x - h\sigma_y}{2}; K_2 = \sqrt{\left(\frac{\sigma_x - h\sigma_y}{2}\right)^2 - p^2\tau_{xy}^2}$, and a, c, and h are functions of anisotropy ratio and they are expressed as

$$\begin{aligned} a &= 2 - 2\sqrt{\frac{\mathrm{R}_{00}\mathrm{R}_{90}}{(1+\mathrm{R}_{00})(1+\mathrm{R}_{90})}} \\ h &= \sqrt{\frac{\mathrm{R}_{00}(1+\mathrm{R}_{90})}{\mathrm{R}_{90}(1+\mathrm{R}_{00})}} \\ &\text{and } c = 2 - a \end{aligned} \tag{13}$$

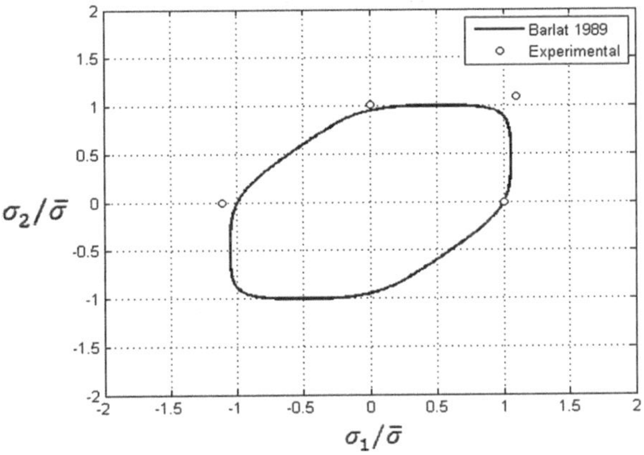

Fig. 3 Barlat 1989 yield locus for Ti–6Al–4V alloy at 400 °C

where r_θ is anisotropy ratio measured for θ orientation from rolling direction p is found out iteratively. Equation (14) is used to find out the value of p iteratively.

$$r_\theta = \frac{2\sigma_y^m}{\left(\frac{\partial\phi}{\partial\sigma_x} + \frac{\partial\phi}{\partial\sigma_y}\right)\sigma_\theta} - 1 \tag{14}$$

The representative Barlat 1989 yield locus for Ti–6Al–4V alloy at 400 °C is shown in Fig. 3.

In sheet metal forming industries, Hill 1948 and Barlat 1989 model are quite popular because it is very easy to determine the material constant by using only uniaxial tensile test. However, most of the metal forming operations is carried out under biaxial states of stress; the stress–strain formability parameters obtained by uniaxial tensile testing are inadequate (Hannon and Tiernan 2008). In tensile tests, the maximum attainable strain is reported to be limited to approximately 30 %. This leads to interpolations of the test data that are necessary beyond the fracture point (Gutscher et al. 2004). Limitations of uniaxial tensile test have been described by several researchers (Kaya et al. 2008; Atkinson 1996). Subsequently, biaxial testing has become of great importance in establishing the formability characteristics of sheet metal, whereas the material is subjected to deformation in more than one plane or axis. There are numerous methods of producing biaxial stresses in a material for different types of specimens. These include the bulge test (Demmerle and Boehler 1993; Vlassak and Nix 1992), combined tension–torsion test (Keefe et al. 1998) and biaxial testing of sheet metal (Hannon and Tiernan 2008).

Biaxial testing with various types of cruciform specimens is the most promising method to realize various stress states of biaxial tension by changing the proportion of load or displacement of the two axes (Xiang-Dong et al. 2005). Presently, there is no any particular standard available in cruciform specimen dimensions (Hannon and Tiernan 2008). Another popular method to determine the biaxial state of stress is the hydraulic bulge test (Bong et al. 2013). However, since the hydraulic bulge

Table 2 Material properties for Ti–6Al–4V alloy at 400 °C

Temperature (°C)	σ_0 (MPa)	σ_{45} (MPa)	σ_{90} (MPa)	σ_C (MPa)	σ_b (MPa)	r_0	r_{45}	r_{90}	r_b
400	681.0	591.0	691.0	751.0	744.9	0.6	1.26	0.512	1.02

test ensures fluid pressure, it is usually very challenging to measure the flow curve at warm forming temperature (Lee et al. 2013).

The development of the latest yield criteria usually needs biaxial yield stress and biaxial anisotropic coefficients. In this study, biaxial data were taken from previous literature (Odenberger et al. 2013). Table 2 indicates the material properties required for determination of various yield criteria parameters at 400 °C.

3.4 Barlat 1996 Yield Criterion

The yield criterion proposed by Barlat et al. (2003) is expressed in Eq. (15)

$$\phi = \propto_1 |S_2 - S_3|^a + \propto_2 |S_3 - S_1|^a + \propto_3 |S_1 - S_2|^a = 2\bar{\sigma}^a \tag{15}$$

where $\bar{\sigma}$ is equivalent stress and a is a material parameter which is the same as Barlat 1989 yield criterion. S_i corresponds to principal values of Cauchy stress deviator.

The equivalent stress deviator is defined as

$$[s] = [L][\sigma]$$
$$\text{where, } L = \begin{bmatrix} (c_2 + c_3)/3 & -c_3/3 & -c_2/3 & 0 \\ -c_3/3 & (c_3 + c_1)/3 & -c_1/3 & 0 \\ -c_2/3 & -c_1/3 & (c_1 + c_2)/3 & 0 \\ 0 & 0 & 0 & c_4 \end{bmatrix} \tag{16}$$

c_1, c_2, c_3, and c_4 are material constants and $\propto_{i=1,2,3}$ depends upon the orientation of frame of the principal value of S and anisotropy axis (Barlat and Lian 1989).

$$\propto_i = \propto_x P_{1i}^2 + \propto_y P_{2i}^2 + \propto_z P_{3i}^2 \tag{17}$$

For a plane stress condition when shear stress is zero, the yield function becomes

$$\phi = \propto_x |S_y - S_z|^a + \propto_y |S_z - S_x|^a + \propto_{z0} |S_x - S_y|^a = 2\bar{\sigma}^a \tag{18}$$

where

$$S_x = \frac{c_3 + c_2}{3}\sigma_x - \frac{c_3}{3}\sigma_y$$
$$S_y = \frac{-c_3}{3}\sigma_x + \frac{c_{11} + c_3}{3}\sigma_y$$
$$S_z = \frac{-c_2}{3}\sigma_x - \frac{c_1}{3}\sigma_y$$

For instance, consider $\propto_{z0} = 1$. When the shear stress is not equal to zero, other constants c_6 and $\propto_{z1}$ are used to introduce anisotropy. Therefore, seven tests are needed to be performed for determining the material constants.

For uniaxial tension in rolling direction, the components of the stress deviator are

$$\begin{aligned} S_x &= \frac{c_3 + c_2}{3}\sigma_0 \\ S_y &= \frac{-c_3}{3}\sigma_0 \\ S_z &= \frac{-c_2}{3}\sigma_0 \end{aligned} \tag{19}$$

By substituting these components into Eq. (18)

$$F = \propto_x |c_2 - c_3|^a + \propto_y |2c_2 + c_3|^a + \propto_{z0} |c_2 + 2c_3|^a - 2(3\bar{\sigma}/\sigma_0)^a \tag{20}$$

Similarly, for uniaxial tension in 90° orientation and biaxial stress state, the equation reduces in the form of Eq. (21)

$$\begin{aligned} F &= \propto_x |c_3 - 2c_1|^a + \propto_y |c_3 - c_1|^a + \propto_{z0} |c_1 + 2c_3|^a - 2(3\bar{\sigma}/\sigma_{90})^a \\ F &= \propto_x |2c_1 + c_2|^a + \propto_y |c_1 + 2c_2|^a + \propto_{z0} |c_1 - c_2|^a - 2(3\bar{\sigma}/\sigma_b)^a \end{aligned} \tag{21}$$

Other three expressions are obtained by a function expressed in Eq. (22) considering r value's directionality requirement.

$$\begin{aligned} r_0 &= \frac{2\bar{\sigma}^a}{(\frac{\partial\varphi}{\partial S_x} + \frac{\partial\varphi}{\partial S_y})\sigma_0} - 1 \\ r_{90} &= \frac{2\bar{\sigma}^a}{(\frac{\partial\varphi}{\partial S_x} + \frac{\partial\varphi}{\partial S_y})\sigma_{90}} - 1 \\ r_b &= \frac{2\bar{\sigma}^a}{(\frac{\partial\varphi}{\partial S_x} + \frac{\partial\varphi}{\partial S_y})\sigma_b} - 1 \end{aligned} \tag{22}$$

Therefore, six material constants are calculated by solving above six nonlinear equations. The material constants for Barlat 1996 yield criteria are presented in Table 3. The yield locus is plotted as shown in Fig. 4.

Table 3 Barlat 1996 model material parameters for Ti–6Al–4V alloy

Temperature (°C)	C_1	C_2	C_3	C_4	α_x	α_y	α_{z0}	a
400	1.059	0.939	1.233	1	0.992	0.992	1	8

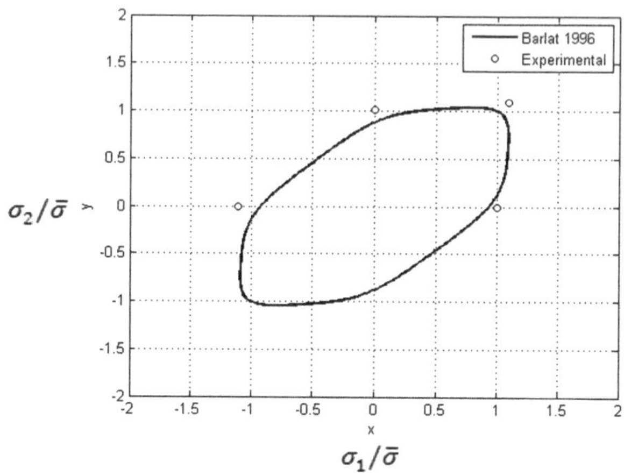

Fig. 4 Barlat 1996 yield locus for Ti–6Al–4V alloy at 400 °C

3.5 *Barlat 2000 Yield Criterion*

Barlat and Lian (1989) proposed an anisotropic yield criterion which is made of two convex functions. The equation for the yield locus is given by Eq. (23)

$$\begin{gathered}\phi' = |X'_1 - X'_2|^a;\ \phi'' = |2X''_2 + X''_1|^a + |2X''_1 + X''_2|^a \\ \phi = |X'_1 - X'_2|^a + |2X''_2 + X''_1|^a + |2X''_1 + X''_2|^a = 2\sigma_y^a\end{gathered} \tag{23}$$

$$\begin{gathered}\{X'\} = [C']\{s\} = [C'][T][\sigma] = [L'][\sigma] \\ \{X''\} = [C'']\{s\} = [C''][T][\sigma] = [L''][\sigma] \\ \text{where} \quad [T] = \begin{bmatrix} 2/3 & -1/3 & 0 \\ -1/3 & 2/3 & 0 \\ 0 & 0 & 1 \end{bmatrix}\end{gathered} \tag{24}$$

in which $X'_{1,2}$ and $X''_{1,2}$ are the principal values of the linearly transformed stress deviator matrices $\{s\}$,

When the matrices C' and C'' are taken as identity matrix, the above criteria reduce to isotropic case. For the simplicity of calculation, anisotropy parameters and the coefficients of L' and L'' are related as,

Table 4 Barlat 2000 model material parameters for Ti–6Al–4V alloy

Temperature (°C)	α_1	α_2	α_3	α_4	α_5	α_6	α_7	α_8	a
400	0.982	0.853	0.787	0.976	0.986	0.787	1.132	1.426	8

$$\begin{bmatrix} L'_{11} \\ L'_{12} \\ L'_{21} \\ L'_{22} \\ L'_{66} \end{bmatrix} = \begin{bmatrix} 2/3 & 0 & 0 \\ -1/3 & 0 & 0 \\ 0 & -1/3 & 0 \\ 0 & 2/3 & 0 \\ 0 & 0 & 1 \end{bmatrix} \begin{bmatrix} \alpha_1 \\ \alpha_2 \\ \alpha_7 \end{bmatrix} \tag{25}$$

$$\begin{bmatrix} L''_{11} \\ L''_{12} \\ L''_{21} \\ L''_{22} \\ L''_{66} \end{bmatrix} = \begin{bmatrix} -2 & 2 & 8 & -2 & 0 \\ 1 & -2 & -4 & -4 & 0 \\ 4 & -4 & -4 & 1 & 0 \\ -2 & 8 & 2 & -2 & 0 \\ 0 & 0 & 0 & 0 & 9 \end{bmatrix} \begin{bmatrix} \alpha_3 \\ \alpha_4 \\ \alpha_5 \\ \alpha_6 \\ \alpha_8 \end{bmatrix} \tag{26}$$

Exponent a is the same parameter as in the three parameter Barlat's criterion. Matrices C' and C'' are expressed in terms of eight anisotropy coefficients $\propto_i$, for isotropic case $\propto_i$ is unity.

For determination of material parameters $\propto_1$ to $\propto_8$, eight tests are needed to be performed. In this study, yield stresses in 0°, 45°, and 90° orientation and Lankford coefficients in the same direction are obtained from experiments. In addition to that, biaxial yield stress and biaxial anisotropy coefficient are considered from previous literature (Odenberger et al. 2013).

Deviatoric components are defined for loading each stress of state as Eq. (27)

$$s_x = \gamma\sigma; \quad s_y = \delta\sigma \tag{27}$$

The material parameters are determined by solving two equations given below (28) for each state of stress.

$$\begin{aligned} F &= |\propto_1\gamma - \propto_2\delta|^a + |\propto_3\gamma - 2\propto_4\delta|^a + \propto_{z0}|2\propto_5\gamma - \propto_6\delta|^a - 2(\bar{\sigma}/\sigma)^a \\ G &= q_x\frac{\partial\phi}{\partial s_x} - q_y\frac{\partial\phi}{\partial s_y} = 0 \end{aligned} \tag{28}$$

The parameter descriptions were given in previous literature (Odenberger et al. 2013). The material constants for Barlat 2000 criterion are presented in Table 4. The yield locus plotted by Barlat 2000 model is as shown in Fig. 5.

Fig. 5 Barlat 2000 yield locus at 400 °C

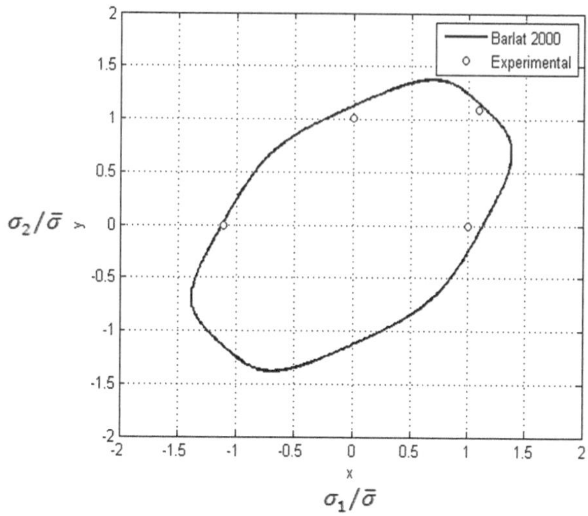

3.6 *Cazacu–Barlat Yield Criterion*

Cazacu et al. (2006) proposed an anisotropic yield criterion which consists of both tension and compression asymmetry. For extending this criterion, stress deviator *s* is linearly transformed and the principle values of Cauchy stress deviator in the yield function are replaced by transformed tensor. The proposed anisotropic yield function is given as Eq. (28)

$$\left[\left|\sum_1\right| - k\sum_1\right]^a + \left[\left|\sum_2\right| - k\sum_2\right]^a + \left[\left|\sum_3\right| - k\sum_3\right]^a = F \tag{29}$$

where $\Sigma = C[s]$

$$C = \begin{bmatrix} C_{11} & C_{12} & C_{13} & & & \\ C_{12} & C_{22} & C_{23} & & & \\ C_{13} & C_{23} & C_{33} & & & \\ & & & C_{44} & & \\ & & & & C_{55} & \\ & & & & & C_{66} \end{bmatrix}$$

where C is a fourth-order transformation tensor with reference to orthotropic (x, y, z) axes and F is the size of yield locus, k is based on yield stress in tension/compression and material parameter a in the yield function.

$$\frac{\sigma_T}{\sigma_C} = \left\{\frac{(2/3(1+k))^a + 2(1/3(1-k))^a}{(2/3(1-k))^a + 2(1/3(1+k))^a}\right\}^{1/a} \tag{30}$$

Material parameter a is considered as 8 for Ti–6Al–4V. The material constants in tensor C are determined by performing uniaxial yield stress in tension and compression along with balanced biaxial test and anisotropy coefficients (Cazacu et al. 2006).

Consider plane stress condition for thin sheet, the nonzero components are the $(\sigma_{xx}, \sigma_{yy}, \tau_{xy})$. Hence, the principal values of Σ are as follows:

$$\begin{aligned}\sum_{1} &= \frac{1}{2}\left[\sum_{xx} + \sum_{yy} + \sqrt{\left(\sum_{xx} - \sum_{yy}\right)^{2} + 4\sum_{xy}^{2}}\right]\\ \sum_{2} &= \frac{1}{2}\left[\sum_{xx} + \sum_{yy} - \sqrt{\left(\sum_{xx} - \sum_{yy}\right)^{2} + 4\sum_{xy}^{2}}\right]\\ \sum_{3} &= \sum_{zz}\end{aligned} \tag{31}$$

where

$$\begin{aligned}\sum_{xx} &= \left(\frac{2}{3}C_{11} - \frac{1}{3}C_{12} - \frac{1}{3}C_{13}\right)\sigma_{xx} + \left(-\frac{1}{3}C_{11} + \frac{2}{3}C_{12} - \frac{1}{3}C_{13}\right)\sigma_{yy}\\ \sum_{yy} &= \left(\frac{2}{3}C_{12} - \frac{1}{3}C_{22} - \frac{1}{3}C_{23}\right)\sigma_{xx} + \left(-\frac{1}{3}C_{12} + \frac{2}{3}C_{22} - \frac{1}{3}C_{23}\right)\sigma_{yy}\\ \sum_{zz} &= \left(\frac{2}{3}C_{13} - \frac{1}{3}C_{23} - \frac{1}{3}C_{33}\right)\sigma_{xx} + \left(-\frac{1}{3}C_{13} + \frac{2}{3}C_{23} - \frac{1}{3}C_{33}\right)\sigma_{yy}\\ \sum_{xy} &= C_{66}\sigma_{xy}\end{aligned} \tag{32}$$

Let σ_0^T and σ_0^C be yield stresses in tension and compression, respectively, along the 0° orientation. The proposed orthotropic yield criterion is written as Eq. (33)

$$\begin{aligned}\sigma_0^T &= \left\{\frac{F}{[|\phi_1| - k\phi_1]^a + [|\phi_2| - k\phi_2]^a + [|\phi_3| - k\phi_3]^a}\right\}^{\frac{1}{a}}\\ \sigma_0^C &= \left\{\frac{F}{[|\phi_1| + k\phi_1]^a + [|\phi_2| + k\phi_2]^a + [|\phi_3| + k\phi_3]^a}\right\}^{\frac{1}{a}}\end{aligned} \tag{33}$$

$$\begin{aligned}\phi_1 &= \left(\frac{2}{3}C_{11} - \frac{1}{3}C_{12} - \frac{1}{3}C_{13}\right)\\ \phi_2 &= \left(\frac{2}{3}C_{12} - \frac{1}{3}C_{22} - \frac{1}{3}C_{23}\right)\\ \phi_3 &= \left(\frac{2}{3}C_{13} - \frac{1}{3}C_{23} - \frac{1}{3}C_{33}\right)\end{aligned} \tag{34}$$

Similarly for transverse direction,

$$
\begin{aligned}
\sigma_{90}^{T} &= \left\{\frac{F}{[|\Psi_1| - k\psi_1]^a + [|\psi_2| - k\psi_2]^a + [|\psi_3| - k\psi_3]^a}\right\}^{\frac{1}{a}} \\
\sigma_{90}^{C} &= \left\{\frac{F}{[|\Psi_1| + k\psi_1]^a + [|\psi_2| + k\psi_2]^a + [|\psi_3| + k\psi_3]^a}\right\}^{\frac{1}{a}}
\end{aligned}
\tag{35}
$$

where

$$
\begin{aligned}
\psi_1 &= \left(-\frac{1}{3}C_{11} + \frac{2}{3}C_{12} - \frac{1}{3}C_{13}\right) \\
\psi_2 &= \left(-\frac{1}{3}C_{12} + \frac{2}{3}C_{22} - \frac{1}{3}C_{23}\right) \\
\psi_3 &= \left(-\frac{1}{3}C_{13} + \frac{2}{3}C_{23} - \frac{1}{3}C_{33}\right)
\end{aligned}
\tag{36}
$$

Similarly for biaxial direction,

$$
\sigma_{b}^{T} = \left\{\frac{F}{[|\Omega_1| + k\Omega_1]^a + [|\Omega_2| + k\Omega_2]^a + [|\Omega_3| + k\Omega_3]^a}\right\}^{\frac{1}{a}}
\tag{37}
$$

where

$$
\begin{aligned}
\Omega_1 &= \left(\frac{1}{3}C_{11} + \frac{1}{3}C_{12} - \frac{2}{3}C_{13}\right) \\
\Omega_2 &= \left(\frac{1}{3}C_{12} + \frac{1}{3}C_{22} - \frac{2}{3}C_{23}\right) \\
\Omega_3 &= \left(\frac{1}{3}C_{13} + \frac{1}{3}C_{23} - \frac{2}{3}C_{33}\right)
\end{aligned}
\tag{38}
$$

Let r_θ is Lankford coefficient under the uniaxial tension case, according to proposed orthotropic criterion

$$
\begin{aligned}
r_0^T &= -\frac{(1-k)^a \Phi_1^{a-1}\Psi_1 + (-1-k)^a\left(\Phi_2^{a-1}\Psi_2 + \Phi_3^{a-1}\Psi_3\right)}{(1-k)^a \Phi_1^{a-1}(\Psi_1 + \Phi_1) + (-1-k)^a\left(\Phi_2^{a-1}\Psi_2 + \Phi_3^{a-1}\Psi_3 + \Phi_2^a + \Phi_3^a\right)} \\
r_{90}^T &= -\frac{(1-k)^a \Psi_2^{a-1}\Phi_2 + (-1-k)^a\left(\Psi_1^{a-1}\Phi_1 + \Psi_3^{a-1}\Phi_3\right)}{(1-k)^a \Psi_2^{a-1}(\Psi_2 + \Phi_2) + (-1-k)^a\left(\Psi_1^{a-1}\Phi_1 + \Psi_3^{a-1}\Phi_3 + \Psi_1^a + \Psi_3^a\right)}
\end{aligned}
\tag{39}
$$

Table 5 Cazacu–Balart yield model material parameters for Ti–6Al–4V alloy

Temperature (°C)	C_{11}	C_{22}	C_{33}	C_{12}	C_{13}	C_{23}	C_{44}	a	k
400	1.959	1.837	1.955	0.684	0.480	0.451	1.880	8	−0.085

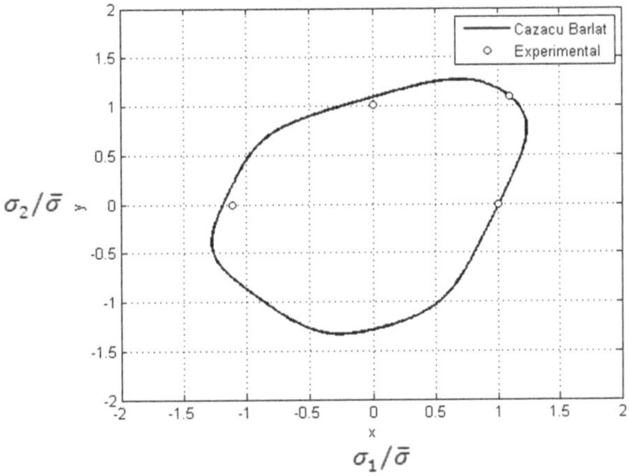

Fig. 6 Cazacu–Barlat yield locus for Ti–6Al–4V alloy at 400 °C

The material constants have been calculated by solving a set of nonlinear equations using MATLAB. The material constants are presented in Table 5. The yield locus plotted by Cazacu–Barlat function is as shown in Fig. 6.

4 Important Points About Finite Element Analysis

The considerable increase in computer processing speeds makes it possible to include the FEM for investigating various aspects of metal forming processes, therefore overcoming most of the obstructive assumptions related with analytical models. Many investigators have adopted FEM to gain better understanding of sheet metal forming processes (Van den Boogaard and Huétink 2006; Abedrabbo et al. 2007). Numerous FE codes such as DYNAFORM/LSDYNA/ABAQUS/ANSYS/AUTOFORM are available for sheet metal forming simulations.

4.1 FE Analysis Method

There are generally two types of methods for solving the ordinary differential equations which are derived from the problem: (a) Implicit method and (b) Explicit method.

Implicit method is not suitable for highly nonlinear problems because of the convergence issue. Sheet metal forming is characterized by many nonlinearity not only due to structural and material nonlinearity, but also due to contact between the bodies makes the problem highly nonlinear. However, by using explicit method, these nonlinearities can be treated without any problem (Nielsen 2000). Few advantages of the explicit formulation are listed below:

- Few computations are required per time step
- The algorithm is simple in logic and structure, so complex nonlinearities are easily handled.
- It is reliable in accuracy and completion of computation.

4.2 Selection of Element Type

One of the major choices when developing a proper finite element model is the choice of element formulation. Most of sheet metal forming processes is considered to be in a plane stress condition. For plane stress condition, generally shell element has better choice. The following are some important shell elements available in LSDYNA.

- Selective reduced Hughes–Liu shell (SRHL)
- Selective reduced co-rotational Hughes–Liu shell (SRCOHL)
- Hughes–Liu shell (HL)
- Englemann–Whirley shell (EW)
- Co-rotational Hughes–Liu shell (COHL)
- Belystchko–Lin–Tsay shell (BLT)
- Belystchko–Wong–Chiang shell (BWC)

The detailed description about each element is available in LSDYNA user manual (LS-DYNA Keyword User's Manual 2007). For the present FE analysis, the blank and the tool components were meshed using Belystchko–Tsay shell elements as it takes less computational time, around 30–50 % less than other elements like Hughes and Liu (Kotkunde et al. 2014a; Nielsen 2000).

4.3 Mesh Adaptivity

Metal forming simulations are considered as large deformation problems. Therefore, the elements may undergo excessive deformation which leads to abrupt increase in aspect ratio of the element. Such increase in aspect ratio of the element increases the stiffness in one direction and decreases in another which further affects the accuracy of the simulation. Also, mesh size affects the time step calculation by taking smallest length into account and as the time step decreases the time required to solve the nonlinear problem increases to a large extent. Therefore, in order to resolve such a problem adaptive meshing is used. Adaptive meshing divides the element into a number of parts when the aspect ratio becomes excessively large and

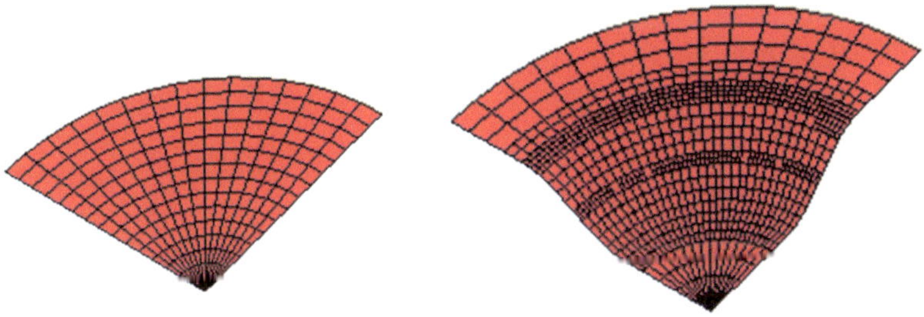

Fig. 7 Mesh adaptivity for blank

maintains the stiffness properties balanced in both direction and thereby improving the results of finite element simulations. Figure 7 shows mesh adaptivity after certain deformation in the blank.

4.4 Mass Scaling

Sheet metal forming simulations are highly nonlinear in nature, and it takes a lot of computational time to get the desired output. In industries, it is desirable to reduce the simulation time to get the required results. Therefore, few methods are available in the commercial FE analysis software for increasing of simulation speed. Specifically, mass scaling method is popularly used to speed the FE analysis (Olovsson et al. 2005).

For a shell element the time step size is calculated using Eq. (40). From this relation, it is clear that the time step size can be changed by changing the element length (*L*), Young's modulus (*E*), and density (ρ). But changing *L* will change the mesh density and deteriorates the results; changing E is not recommended as it greatly affects the material stiffness. The last option is changing the density which may not have a significant effect on analysis results. This process of changing density to increase the time step and to minimize the computational time is called as mass scaling.

$$\Delta t_c = \frac{L}{c} \qquad \text{where} \quad c = \sqrt{\frac{E}{\rho(1-\vartheta^2)}} \tag{40}$$

Mass scaling is done manually or automatically in commercial software packages. In manual mass scaling, the density of material is changed in the material card. The excess mass scaling also induces dynamic effects; thus, it is important to monitor the kinetic energy of the system to understand about the effect of mass

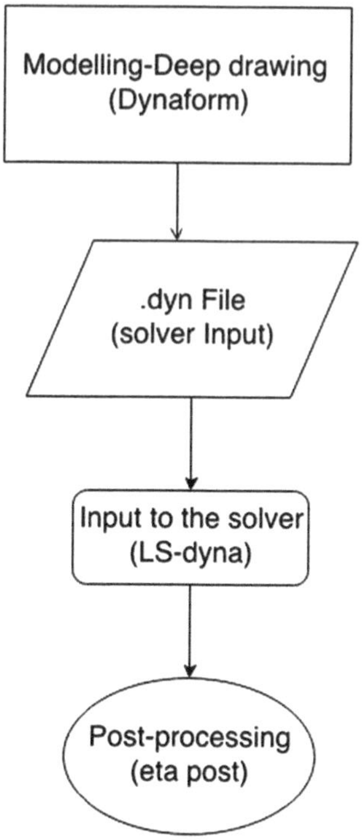

Fig. 8 Workflow of DYNAFORM software program

scaling on the results. These are a few salient points one needs to take into account before a FE analysis. Figure 8 shows the work flow of DYNAFORM software program.

5 Finite Element Analysis of Warm Forming for Ti–6Al–4V Alloy

5.1 Experimental Details

Ti–6Al–4V alloy sheet of 0.9 mm thickness is used. The drawability range of Ti–6Al–4V alloy was found by performing experiments at various temperatures. Temperature higher than 400 °C increases the oxygen contamination in Ti–6Al–4V alloy, and with oxygen, the material becomes more brittle due to formation of α-scale. Therefore, it is preferred to perform warm forming of Ti–6Al–4V alloy in an inert and protective environment (Kotkunde et al. 2014a). Considering the limitations of experimental facility at higher temperatures with inert environment, the experiments were performed up to 400 °C at interval of 50 °C on different diameter blanks. Molykote was used for forming process at elevated temperatures as an effective lubricant (Singh et al. 2010). Blanks were kept at particular temperature for certain duration (approximately 3–5 min.) for uniform heating of the sheet. Deep drawing operation was performed when the blank reaches the required temperature.

Based on the reported study, it is seen that Ti–6Al–4V alloy is difficult to form at room temperature (Odenberger et al. 2013). The similar results are observed in the present investigation for temperature up to 150 °C. A maximum of 54-mm-diameter blank was drawn successfully at 400 °C; hence, the maximum limiting draw ratio (LDR) of 1.8 is achieved which is lesser than other metallic materials such as steel and aluminum (Singh et al. 2010; Hussaini et al. 2014).

5.2 Preprocessing for FE Analysis

The input models such as die, blank, blank holder, and punch were constructed in preprocessor. For the present investigation, isothermal model was used. After the surface was generated, fine meshing was done on the surface of the tool components and on the blank. Adaptive meshing was used on the blank to obtain accurate results. The complete model in the preprocessor is shown in Fig. 9.

The blank and the tool components were meshed using Belytschko–Tsay shell elements as it takes less computational time, around 30–50 % less than others (Kotkunde et al. 2014a). Material properties were calculated using uniaxial tensile test performed at room temperature to 400 °C at interval of 50 °C and a constant velocity of 2 mm/s. Table 1 indicates the material properties for Ti–6Al–4V alloy at warm conditions. Selective mass scaling was used for finite element (FE) simulations. von-Mises, Hill 1948, and Barlat 1989 yield criteria with Hollomon power law are used for FE analysis. The evaluated material properties used for the simulations are shown in Table 1. The coefficient of friction under the forming conditions was calculated by inverse approach. The methodology was adopted by varying the coefficient of friction values and superimposing experimental and simulated plot of maximum punch force. Figure 10 shows representative load

Fig. 9 Axisymmetric model of deep drawing setup

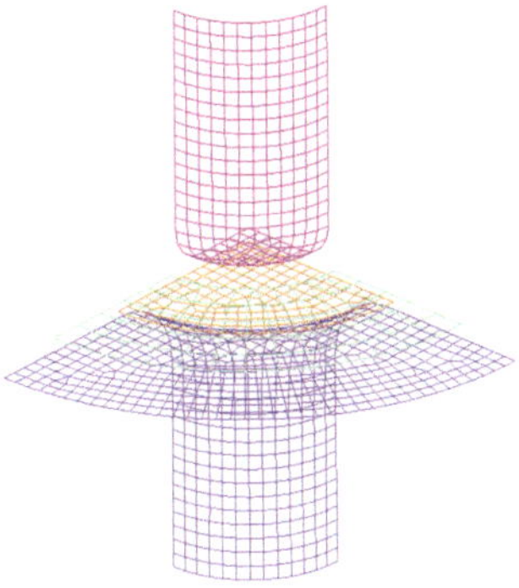

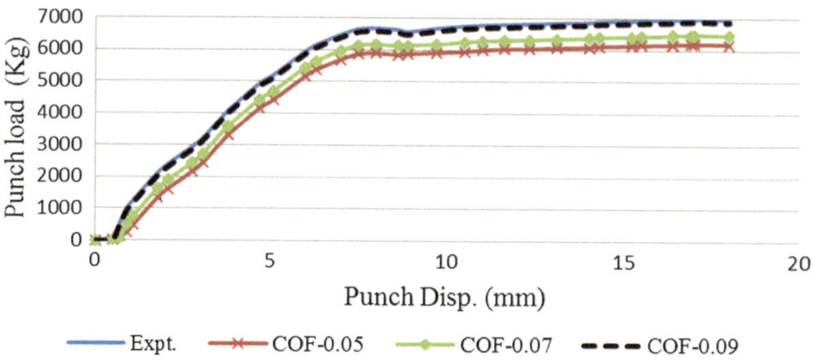

Fig. 10 Representative experimental and FE validation of punch load versus punch disp. at 400 °C

Table 6 Variation of coefficient of friction value with temperature

Temperature (°C)	150	200	250	300	350	400
COF	0.05	0.05	0.07	0.07	0.08	0.09

versus displacement graph for the punch at 400 °C. Table 6 shows variation of coefficient of friction value with temperature. As expected, as temperature increases, coefficient of friction value increases from 0.05 to 0.09. The variation in the value is negligible, it indicates the effectiveness of Molkote lubricant during the experimentation at high temperature (Singh et al. 2010).

5.3 Failure Study

The fracture can be distinguished in two ways, initial fracture and final fracture, shown in Figs. 11a, b and 12a. Similar fractures were reported in the previous literature (Djavanroodi and Derogar 2010; Zhang et al. 2007). Initial fracture appears in the punch corner region since the material possesses very low strain-hardening

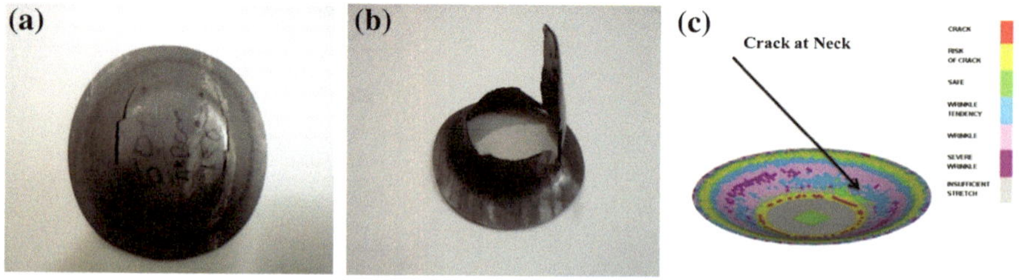

Fig. 11 Comparison of neck failure at experimentation and FE analysis using Barlat 1989 yield criterion

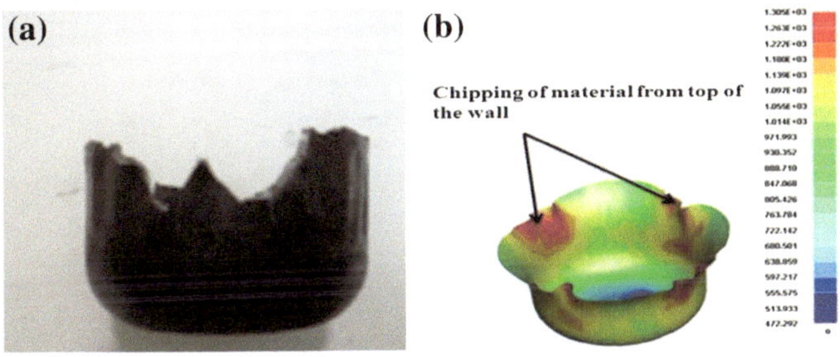

Fig. 12 Comparison of chipping of material at experimentation and FE analysis using Barlat 1989 yield criterion

exponent at room temperature. For larger diameter blanks, due to more punch load requirement the extent of thinning will be more; hence, failure will be appearing in the neck region.

Another type of fracture observed is final fracture, i.e., chipping of wall materials from upper region. This type of fracture is seen at the temperature of 400 °C with larger blank diameter (above 56 mm). Bauschinger effect is more pronounced in Ti–6Al–4V alloy upto lower elevated temperature (Odenberger et al. 2013). Heavy compressive hoop stresses and radial tensile stresses are developed in the flange region. Since the material loses its compressive strength due to Bauschinger effect therefore, even a slight tendency of wrinkling will lead to fracture. Moreover, at this temperature work-hardening tendency of the material is almost insignificant as mentioned in Table 1. As the material is moving from flange toward die curvature, there is possibility of increased friction at higher temperatures which is evidence from Table 6. Therefore, excessive shear stresses will be developed on the outer surface of material in the bend region of die. In addition to this, compressive stresses will be developed on inner curvature of the blank. Since the material is weak in compression due to Bauschinger effect, the inner curvature of blank may become the site for the initiation of failure. As discussed before, this material does not exhibit much work hardening even at higher temperatures; hence, drawing may lead to shear fracture in the upper region of the wall. One of the possible solutions to overcome this problem is increasing the die curvature which decreases the intensity of the compressive stresses on the inner surface of the sheet metal. When large-diameter blank is being drawn, due to the excessive shear stresses on the bend region of die, the initiated crack has sufficient time to grow and fracture the material. On the other hand for smaller diameter blanks, the initiated crack does not have sufficient time to propagate and the cups can be successfully drawn before failure takes place. These are the major reasons that in the drawn cups, the fracture occurs near the region of upper wall. The fracture at upper wall portion (chipping of material) is shown in Fig. 13a.

Strain data point in the simulation shows fracture in the punch corner as shown in Fig. 12c; this indicates that the fracture may appear at the beginning of an experimental drawing process. Simulation of 56-mm-and-above-diameter cups at

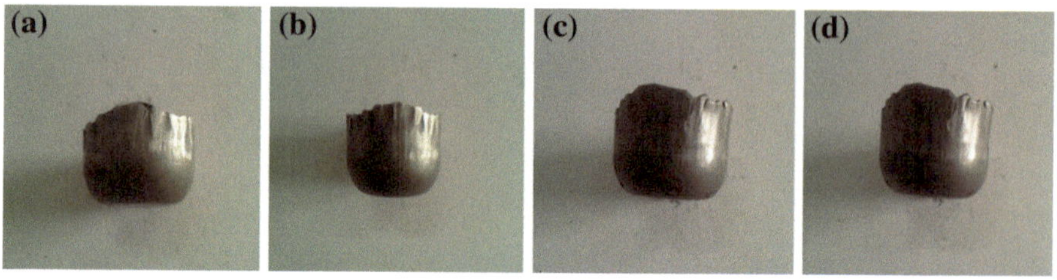

Fig. 13 Representative drawn cups at **a** 50-mm-diameter blank and 150 °C, **b** 52-mm-diameter blank and 250 °C, **c** 54-mm-diameter blank and 250 °C, **d** 54-mm-diameter blank and 400 °C

400 °C shows excessively high von-Misses stresses in the upper region of the cup, and it is shown in Fig. 13b. The von-Misses stresses should not exceed mean flow stress as it may lead to fracture in the material. Hence, it can be interpreted that the fracture can take place in this area. The detailed study of failure requires incorporating damage model in FE analysis. The most popular damage models are Johnson–Cook damage model and Gurson damage model (Wierzbicki et al. 2005). For incorporating these models in FE analysis, one need to find out parameters for these models using experimental data. For present FE analysis, we were not considering any damage model. Our intention is to show the crack initiation and possibility of failure by FE analysis.

5.4 Thickness Distribution and Earing Profile Prediction

It can be seen from Table 1 that the strain-hardening exponent (n) is very low at room temperature. Similar range of n values was reported in the previous literature (EI-Domiaty 1992). With increase in temperature, there is gradual increase in the strain-hardening exponent which implies the material has more scope for elongation before necking. That may be the reason Ti–6Al–4V alloy could be drawn above 150 °C. Representative drawn cups are shown in Fig. 13.

Drawn cups were cut using wire cut EDM, and the thickness was measured from the center toward the cup wall. Thickness plots at various temperatures for different blank diameters are shown in Fig. 14. As expected, neck formation appeared near the punch corner radius and increase in thickness was observed in wall region. As discussed before, at higher temperature between 150 and 400 °C, there is a consistent increase in the strain hardening of material compared to room temperature. Hence, reduced necking tendency can be seen in all thickness plots shown in Fig. 14. As the temperature of blank increases, there is consistent decrease in ultimate tensile strength and yield strength of material as mentioned in Table 1; this decreases the mean flow stress at which the material is being drawn. Hence, lesser amount of forces will be appeared on the punch; this will further decrease the variation of the thickness in drawn cup. Therefore, at higher temperature, more

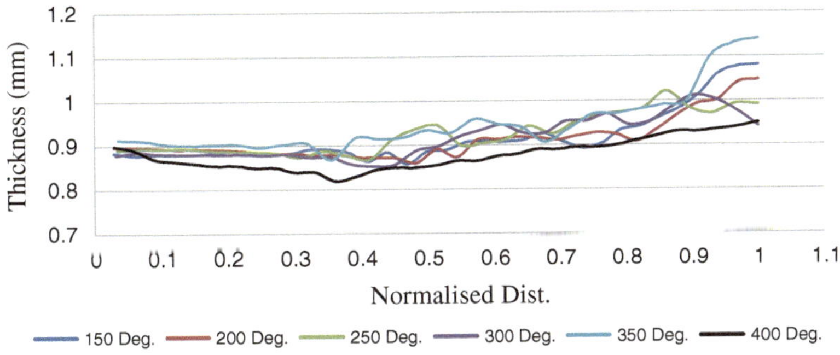

Fig. 14 Representative thickness distribution at 52-mm-diameter blank with variation of temperature

uniform thickness can be observed due to lower mean flow stress and slightly larger work-hardening exponent. Alternatively, for larger blank diameter since a larger area of flange has to be drawn into die, there will be increase in punch load, and hence, increased tendency of thinning will appear at the punch corner. Based on this observation, it can be concluded that as temperature increases, thickness is more uniform which means better quality of cups will be obtained with optimum blank diameter. FE analysis has been carried out for all the possible blank diameters and variation of temperature from 150 to 400 °C at interval of 50 °C. For present analysis von-Mises, Hill 1948 and Barlat 1989 yield criteria have been used. Figure 15 shows representative thickness distribution comparison at 400 °C and 52-mm-diameter blank. It has been clearly indicated that von-Mises yield criterion thickness prediction is very poor compared to other two yield criteria. Even tendency of necking prediction at punch corner region is poorly predicted by von-Mises yield criterion. The average absolute error is almost 10 % which indicate very poor prediction of thickness distribution. However, Hill 1948 and Barlat 1989 yield criteria follow the trend of experimental thickness distribution. Barlat 1989 yield

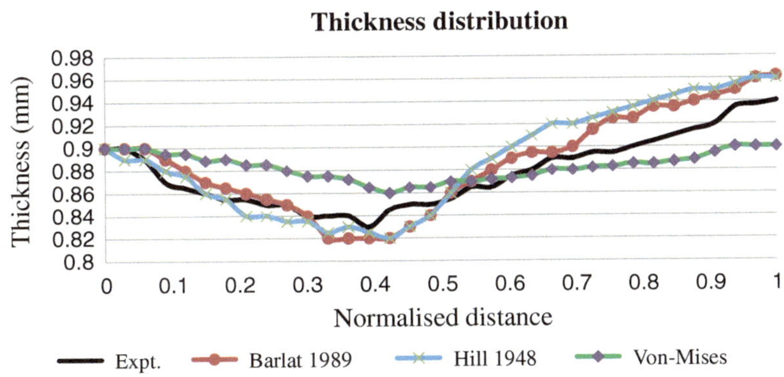

Fig. 15 Representative thickness distribution at 400 °C and 52-mm-diameter blank

Table 7 Average absolute error for experimental thickness

Blank diameter (mm)	Temperature (°C)	von-Mises error (%)	Hill 1948 error (%)	Barlat 1989 error (%)
50	150	9.25	5.21	5.00
50	200	8.57	6.2	5.13
50	250	9.34	5.37	5.77
50	300	9.75	5.57	5.24
50	350	8.89	6.25	5.77
50	400	8.78	5.2	4.44
52	150	9.15	5.2	4.65
52	200	9.25	4.5	3.62
52	250	8.75	5.2	4.13
52	300	8.5	5.0	4.35
52	350	10.25	7.0	6.75
52	400	10.50	8.25	7.16
54	150	8.75	6.25	5.05
54	200	8.0	4.0	2.74
54	250	9.25	4.25	2.93
54	300	8.75	5.25	5.98
54	350	9.25	5.25	4.22
54	400	9.85	5.0	4.24

model prediction is better than that of Hill 1948 criterion, specifically in the wall region thickness prediction. The average relative error is almost 5 % in both the cases. Table 7 indicates relative error between experimental thickness distribution and all above-mentioned three yield criteria for all the possible blanks with variation of temperature.

Earing in deep drawing for anisotropic metals is a pronounced phenomenon. Accurate prediction of earing profile is one of the main criteria for appropriate selection of yield criterion in FE analysis. Figure 16 shows earing profile of experimental and simulated cups. It clearly indicates that earing is poorly predicted using all the yield criteria. However, Barlat 1989 model cup height prediction is better than all other models. Therefore, considering a thickness distribution and earing profile, Barlat 1989 model is better than Hill 1948 and von-Mises yield criteria.

Deviation between experimental and FE results of thickness distribution and earing profile can be further minimized by using advanced yield criteria. But these advanced yield models have more number of material parameters which further need more number of tests such as bulge test, shear test, compression test, etc. For present study, biaxial yield data have been taken from previous literature (Odenberger et al. 2013). Barlat 1996, Barlat 2000 and Cazacu–Barlat yield criteria are chosen for further analysis.

Representative thickness distribution for the 54-mm-diameter blank is as shown in Fig. 17. All these three yield criteria prediction is better than that of Barlat 1989

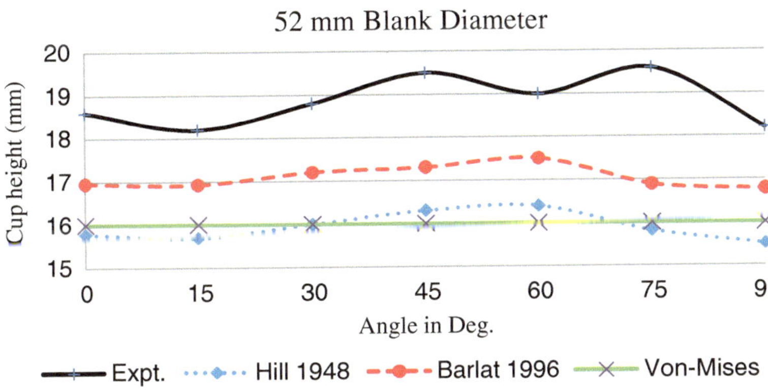

Fig. 16 Earing profile of experimental and simulated cups

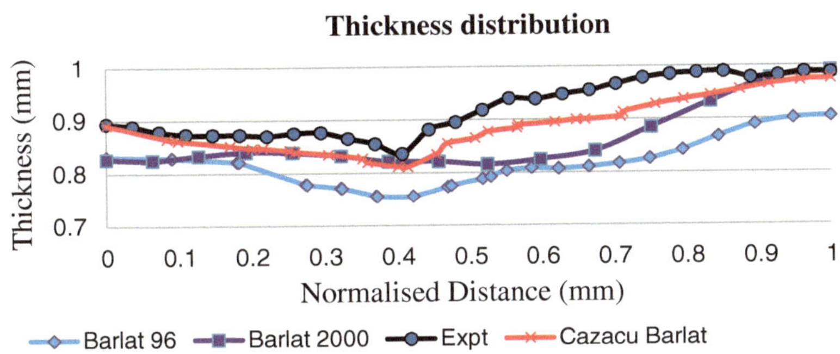

Fig. 17 Thickness distribution of 54-mm-diameter blank at 400 °C

model. The calculation of relative error of thickness along deep drawn cup is presented in Table 8. Average percentage error of thickness distribution and its standard deviation are chosen as statistical measures for comparing anisotropic yield criteria. It is clear from Table 8 that Cazacu–Barlat yield has the least error in prediction compared to other models.

Figure 18 shows earing profile of experimental and simulated cups using advanced yield criteria. Earing is poorly predicted by Barlat 1996 and Barlat 2000 yield models. However, Cazacu–Barlat yield criterion shows a better result in earing prediction. Therefore, considering thickness distribution and earing profile are qualitative parameters of deep drawn cups, Cazacu–Barlat yield model is well

Table 8 Comparison of average relative error (%) in thickness distribution

Material model	Barlat 1996	Barlat 2000	Cazacu–Barlat
Average relative error (%)	3.6	3.0	1.6
Standard deviation of error	0.0161	0.0232	0.0112

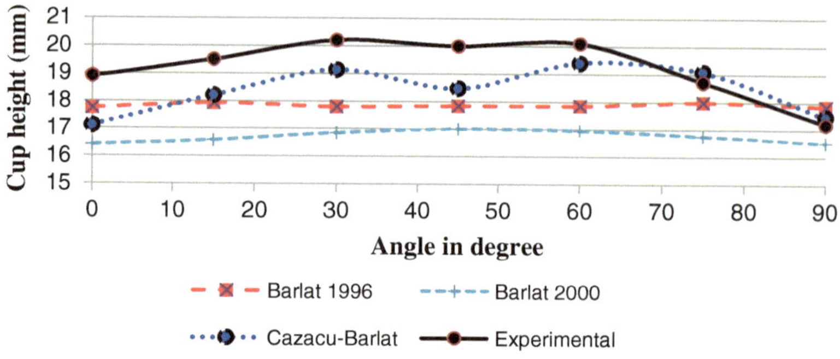

Fig. 18 Earing profile of experimental and simulated cup

suited for warm deep drawing of Ti–6Al–4V alloy. Table 9 indicates mechanical parameters required for all above-mentioned yield criteria.

For the simplicity in implementation of yield criteria, it is important to consider a number of required material parameters and experimental tests to be carried out for their evaluation. Table 9 shows different yield models and mechanical tests required for their determination. Hill 1948 has only four material parameters which are determined by performing simple uniaxial tension tests. Also, the simulation time taken for these models is relatively lesser than that of other models. Table 10 shows comparison of simulation time required for different anisotropic yield criteria. On the other hand, advanced yield models such as Barlat 1996 and Barlat 2000 and Cazacu–Barlat yield models need different mechanical tests to be carried out such as biaxial tensile test and compression test. which makes them complicated for their application in finite element simulation. Moreover, the simulation time required for these models are very large compared to Hill 1948 and Barlat 1989 yield models. It should be noticed that the simulation time for Cazacu–Barlat yield model is less as compared to Barlat 1996 and Barlat 2000 with adequate accuracy along with prediction of thickness distribution and earing profile in the deep drawn cup. Hence, Cazacu–Barlat yield model is most suitable for deep drawing simulation of Ti–6Al–4V alloy.

Table 9 Mechanical parameters required for yield criteria

Yield model	Parameters	Tests required								
		σ_0	σ_{45}	σ_{90}	σ_b	r_0	r_{45}	r_{90}	r_b	σ_c
von-Mises	1	*Y*								
Hill 1948	4	*Y*				*Y*	*Y*	*Y*		
Barlat 1989	3	*Y*				*Y*	*Y*	*Y*		
Barlat 1996	8	*Y*	*Y*	*Y*	*Y*	*Y*	*Y*	*Y*	*Y*	
Barlat 2000	8	*Y*	*Y*	*Y*	*Y*	*Y*	*Y*	*Y*	*Y*	
Cazacu–Barlat	8	*Y*	*Y*	*Y*	*Y*	*Y*	*Y*	*Y*		*Y*
Y = Test required										

Table 10 Total CPU time taken for simulation

Yield model	Time (s)
von-Mises	6578
Hill 1948	7166
Barlat 1989	8769
Barlat 1996	82,722
Barlat 2000	76,014
Cazacu–Barlat	37,398

6 Conclusions

This work involves experimentations of deep drawing on Ti–6Al–4V alloy, its failure, and formability studies at warm condition. In addition to that, determination of various anisotropic yield criteria parameters and its implementation in FE analysis is also carried out. The results obtained from FE analysis is validated with experimental data. The important conclusions are as follows:

(a) Ti–6Al–4V alloy is difficult to draw from room temperature to 150 °C. Maximum LDR of 1.8 was achieved at 400 °C, but it is still lesser than other structural alloy which is an indication of poor formability at lower elevated temperatures.
(b) Failure is seen in two regions, namely the neck and upper wall. FE simulations also show failures in the same region which promises the suitability of FE simulations for deep drawing.
(c) Cazacu–Barlat criterion is the most suited for Ti–6Al–4V alloy among other anisotropic yield criteria since anisotropy in yielding and stress asymmetry resulted in excellent validation of yield function with experimental thickness distribution and earing profile.
(d) Although a similar complexity is involved in the material parameter identification of Barlat 1996, Barlat 2000, and Cazacu–Barlat yield model, considering the computational time, Cazacu–Barlat is more efficient in finite element analysis for deep drawing of Ti–6Al–4V alloy.

Acknowledgments The financial support received for this research work from Department of Science and Technology (DST), Government of India, SERB-DST, SR/FTP/ETA-0056/2011 is gratefully acknowledged.

References

Abedrabbo N, Pourboghrat F, Carsley J (2007) Forming of AA 5182-O and AA 5754-O at elevated temperatures using coupled thermo-mechanical finite element models. Int J Plast 23:841–875

Altan T, Oh SI, Gegel HL (1983) Metal forming: fundamentals and applications. Am Soc Met 1:4–10

Atkinson M (1996) Accurate determination of biaxial stress–strain relationships from hydraulic bulging tests of sheet metals. Int J Mech Sci 39(7):761–769

Banabic D (2010) Sheet metal forming process: constitutive modeling and numerical simulations. Springer, Heidelberg

Barlat F, Lian J (1989) Plastic behaviour and strechability of sheet metals Part I: a yield function for orthotropic sheets under plane stress condition. Int J Plast 5:51–56

Barlat F, Brem JC et al (2003) Plane stress yield function for aluminium alloy sheet. Part 1 theory. Int J Plast 19:1297–1319

Beal JD, Boyer R, Sanders D (2006) The boeing company forming of titanium and titanium alloys. Metalworking Sheet Form 14 (ASM Handbook): 656–669

Bergström Y (1983) The plastic deformation of metals—a dislocation model and its applicability. Rev Powder Metall Phys Ceram 2:105–115

Bolt PJ, Lamboo NAPM, Rozier PJCM (2001) Feasibility of warm drawing of aluminum products. J Mater Process Technol 115:118–121

Bong HJ, Barlat F et al (2013) Formability of austenitic and ferritic stainless steels at warm forming temperature. Int J Mech Sci 75:94–109

Cazacu O, Plunkett B, Barlat F (2006) Orthotropic yield criterion for hexagonal closed packed metals. Int J Plast 22:1171–1194

Cole GS, Sherman AM (1995) Lightweight materials for automotive applications. Mater Charact 35:3–9

Davis JR (ed) (2004) Tensile testing, 2nd edn. ASM International, Ohio

Demmerle S, Boehler JP (1993) Optimal design of biaxial tensile cruciform specimens. J Mech Phys Solids 41(1):143–181

Djavanroodi F, Derogar A (2010) Experimental and numerical evaluation of forming limit diagram for Ti-6Al-4V titanium and Al6061-T6 aluminum alloys sheets. Mater Des 31:4866–4875

EI-Domiaty A (1992) The effect of strain, strain rate and temperature on formability of Ti-6Al-4V alloy. J Mater Process Technol 32:243–251

Güner A, Brosius A, Tekkaya AE (2009) Analysis of the hydraulic bulge test with FEA concerning the accuracy of the determined flow curves. Key Engineering Materials 410:439–447

Gupta AK, Anirudh VK, Singh SK (2013) Constitutive models to predict flow stress in austenitic stainless steel 316 at elevated temperatures. Mater Des 43:410–418

Gutscher G, Wu HC, Ngaile G, Altan T (2004) Determination of flow stress for sheet metal forming using the viscous pressure bulge (VPB) test. J Mater Process Technol 146:1–7

Hannon A, Tiernan P (2008) A review of planer biaxial tensile test system for sheet metal. J Mater Process Technol 198:1–13

Hill R (1950) The mathematical theory of plasticity. Clarendon, Oxford

Hollomon JH (1945) Trans AIME 162:268–290

Hosford W, Canddell R (2014) Metal forming mechanics and metallurgy, 3rd edn. Cambridge University Press, Cambridge, pp 52–75

Hussaini SM, Singh SK, Gupta AK (2014) Experimental and numerical investigation of formability for austenitic stainless steel 316 at elevated temperatures. J Mater Res Technol 3 (1):17–24

Kaya S, Altan T, Groche P, Klöpsch C (2008) Determination of the flow stress of magnesium AZ31-O sheet elevated temperatures using hydraulic bulge test. Int J Mach Tools Manuf 48:550–557

Keefe AC, Carman GP, Jardine P (1998) Torsional behavior of shape memory alloys. Smart Struct Mater 3324:58–67

Keum YT, Ghoo BY, Waginer RH (2001) 3-Dimensional finite element analysis of non-isothermal forming processes for non-ferrous sheets. Simulation of materials processing: theory, methods and applications. A.A. Balkema, Lisse, pp 813–818

Koç Muammer, Billur Eren, Cora Ömer Necati (2011) An experimental study on the comparative assessment of hydraulic bulge test analysis methods. Mater Des 32:272–281

Kotkunde N, Deole AD et al (2014a) Failure and formability studies in warm deep drawing of Ti–6Al–4V alloy. Mater Des 60:540–547

Kotkunde N, Deole AD et al (2014b) Experimental and numerical investigation of anisotropic yield criteria for warm deep drawing of Ti–6Al–4V alloy. Mater Des 63:336–344

Kotkunde N, Krishnamurthy HN et al (2014c) Microstructure study and constitutive modeling of Ti–6Al–4V alloy at elevated temperatures. Mater Des 54:96–103

Kotkunde N, Deole AD et al (2014d) Comparative study of constitutive modeling for Ti–6Al–4V alloy at low strain rates and elevated temperatures. Mater Des 55:999–1005

Kurukuri S, van den Boogaard AH et al (2009) Warm forming simulation of Al–Mg sheet. J Mater Process Technol 209:5636–5645

Lange K (1985) Handbook of metal forming. McGraw-Hill, New York

Lee JY, Xu L, Barlat F et al (2013) Balanced biaxial testing of advanced high strength steels in warm conditions. J Exp Mech 53(9):1681–1692

Lin YC, Chen X-M (2011) A critical review of experimental results and constitutive descriptions for metals and alloys in hot working. Mater Des 32:1733–1759

LS-DYNA Keyword User's Manual (2007) Volume 1, Livermore Software Technology Corporation (LSTC). (http://lstc.com/pdf/ls-dyna_971_manual_k.pdf)

Ludwik P (1909) Elemente der Technologischen Mechanik. Verlag Von Julius Springer, Leipzig, p 32

Marciniak Z (2002) Mechanics of sheet metal forming. Butterworth-Heinemann, Oxford

Marciniak Z, Kuczyński K, Pokora T (1973) Influence of the plastic properties of a material on the forming limit diagram for sheet metal in tension. Int J Mech Sci 15(10):789–800

Mishra NS, Mishra S, Ramaswamy V (1989) Analysis of the temperature dependence of strain-hardening behavior in high-strength steel. Metall Trans A 20:2819

Naka T, Yoshida F (1999) Deep drawability of type 5083 aluminum–magnesium alloy sheet under various conditions of temperature and forming speed. J Mater Process Technol 89(90):19–23

Nielsen KB (2000) sheet metal forming simulation using explicit finite element methods, 3rd edition, Aalborg University, Denmark

Odenberger E-L, Schill M, Oldenburg M (2013) Thermo-mechanical sheet metal forming of aero engine components in Ti–6Al–4V—part 2: constitutive modelling and validation. Int J Mater Form 6:403–416

Olovsson L, Simonsson K, Unosson M (2005) Selective mass scaling for explicit finite element analyses. Int J Numer Meth Eng 63(10):1436–1445

Pearce R (1968) Some aspects of anisotropic plasticity in sheet metals. Int J Mech Sci 10(12):995–1004

Singh SK, Mahesh K et al (2010) Understanding formability of extra-deep drawing steel at elevated temperature using finite element simulation. Mater Des 31:4478–4484

Swift HW (1952) Plastic instability under plane stresses. J Mech Phys Solids 1:1–18

Takuda H, Mori K et al (2003) Finite element analysis of the formability of an austenitic stainless steel sheet in warm deep drawing. J Mater Process Technol 143:242–248

Toros S, Ozturk F, Kacar I (2008) Review of warm forming of aluminum–magnesium alloys. J Mater Process Technol 207:1–12

Van den Boogaard AH, Huétink J (2006) Simulation of aluminium sheet forming at elevated temperatures. Comput Methods Appl Mech Eng 195:6691–6709

Vlassak JJ, Nix WD (1992) A new bulge test technique of Young's modulus and Poisson's ratio of thin films. J Mater Res 7(12):3242–3249

Voce E (1948) J Inst Met 74:537–562

Wessel JK (2004) Handbook of advanced materials Hoboken. Wiley, Hoboken

Wierzbicki T, Bao Y et al (2005) Calibration and evaluation of seven fracture models. Int J Mech Sci 47(4–5):719–743

Xiang-Dong W, Wan M, Xian-Bin Z (2005) Biaxial tensile testing of cruciform specimen under complex loading. J Mater Process Technol 168:181–183

Zhang SH et al (2007) Deep-drawing of magnesium alloy sheets at warm temperatures. J Mater Process Technol 185:147–151

Springback and Formability Studies on Friction Stir Welded Sheets

Sudhindra Katre, Siddhartha Karidi, B. Durga Rao, P. Janaki Ramulu and R. Ganesh Narayanan

Abstract In the present work, the effect of selected welding conditions on the temperature evolution, springback behaviour and formability of friction-stir welding (FSW) sheets is investigated. The Al6061T6-Al6061T6 combination is used for temperature evolution studies, while dissimilar Al combination made of Al6061T6 and Al5052H32 is used in springback and formability tests. The springback experiments are performed with a V-bending set-up without tension, and formability experiments are conducted using a limiting dome height (LDH) set-up. By increasing the shoulder diameter, plunge depth and tool rotation speed, the peak temperature is increased, and it is vice versa for welding speed. In the case of springback, the performance of FSW sheets is in between that of Al6061T6 and Al5052H32 base materials. Reduction in springback of FSW sheets is observed at higher rotational speed and welding speed, and this is correlating well with the modifications in tensile properties such as yield strength to elastic modulus ratio and strain hardening exponent of the weld zone (WZ) with varying FSW conditions. Three different modes of failures, normal and parallel to the WZ, are observed during formability testing of FSW sheets made of dissimilar Al grades at two different strain-paths. It is better to evaluate the forming limit of FSW sheets at individual strain-paths, rather than in all the strain-paths using a forming limit curve (FLC), because of changing failure pattern with changing strain-paths.

Keywords Friction-stir welding · Springback · Formability · Aluminium alloys · Forming limit

---

S. Katre · S. Karidi · R.G. Narayanan (✉)
Department of Mechanical Engineering, IIT Guwahati, Guwahati 781039, India
e-mail: ganu@iitg.ernet.in

B. Durga Rao
Parala Maharaja Engineering College, Sitalapalli, Berhampur 761003, Orissa, India

P.J. Ramulu
Department of Mechanical Engineering, Vardhaman College of Engineering, Kacheram, Shamshabad 501218, Hyderabad, India

R.G. Narayanan and U.S. Dixit (eds.), *Advances in Material Forming and Joining*, Topics in Mining, Metallurgy and Materials Engineering,
DOI 10.1007/978-81-322-2355-9_7

1 Introduction

Friction-stir welding (FSW) is a solid-state joining process. It involves joining of metals without filler materials. A constantly rotating cylindrical-shouldered tool with a profiled nib is transversely fed at a constant rate into a butt joint between two clamped sheets (Fig. 1). Frictional heat is generated between the wear-resistant welding components and the work pieces. This heat, along with that generated by the mechanical mixing process and the adiabatic heat within the material, cause the stirred materials to soften without melting. Hence, this is called as a solid-state welding process. The resultant plasticized material is transferred from the leading edge of the tool to the trailing edge of the tool probe and is forged together by the intimate contact of the tool shoulder and the pin profile (Mishra and Ma 2005). There are many advantages of FSW including energy savings, exclusion of consumables and shielding gases, good dimensional stability, low distortion of work sheet, enriched metallurgical properties in the weld region, fine microstructure, possibility of welding dissimilar materials and reduced usage of fasteners for joining multiple parts. It is considered as 'green technology' because of the environmental friendliness and energy efficiency.

The important parameters in FSW are tool material and its geometry (both pin and shoulder), welding speed, tool rotational speed, axial force and work piece material quality. An appropriate combination of these parameters determines the excellence of welds including the presence of defects and microstructure and hence as a consequence the mechanical properties and formability of FSW sheets gets affected. The remaining part of the section describes some of the available work performed on springback and formability of unwelded and welded sheets.

During sheet bending, while one side of the sheet metal is subjected to compressive stress, the other surface is subjected to tensile stress. Once the load is released by withdrawal of the punch, an elastic recovery occurs because of the release of the elastic stresses in the bending process. This elastic recovery is called

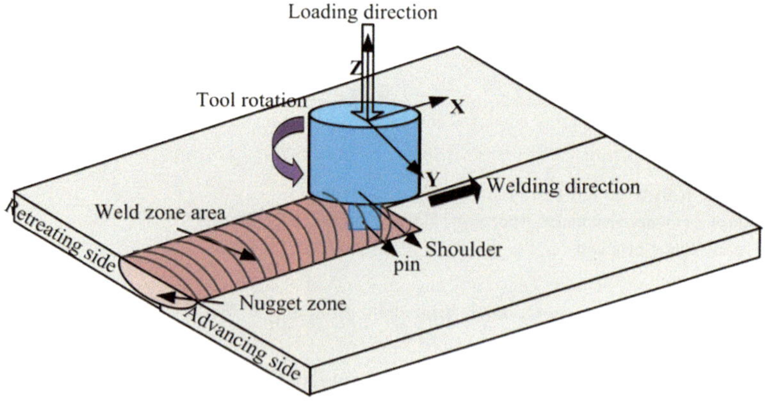

Fig. 1 Schematic of friction-stir welding process

Springback. It is seen as a practical problem to predict the final geometry of the part accurately, after springback, and to design appropriate tooling for compensation. Overbending, bottoming and stretch bending are common methods of controlling springback. There are other methods, such as heat treating the materials, change in working temperature of material and tools, appropriate tool design, tool geometry design and recently the friction-stir processing/welding, that are used for springback control of sheets. It is shown by Moon et al. (2003) that the hot die and cold punch combination can reduce the springback of Al 1050 sheet up to 20 % when compared to conventional bending done at room temperature. The ram speed was also shown to have significant effect on controlling springback. Garcia-Romeu et al. (2007) have shown through experiments that sheet materials, their thicknesses and die widths affect the springback behaviour in a compounding fashion. It has been concluded by them that though small radius is preferred for springback, large radius dies are preferred in consideration of mechanical properties of bent part. Hence, one should optimize various factors governing the springback of sheets. Alves de Sousa et al. (2008) investigated the unconstrained springback of Al–Mg–Si sheets, by considering pre-strain before the actual deformation and later subjected to different sitting times at normal temperatures show that extensive amount of stress is relaxed when the punch is totally released. It is also observed that the springback angle decreases as the sitting time increases.

It is shown by Greze et al. (2010) that the effect of temperature is to decrease the stress gradient in the cup wall, because of which the springback opening of the ring cut from the deep drawn cup decreases. The significant decrease of springback at 773 K for high-strength steel is caused by the increase of high-temperature creep strain just after loading process, resulting in springback reduction. It is also confirmed that the change in flow stress and Young's modulus at elevated temperatures plays a minor role in springback reduction. The study of Wang et al. (2013) reveals that the springback of AZ31B Mg alloy decreased with increase of forming temperature and decrease of punch radii during V-bending.

The study of Wang et al. (2004) and Lim et al. (2012) revealed that the time at which springback is measured also influences the accuracy of springback evaluation, and hence, the controlling and compensating strategies depend on this. A study of time-dependent springback on Al alloys such as 2008-T4, 5182-O, 6022-T4 and 6111-T4, after draw-bend tests was performed by Wang et al. It is demonstrated that the springback is proportional to log (time) up to few months, in a span of 15 months, after which no effect was observed. The creep driven by residual stress and anelasticity is proposed to be responsible for time-dependent springback. But finally, it was concluded that anelasticity is improbable to play large role in long-term time-dependent springback, but it can contribute to short-term response. Later, the workdone by Lim et al. (2012) showed that advanced high-strength steels (AHSS) such as dual-phase (DP) steel, transformation induced plasticity (TRIP) steel and conventional steels also show time-dependent springback. It was even shown that the springback found in AHSS is approximately 1/3rd of that for aluminium alloys mentioned in Wang et al. (2004) work.

In the case of tailor-welded sheets made by FSW, the selection of welding and tool conditions decides the final mechanical properties and weld microstructures and hence the overall forming performance. In this context, Park et al. (2008) have shown through unconstrained bending tests that in the case of FSW sheets made of 5052-H32 of 1.5 mm thickness, with increase in friction-stir pin diameter from 5 to 10 mm, the springback have reduced by about 3°–5°, when the weld zone (WZ) is placed longitudinally. Similarly, about 4° decrease in springback was found in the case of transverse weld, with respect to base material. The springback tests such as the unconstrained cylindrical bending, 2D draw bending and draw-bend tests were simulated by using the combined isotropic–kinematic hardening law based on the modified Chaboche model and Yld2000-2d yield function, and the predictions agree reasonably well with the experimental results (Kim et al. 2011). The work of Miles et al. (2006) also highlights the importance of friction-stir process in increasing the bending limit of welded plates made of 6061-T6 and 7075-T7451. Though the failure models that are used during finite element analysis predicted the necking limits accurately in the case of unprocessed 6061 Al plate, their prediction for friction-stir processed plate was moderately accurate. The prediction accuracy has improved once the gradient of mechanical properties in through thickness direction was captured properly by the stress–strain behaviour. Chang et al. (2002) investigated the springback characteristics of tailor-welded strips of dissimilar thicknesses made by laser welding during U-bending. The springback encountered by longitudinally welded strip was same as that of un-welded base sheets. In the case of the transverse welded strips, a significant reduction of the springback was observed in the thinner side compared with the same thickness of un-welded strip.

Ramulu et al. (2013a) analysed the effect of welding speed, rotation speed, plunge depth and shoulder diameter on the formation of internal defects, axial force and torque during FSW of 6061T6 sheets. At higher welding speed, higher rotation speed and higher plunge depth, internal defect-free weld joints are produced. The axial force and torque were not constant, and a large variation was seen with respect to FSW parameters that produced defective welds. In the case of defect-free welds, the axial force and torque were relatively constant. Recently, Ramulu et al. (2013b) studied about the effect of the welding speed and tool rotation speed on the forming limit of FS welded sheets made of AA6061-T6 with thickness of 2.1 mm using limit dome height test. With weld oriented along major straining direction, the results showed that the formability has improved by decreasing welding speed and increasing tool rotation speed. Moreover, the forming limits of FSW sheets are better than that of un-welded base materials. The strain hardening exponent of weld (n) increases with increase in tool rotation speed and it decreases with increase in feed rate. Lee et al. (2009) evaluated the formability of FSW sheets in longitudinal weld direction, using numerical simulation and experiments. FSW sheets made of 6111-T4, 5083-H18, 5083-O, dual-phase steel (DP590) and AZ31 magnesium alloy were considered. It is observed that the improved ductility of the weld promoted the formability performance of AA5083-H18, AA5083-O and AZ31 TWB sheets. Zadpoor et al. (2008, 2009) studied the effects of the employment of the weld details on the accuracy of the failure prediction of FSW blanks, for the limiting dome height (LDH) test and S-rail. The Marciniak–Kuczynski theory is used for the prediction of forming limits diagrams of the different zones of the FSW blanks. The WZ was modelled as a double zone, i.e. WZ and heat affected zone (HAZ) considered separately and finally recommended that implementation of the nugget zone properties is sufficient for reasonably accurate predictions of the springback behaviour.

The major objectives of the present work are (i) demonstrating the possibilities of controlling the springback of Aluminium alloys by FSW, (ii) studying the impact of a few chosen welding parameters on the springback and formability of FSW sheets and (iii) prediction of springback of FSW sheets using finite element simulations. A brief discussion has been made on the influence of few FSW parameters on the temperature evolution. A dissimilar aluminium grade combination (Al6061T6-Al5052H32) is used for springback and formability evaluation, while similar aluminium grade combination (Al6061T6-Al6061T6) is used for temperature measurement.

2 Methodology

2.1 Temperature Measurement

The base material used for temperature measurement is AA6061-T6 alloy with 2.1 mm thickness, i.e. a similar Al grade combination. The nominal chemical composition (in weight percentage) of the base material is Si: 0.65, Fe: 0.3, Cu: 0.27, Mn: 0.02, Mg: 1.0, Cr: 0.17, Zn: 0.04, Ti: 0.02 and Al: remaining. The friction-stir welding was carried out on a machine designed and developed by IISc and ETA Technologies. The process parameters varied were tool rotational speed, welding speed, plunge depth and shoulder diameter, while the tool tilt angle was kept constant at 25 °C throughout the temperature measurement. Three types of tools with flat shoulder having different diameters, and pin with equal lengths were used. These three tools were made of hot die steel with shoulder diameter 12, 15 and 18 mm, pin of frustum shape with base diameter 6 mm and top diameter 4 mm and length 1.7 mm. The welding trials were conducted at two levels of chosen parameters (Table 1) (Ramulu et al. 2013a). A third level was also selected with 15 mm shoulder diameter, at plunge depth of 1.95 mm, welding speed of 120 mm/min and tool rotation speed of 1500 rpm.

Table 1 Optimum range of process parameters for FSW without defect formation

Shoulder diameter (mm)	Plunge depth (mm)	Tool rotation speed (rpm)	Welding speed (mm/min)
12, 18	1.85, 1.9	1300, 1400	90, 100
15	1.95	1500	120

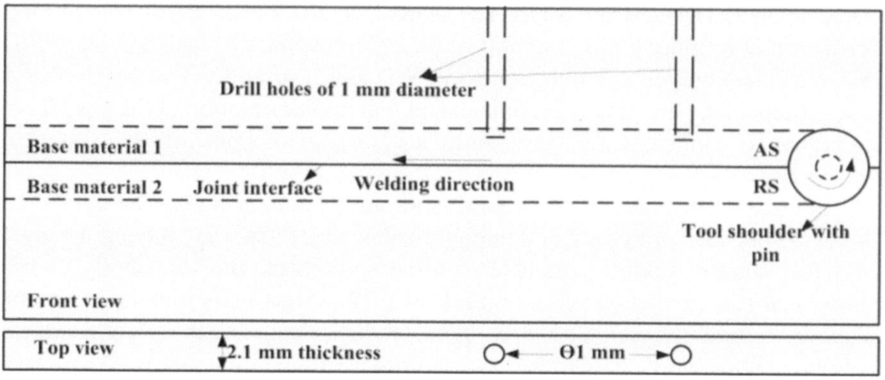

Fig. 2 Schematic of FSW process for measuring temperature; Here, *AS* advancing side; *RS* retreating side

The drill holes were made using a conventional high-speed drilling machine with 1-mm diameter tool bit. The location of drill holes were decided by the linear interpolation of parameters for different tools as shown schematically in Fig. 2. Temperature was recorded with the help of K-type thermocouples which was inserted on end to the drilled holes of 1 mm diameter of base material in the advancing side (AS) and to the data acquisition (DAQ) system during welding. The positions of the thermocouples are such that they are 3–4 mm away from the pin for all the experiments. The temperature is recorded in the shoulder influenced region. The experimental FSW parameters for which temperature measurements were performed are tabulated in the Table 2. With tool of 12 and 18 mm shoulder diameters, three levels of experiments were performed, and with 15 mm shoulder diameter, only one experiment was performed. Temperature was recorded in all the experiments.

Table 2 Range of FSW parameters for temperature measurement

Shoulder diameter (mm)	FSW varied parameters for experiments			Conditions for temperature measurement
	PD (mm)	RS (rpm)	WS (mm/min)	
12	1.8–1.95	1400	90	1.85 and 1.9 mm PD
	1.9	1200–1500	90	1300 and 1400 rpm RS
	1.9	1400	80–120	90 and 100 mm/min WS
18	1.8–1.95	1400	90	1.85 and 1.9 mm PD
	1.9	1200–1500	90	1300 and 1400 rpm RS
	1.9	1400	80–120	90 and 100 mm/min WS
15	1.95	1500	120	Single data

PD Plunge diameter, *RS* Transverse rotational speed, *WS* Welding speed

2.2 Springback Experiments and Prediction

The first step was to produce samples of friction-stir welded sheets made of Al6061T6 and Al5052H32 (a dissimilar Al grade combination) by varying the tool rotational and translational (welding) speed. The tool rotational speed was varied from 600 to 800 rpm, in three steps, and the tool translational speed was varied from 80 to 120 mm/min in three steps. AA6061T6 was placed on the AS and AA5052H32 on the retreating side (RS) of the weld during welding trials to fabricate FS welded sheets. The material properties and Lankford coefficients (r) of the base materials (Tables 3 and 4) and weld regions (Table 5) were determined by tensile tests by using ASTM standard sub-size 1 specimen dimensions. Three trials were conducted to evaluate the tensile properties. Later, V-bending tests were performed on all the welded blanks and base materials of dimensions 160 mm × 50 mm. The weld was oriented longitudinally during bending experiments. Since time-dependent springback was reported by Wang et al. (2004) and Lim et al. (2012), the springback of all the bent sheets in the present work (included angle) were measured within an hour, after V-bending was completed. Two trials were conducted to evaluate the springback in each case. A schematic of the

Table 3 Tensile properties of Al6061T6 base sheet

Rolling direction (degrees)	E (GPa)	Yield stress (MPa)	Ultimate stress (MPa)	Uniform elongation	n	K	r
0	23	206	282	0.096	0.141	397.6	0.44
15	23	206	277	0.085	0.14	392.2	0.47
30	21	205	272	0.08	0.139	386.9	0.64
45	26	187	256	0.08	0.142	368.4	0.64
60	29	206	282	0.095	0.134	386.9	0.65
75	22	200	275	0.093	0.143	387.5	0.66
90	22	196	276	0.093	0.148	395.1	0.66

E Elastic modulus, K Strength coefficient, n Strain hardening exponent and r Plastic strain ratio

Table 4 Tensile properties of Al5052H32 base sheet

Rolling direction (degrees)	E (GPa)	Yield stress (MPa)	Ultimate stress (MPa)	Uniform elongation	n	K	r
0	23	128	232	0.132	0.26	408.1	0.53
15	26	130	226	0.12	0.192	344.4	0.98
30	24	129	225	0.139	0.453	600.5	0.58
45	24	129	217	0.116	0.185	326.9	0.45
60	25	131	218	0.116	0.184	326.5	0.4
75	24	131	220	0.115	0.187	330.7	0.43
90	24	133	224	0.114	0.189	335.7	0.87

Table 5 Tensile properties of weld region

Rotational speed (rpm)	Translational speed (mm/min)	E (GPa)	Yield stress (MPa)	Ultimate stress (MPa)	Uniform elongation	n	K	r
600	80	29.9	110	226.3	0.167	0.211	324	0.49
600	120	30.1	100.9	226.3	0.167	0.260	359	1.68
700	80	25.5	105.3	235.3	0.181	0.255	363	0.36
700	100	30.4	113.8	252.5	0.191	0.249	380	0.54
700	120	27.5	104.6	236.4	0.175	0.256	369	0.21
800	80	30.2	104.6	245.2	0.190	0.256	374	1.11
800	100	28.8	103.9	246.5	0.186	0.259	368	0.86
800	120	26.1	86.5	217.8	0.196	0.274	340	0.73

V-bending set-up that was fabricated to conduct the experimental trials is shown in Fig. 3.

In order to predict the springback, finite element simulations were performed for V-bending (using the dimensions of setup in Fig. 3) of friction-stir welded sheets using a commercially available elasto-plastic, explicit dynamic finite element code. The adaptive meshing that automatically refines the mesh was used for finite element simulations. The meshing was done with quadrilateral shell elements of the Belytschko–Tsay formulation. Five through thickness integration points were used during simulation. The average mesh size of about 2 mm was used throughout the sheet specimen and tools. The tools were modelled as rigid bodies. The base materials properties and WZ properties obtained from experiments were incorporated during FE simulations. Hollomon's strain hardening law was used to describe the stress–strain behaviour of base material and WZ. Hill's 1990 yield criterion was used as the plasticity model. The '*m*' value used in the yield criterion was optimized to have less springback error with experimental values. The sheets were subjected to springback explicit analysis, after bending, and springback angle was evaluated. The different stages of V-bending of friction-stir welded sheets during simulations are shown in Fig. 4.

2.3 Formability Testing of Friction-Stir Welded Sheets

In order to evaluate the influence of shoulder diameter, rotational speed and welding speed on the formability, LDH tests were performed on the FS welded sheets made of dissimilar aluminium grade combination made of Al6061T6 and Al5052H32. AA6061T6 was placed on the AS and AA5052H32 on the RS of the weld during welding trials. The base materials properties are given in Tables 1 and 2. Standard LDH test tool geometry was used for the formability tests. Few different sheet widths (like 100 × 300 mm, 125 × 300 mm etc.) with two lubrication (dry, grease) conditions simulating four different predefined strain-paths in stretching side were

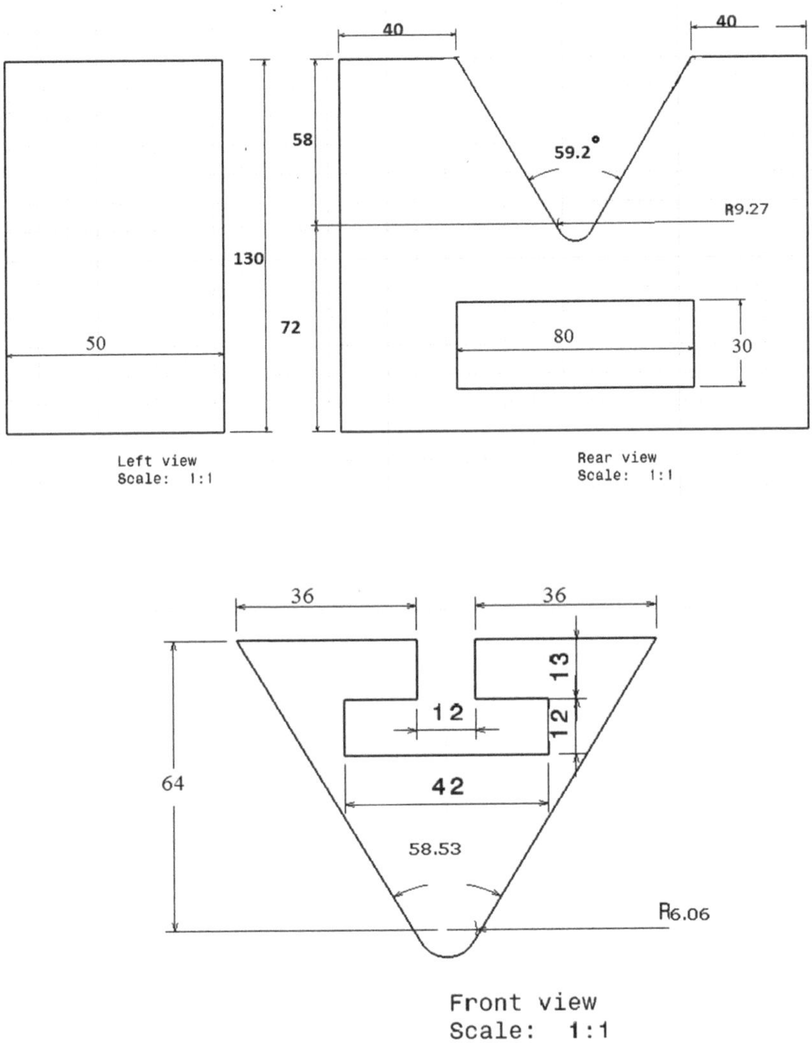

Fig. 3 Schematic of die and punch in V-bending set-up

deformed to construct a Forming limit curve (FLC). During the LDH test, samples were first thoroughly cleaned with acetone to remove oil and dust particles on the sheet surface. The deformation of the welded sheets was analysed through the online monitoring system using spackle pattern. The pixel size in the spackle pattern was fixed at 15 × 6 with a 0.6 deviation before deformation of FSW sheets. The samples were placed on the die surface with sufficient blank holding pressure. All the sheets were deformed until failure (or necking) occurs. The true major

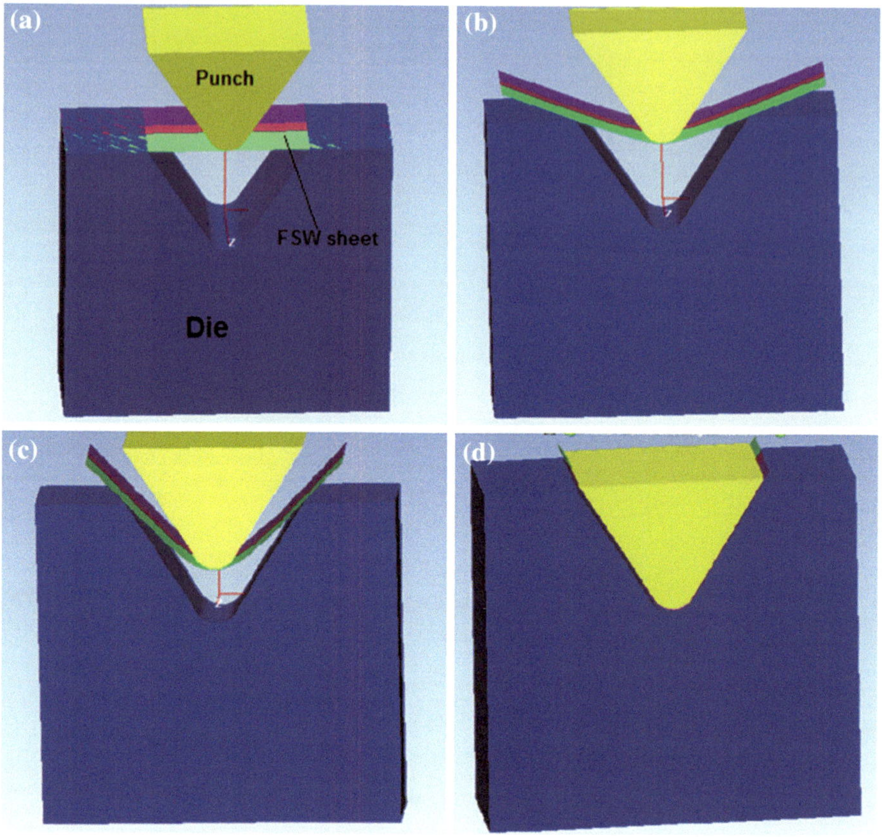

Fig. 4 Stages of V-bending of friction-stir welded sheet during finite element simulations

strains and true minor strains near the failure and away from the failure were measured. These failure strain data were plotted on a major strain (Y-axis), minor strain (X-axis) graph. All the failed strains and safe strains were incorporated in the same graph. The FLCs were plotted such that they separate all the failed strains from the safe strains.

3 Results and Discussion

3.1 Temperature Evolution

The influence of plunge depth, rotational speed, welding speed and shoulder diameter on the temperature evolution is described in the present section. Figures 5 and 6 show the temperature evolution at plunge depths of 1.85 and 1.9 mm, for

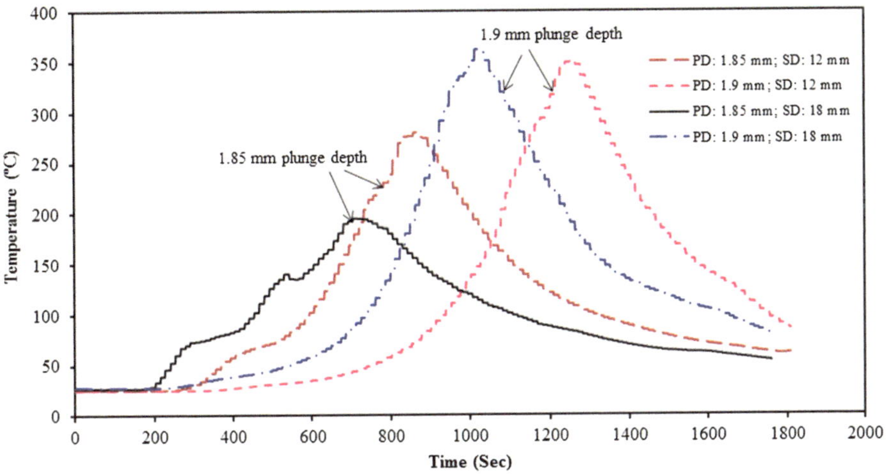

Fig. 5 Influence of plunge depth on the temperature evolution (rotational speed: 1400 rpm, welding speed: 90 mm/min)

shoulder diameters of 12 and 18 mm, at a tool rotational speed of 1400 rpm and 90 mm/min welding speed. It is observed that the peak temperature at 1.9 mm plunge depth is more than at the 1.85 mm irrespective of the tool shoulder diameter. The peak temperatures at 1.85 mm plunge depth are 280 and 195 °C for 12 and 18 mm shoulder diameter, respectively, whereas at 1.9 mm plunge depth, 352 and 362 °C are recorded.

Similarly by increasing the tool rotational speed from 1300 to 1400 rpm, the peak temperature has increased (Fig. 6). The peak temperatures measured are 205

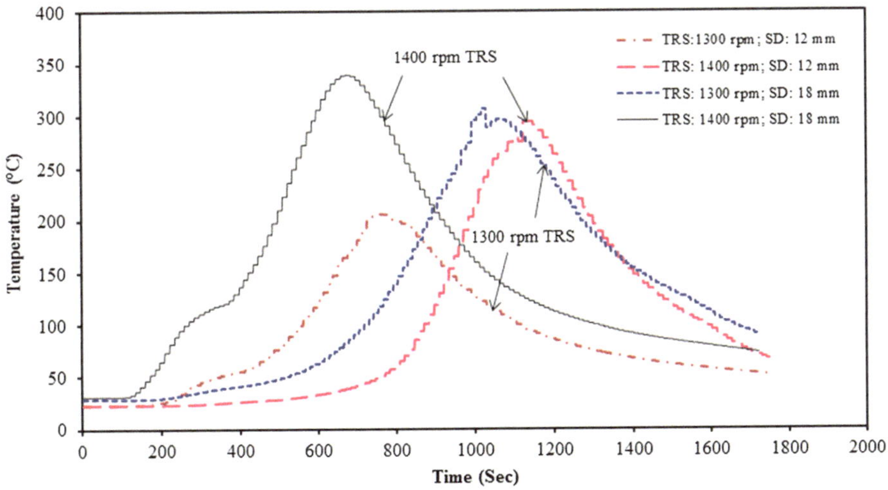

Fig. 6 Influence of rotational speed on the temperature evolution (welding speed: 90 mm/min)

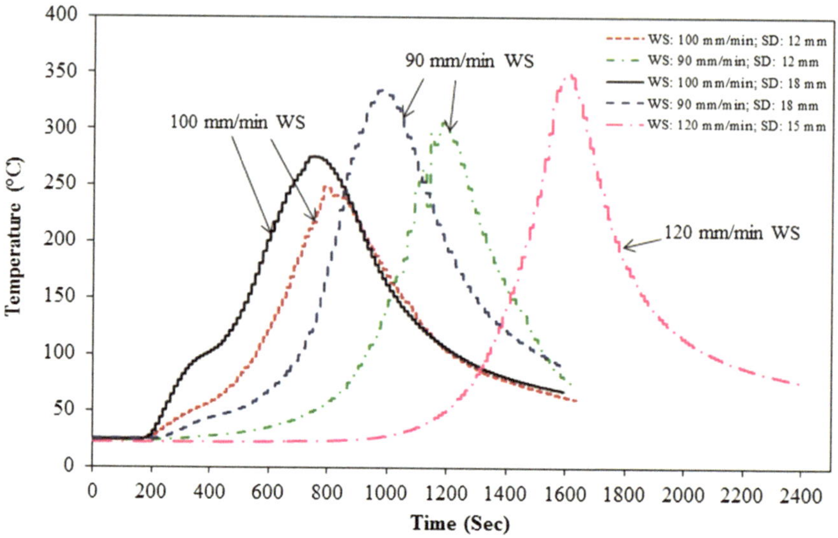

Fig. 7 Influence of welding speed on the temperature evolution

and 307 °C, at 1300 and 1400 rpm tool rotational speed, respectively. In the other case with 18 mm tool shoulder diameter at 1.9 mm plunge depth and 90 mm/min welding speed, the peak temperatures recorded are 295 and 340 °C for 1300–1400 rpm, respectively.

In case of welding speed, the peak temperature is decreased while increasing the welding speed from 90 to 100 mm/min, irrespective of tool shoulder diameter and other parameters. In the case of 12 mm shoulder diameter at 1400 rpm rotational speed and 1.9 mm plunge depth, temperature at 90 mm/min welding speed is 306 °C, and it is 249 °C at 100 mm/min welding speed (Fig. 7). Similarly in the case of 18 mm shoulder diameter, peak temperatures of 335 and 275 °C are seen at 90 and 100 mm/min welding speeds, respectively. For the third level with 120 mm/min welding speed and 15 mm shoulder diameter, the peak temperature recorded is 350 °C. The summary of peak temperature measured for different FSW combinations is given in Table 6.

The overall observation from these temperature measurements are, by increasing the shoulder diameter, plunge depth, tool rotational speed, the peak temperature has increased, while opposite behaviour is seen for welding speed. The shoulder diameter effect is due to the increase in higher temperature region, increasing the peak temperature. The plunge depth influence is due to the forging action of shoulder onto the sheet surface increasing the peak temperature. At high rotational speed, the relative velocity between the tool and sheet is high, and as a result, the heat generation rate and the peak temperature are also high. Finally at higher welding speeds, the heat input per unit length declines and at the same time, heat is dissipated over a wide region of the sheet, reducing the peak temperature.

Table 6 Summary of peak temperature measured for different FSW combinations

FSW parameters				Peak temperature (°C)
SD (mm)	RS (rpm)	WS (mm/min)	PD (mm)	
12	1400	90	1.85	280
12	1400	90	1.9	352
12	1300	90	1.9	205
12	1400	90	1.9	293
12	1400	90	1.9	305
12	1400	100	1.9	249
18	1400	90	1.85	195
18	1400	90	1.9	362
18	1300	90	1.9	307
18	1400	90	1.9	340
18	1400	100	1.9	275
15	1500	120	1.95	350

3.2 SpringBack Evaluation and Prediction

The tensile behaviour of the WZ for different rotational speeds, at a constant translational speed, and for different translational speeds, at a constant rotational speed is shown in Figs. 8 and 9, respectively. It is evident that the weld properties such as yield strength, ultimate tensile strength decreases with the increase in welding speed, at a constant rotational speed of 800 rpm. There is some improvement in the ductility of the WZ with increase in welding speed and rotational speed

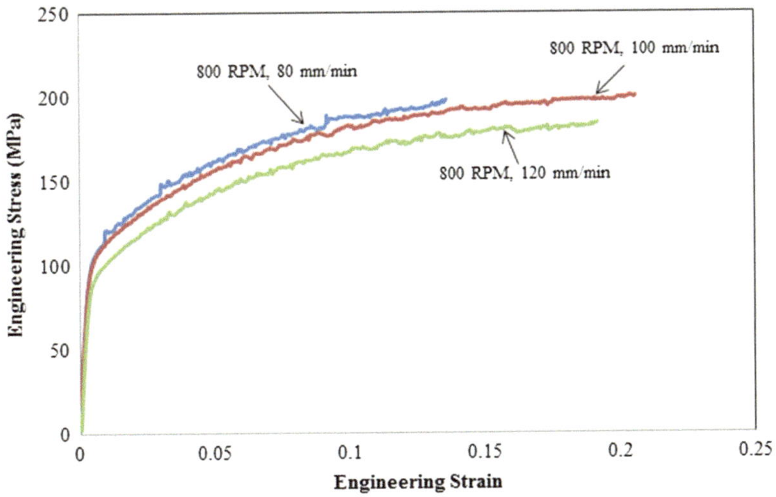

Fig. 8 Tensile behaviour of WZ at different welding speeds

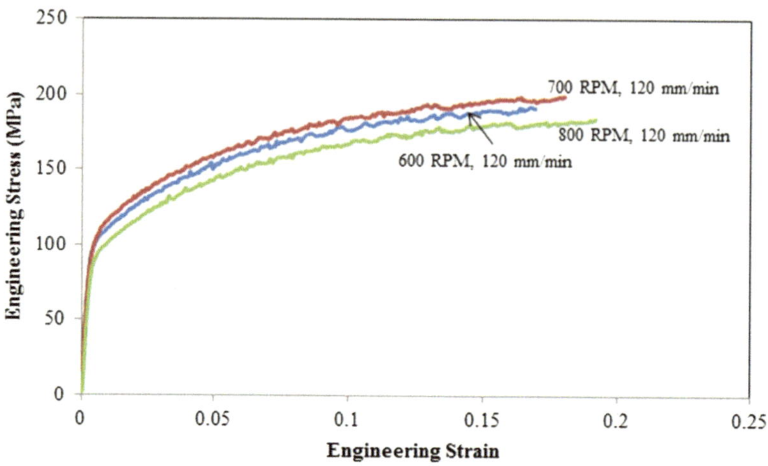

Fig. 9 Tensile behaviour of WZ at different rotational speeds

(Figs. 8 and 9). This improvement is due to the improvement in strain hardening exponent of WZ (Table 5), improving the overall ductility of the FSW sheets. As the rotational speed increases or the welding speed reduces, the temperature in the stirring zone is expected to increase. This means that the final grain size would be larger, and it is also expected that the dislocation density reduces as the temperature increases. This leads to more static recrystallization in the WZ. It is known that the strain hardening behaviour is typically related to the accumulation and interaction of the dislocations. If the dislocation density is lower, more strain can be brought into the material, thus increasing the ductility, and finally, its forming limit has improved. In contrast, in the base material, the saturation of the dislocation density results in the lower strain hardening exponent and poor ductility, resulting in less formability. The effect of strain hardening exponent of HAZ is not considered for discussion, and hence, WZ is nothing but the nugget zone and thermo-mechanically affected zone. The relationship between strain hardening exponent of WZ and the dislocation density has been analysed by Woo et al. (2008) for a 6061-T6 Aluminium alloy, and it has been demonstrated that the low initial dislocation density in the recrystallized region is the main reason of the high strain hardening in the nugget zone.

The results obtained after performing V-bending on base material sheets and welded sheets are tabulated in Tables 7 and 8, respectively. It is evident that the springback of Al5052 base material is less than that of all the welded sheets, which is lesser than the springback of Al6061 base sheets. This is shown schematically in Fig. 10.

Table 7 Springback of base materials

Springback angle (°)	
Al6061T6	Al5052H32
64.77	56.85

Table 8 Springback of friction-stir welded sheets

Rotational speed (rpm)	Translational speed (mm/min)	Springback (°)
600	80	62.86
700	80	58.97
800	80	60.46
600	100	63.10
700	100	62.32
800	100	61.05
600	120	61.52
700	120	60.64
800	120	59.95

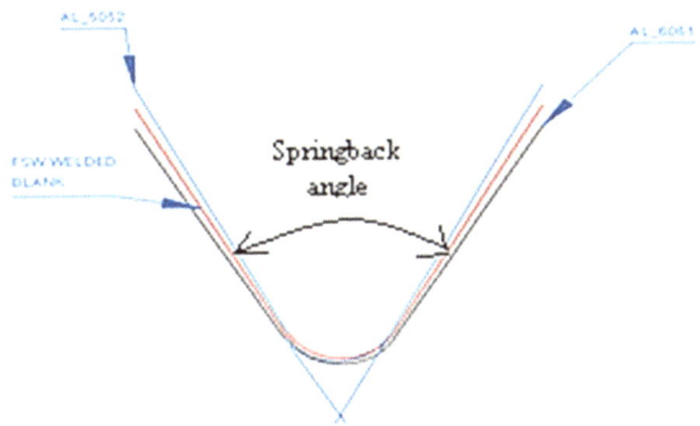

Fig. 10 Representation of relative springback of base materials and welded sheets

The influence of friction-stir welding on the springback is due to the change in WZ mechanical properties as compared to that of base materials. Two important mechanical properties, yield strength to Young's modulus ratio and strain hardening exponent, of WZ are evaluated for varied FSW conditions given in Table 5. Because of the change in rotational speed and welding speed, there is a significant change in yield stress to Young's modulus ratio (σ_y/E) and strain hardening exponent (n) of weld and this affects the overall springback of the welded sheets. Theoretically, it is known that the springback increases with the increase of σ_y/E ratio and decreases with n value. The influence of rotational speed and welding speed on σ_y/E ratio, strain hardening exponent of WZand springback of welded sheets are shown in Figs. 11, 12 and 13.

It is observed that springback decreases with increase in rotational speed and welding speed. With increase in rotational speed, σ_y/E ratio of WZ decreases and 'n' value of WZ increases, decreasing the springback. At the same time, for increasing welding speed from 80 to 120 mm/min, σ_y/E ratio of WZ decreases and 'n' value of WZ increases, decreasing the springback of FSW sheets.

While predicting the springback, as the materials used are aluminium alloys, the yield criterion used is Hill's 1990, and the Hill's non-quadratic yield parameter 'M'

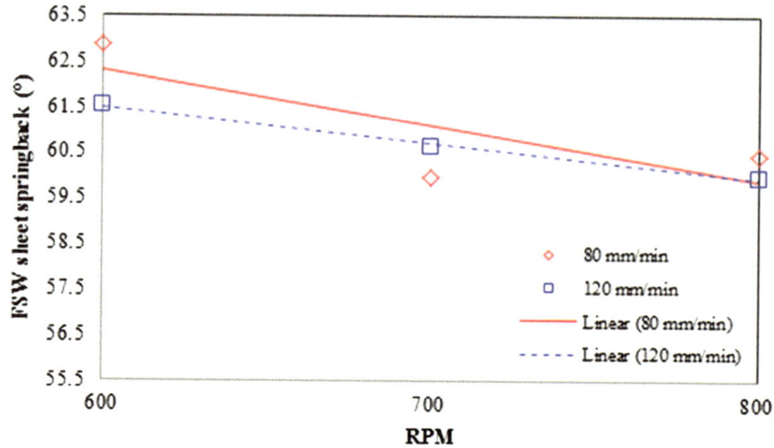

Fig. 11 Effect of rotational speed and welding speed on springback of FSW sheets (error in angle: 1.2°)

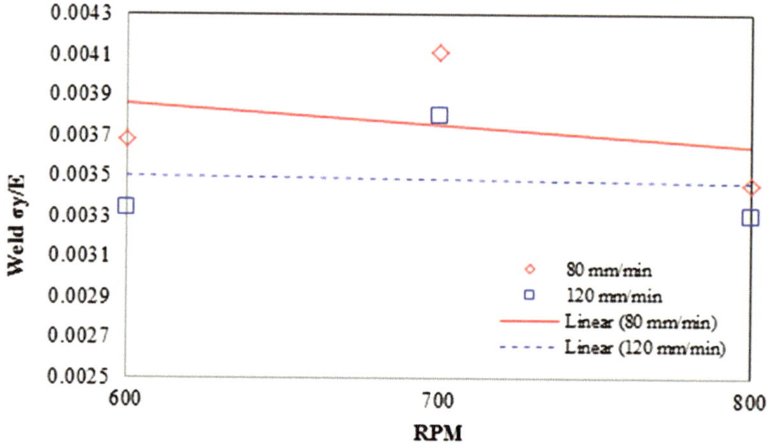

Fig. 12 Effect of rotational speed and welding speed on σ_y/E ratio of WZ (error in σ_y/E: 0.0001)

in the yield function is optimized such that for different *M* values between 1 and 3, the predicted and experimentally evaluated springback angles of base sheets and FSW sheets are compared (Table 9). This parameter can be evaluated through few different experimental methods including balanced biaxial stretch test and through thickness compression tests (Mulder et al. 2012). For Hill's 1990 yield criterion, the inputs given to base material are as follows: yield strength and plastic strain ratios in seven different rolling directions, 0°, 15°, 30°, 45°, 75° and 90°, whereas for WZ, only one yield strength and plastic strain ratio are given corresponding to varied FSW conditions.

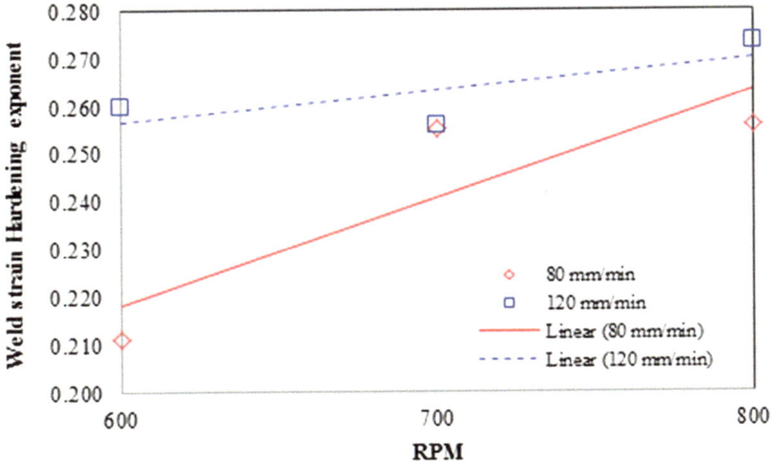

Fig. 13 Effect of rotational speed and welding speed on WZ '*n*' value (error in *n*: 0.002)

Table 9 Optimization of *M* value in Hill's 1990 yield criterion

M	Al6061T6 (°)	Al5052H32 (°)
1.8	63.04	56.24
1.9	63.56	56.43
2	**64.45**	**56.48**
2.1	64.22	56.25
2.2	64.23	55.85
2.3	64.18	55.92

It is observed that the springback values obtained at $M = 2$ closely agree with the experimental springback values of 64.77° and 56.85° for Al6061T6 and Al5052H32 base sheets, respectively. Hence, $M = 2$ was incorporated during FE simulations. The base sheets are considered as anisotropic material, and hence, plastic strain ratios given in Tables 1 and 2 are utilized. The weld region is a mix of two aluminium alloys,; therefore, simulations were performed twice, once by using the plastic strain ratios (r) obtained experimentally (from Table 3), indicating the WZ to be anisotropic, and other by assuming that the weld region as isotropic, i.e. $r = 1$. The values of springback found experimentally, and those predicted by simulations are closely agreeing with each other (Table 10). The overall error between the predictions and the experimental value is 2.33 and 2.26 % in case isotropic and anisotropic WZ assumptions, respectively. Therefore, the assumption of isotropic WZ for springback predictions is acceptable for springback predictions, though anisotropic assumption will yield accurate results. Some of the bent FS welded samples are shown in Fig. 14.

Table 10 Comparison of experimental and predicted springback values of FSW sheets

Rotational speed (rpm)	Translational speed (mm/min)	Experimental springback (°)	From simulation (assuming isotropic weld) (°)	From simulation (assuming anisotropic weld) (°)
600	80	62.86	60.13	60.70
600	120	61.52	59.77	60.08
700	80	58.97	59.54	60.26
700	100	62.32	59.97	59.98
700	120	60.64	60.04	59.49
800	80	60.46	59.72	60.27
800	100	61.05	59.18	59.09
800	120	59.95	59.10	59.40

Fig. 14 Parent friction-stir welded sheet and bent samples

3.3 Formability Evaluation

The LDH tests were conducted to evaluate the forming limit of friction-stir welded sheets made of dissimilar aluminium alloy sheets under two different strain-paths and at different friction conditions (or lubrication). The sheets are formed till failure occurs, and limit strains are measured as described earlier. With this measured strain data, FLC are drawn for FSW sheets at different welding conditions to study their influence. The failure pattern of all the FSW sheets at two different strain-paths—100 (width) × 200 (length) mm (near plane-strain) and 125 × 200 mm (one of the stretching strain-paths) is given in Table 11. The FSW sheets failed under three different modes: (i) failure normal to the WZ, (ii) failure along the WZ, but on the AS, and (iii) failure along the WZ, but at the centre, as shown in Fig. 15a, b (Durga Rao and Narayanan 2014). The first failure mode is seen predominantly in the case

Table 11 Summary of failure patterns during formability testing of FSW sheets

Condition	FSW parameters			Failure pattern	
	SD (mm)	TRS (rpm)	WS (mm/min)	100 × 200 mm strain path	125 × 200 mm strain path
1	12	600	80	Pattern 1	Pattern 2
2	12	600	100	Pattern 1; draw bead failure	Pattern 2
3	12	600	120	Pattern 1	Pattern 2
4	12	700	80	Pattern 1	Pattern 2
5	12	700	100	Pattern 1; draw bead failure	Pattern 2
6	12	700	120	Pattern 1	Pattern 2
7	12	800	80	Pattern 1; draw bead failure	Pattern 2
8	12	800	100	Pattern 1	Pattern 2
9	12	800	120	Pattern 1	Pattern 2
10	16	600	80	Pattern 1; draw bead failure	Pattern 2
11	16	600	100	Pattern 1; draw bead failure	Pattern 2
12	16	600	120	Pattern 1	Pattern 3
13	16	700	80	Pattern 1; draw bead failure	Pattern 2
14	16	700	100	Pattern 1	Pattern 2
15	16	700	120	Pattern 1; draw bead failure	Pattern 2
16	16	800	80	Pattern 1	Pattern 2
17	16	800	100	Pattern 1	Pattern 2
18	16	800	120	Pattern 1	Pattern 2
19	18	600	80	Pattern 1; draw bead failure	Pattern 3
20	18	600	100	Pattern 1; draw bead failure	Pattern 2
21	18	600	120	Pattern 1; draw bead failure	Pattern 2
22	18	700	80	Pattern 1	Pattern 3
23	18	700	100	Pattern 1	Pattern 3
24	18	700	120	Pattern 1	Pattern 2
25	18	800	80	Pattern 1; draw bead failure	Pattern 2
26	18	800	100	Pattern 1	Pattern 2
27	18	800	120	Pattern 1	Pattern 2

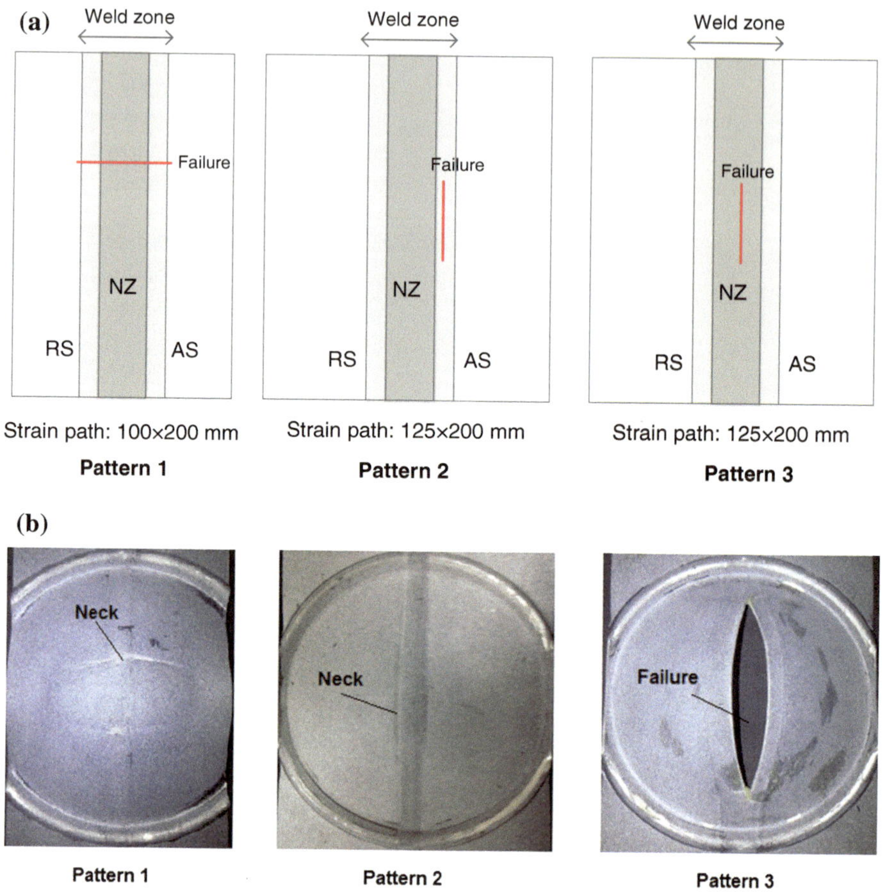

Fig. 15 Failure pattern during LDH test of FSW sheets, **a** schematic of failure pattern at plane-strain and stretching strain-paths, **b** actual FS welded samples after failure (*NZ* nugget zone, *AS* advancing side, *RS* retreating side)

of plane-strain mode of deformation, i.e. for 100 × 200 mm sample. Since deformation is seen only along the longitudinal and in thickness direction of the sheet, and not in width direction, the failure is expected to happen normal to the major straining direction, like in the case of simple tensile test. The second and third failure modes are seen in stretching mode of deformation, i.e. for 125 × 200 mm sample. In this case, there is another pull on the width direction of the sheet, along with the longitudinal pull, making the welded sheet to fail in the weaker zone of the weld on the AS, but along the weld. In very few cases, weld failure is seen at the centre.

The forming limit diagram of FSW sheets made at two different welding conditions is shown in Fig. 16a, b. All the failed and safe strains measured during the LDH test are plotted in the same graph with true major strain in X-axis and true

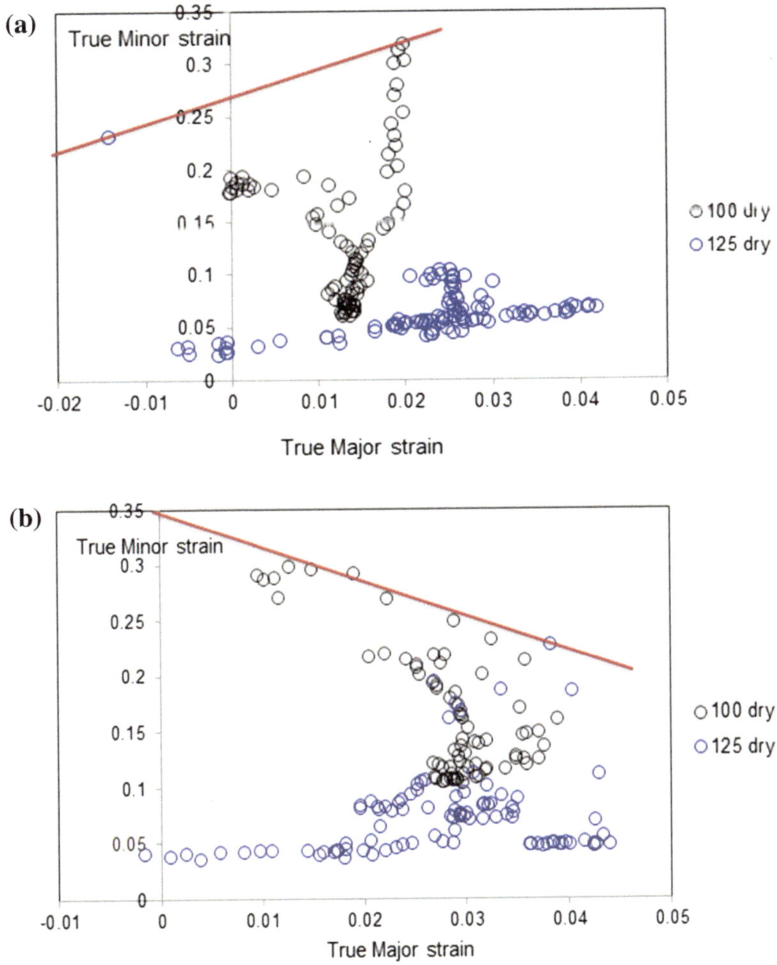

Fig. 16 Forming limit diagram (FLD) of FSW sheets, **a** SD: 12 mm, WS: 120 mm/min, RS: 800 rpm, **b** SD: 16 mm, WS: 120 mm/min, RS: 600 rpm

minor strain in Y-axis. The largest strain defines the limit of useful formability, and hence, FLC is drawn such that all the safe strains are below the curve (Fig. 16). The same procedure is followed for all the FSW sheets, and FLC is plotted. These curves are plotted in the same graph to study the effect of FSW parameters. It should be noted from the Fig. 16b that the forming limit in the stretching side of the FLD is less than that of in plane-strain mode, i.e. FLC of FSW sheet has a negative slope, as opposed to what is generally observed in the case of un-welded sheets. This is mainly due to the change in failure pattern in the plane-strain and stretching strain-paths. In the plane-strain mode, failure occurs normal to the WZ (pattern 1),

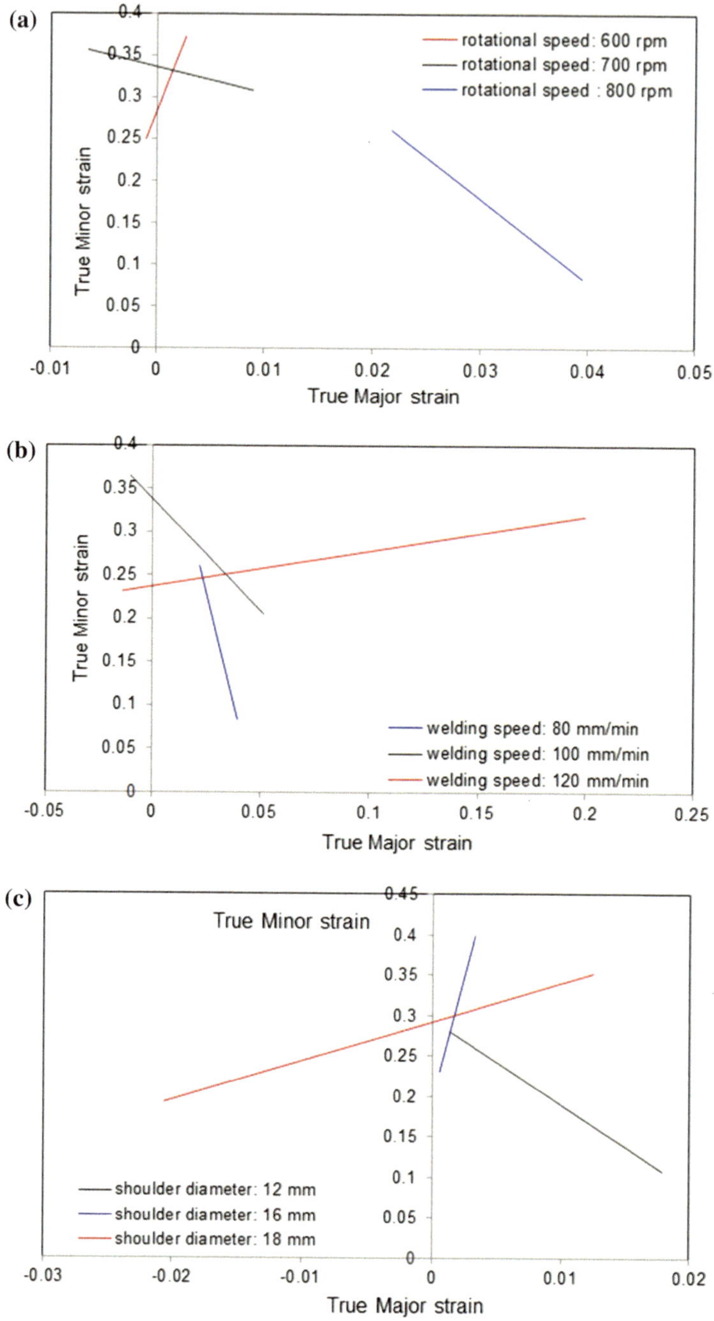

Fig. 17 Effect of rotational speed, welding speed and shoulder diameter on the forming limit curves of FSW sheets

which has better ductility than the base material, while along the WZ in stretching mode (pattern 2 or 3), reducing the overall forming limit because of gauge effect.

The effect of rotational speed, welding speed and shoulder diameter on the FLC of FSW sheets is shown in Fig. 17a–c. It is observed that the FLC exhibit both positive and negative slopes, in an arbitrary manner, because of change in failure patterns from plane-strain to stretching mode. It has been concluded that for studying such effects on the forming limit of FSW sheets made of dissimilar grades, it is better to choose forming limit in any one of the strain-paths, either in plane-strain or tensile or stretching, rather than the whole FLC.

4 Conclusions

The effect of chosen FSW parameters on the temperature evolution, springback and formability of FSW sheets made of similar and dissimilar grades is studied. The following are the conclusions drawn from the present work.

1. By increasing the shoulder diameter, plunge depth and tool rotation speed, the peak temperature is increased, whereas it is decreasing for increasing welding speed. The effect of shoulder diameter is due to the increase in higher temperature region. The influence of plunge depth is due to the forging action of shoulder onto the sheet metal surface. The effect of rotational speed is due to higher relative velocity between the tool and sheet, and the influence of welding speed is due to the lower heat input per unit length and at the same time, dissipation of heat over a wider region of the sheet.
2. The ductility of WZ increases with increase in welding speed and rotational speed. This is due to the improvement in strain hardening exponent of WZ at higher rotational speed and welding speed.
3. The springback of FSW sheets is in between that of Al6061T6 and Al5052H32 base materials.
4. Reduction in springback of FSW sheets is observed at higher rotational speed and welding speed. The modifications in springback have got reasonable correlation with the modifications in σ_y/E ratio and n value of WZ.
5. A good correlation is observed between experimental and predicted springback values at different FSW conditions. The use of isotropic WZ assumption is acceptable during springback prediction of FSW sheets, through anisotropic WZ assumption yield slightly better results.
6. Three different modes of failures, failure normal to the WZ, failure along the WZ, but on the AS, and failure along the WZ, but at the centre, are observed during formability testing of FSW sheets made of dissimilar Al grades at two different strain-paths.

7. It is better to evaluate the forming limit of FSW sheets at individual strain-paths, rather than one full forming limit curve, because of the change in failure pattern when moved from plane-strain to stretching side. The forming limit of FSW sheets on the stretching side is lower than in plane-strain deformation mode.

Acknowledgments R. Ganesh Narayanan acknowledges DST, India and ARDB, India for sponsoring the projects. The present work is part of those projects.

References

Alves de Sousa RJ, Correia JPM, Simoes FJP, Ferreira JAF, Cardoso RPR, Gracio JJ, Barlat F (2008) Unconstrained springback behavior of Al–Mg–Si sheets for different sitting times. Int J Mech Sci 50:1381–1389

Chang SH, Shin JM, Heo YM, Se DG (2002) Springback characteristics of the tailor-welded strips in U-bending. J Mater Process Technol 130–131:14–19

Durga Rao B, Narayanan RG (2014) Springback of friction-stir welded sheets made of aluminium grades during V-bending: an experimental study. ISRN Mech Eng Article ID 681910. doi:10.1155/2014/681910

Garcia-Romeu ML, Ciurana J, Ferrer I (2007) Springback determination of sheet metals in an air bending process based on an experimental work. J Mater Process Technol 191:174–177

Greeze R, Manach PY, Laurent H, Thuillier S, Menezes LF (2010) Influence of the temperature on residual stresses and springback effect in an aluminium alloy. Int J Mech Sci 52:1094–1100

Kim J, Lee W, Chung KH, Kim D, Kim C, Okamoto K, Wagoner RH, Chung K (2011) Springback evaluation of friction-stir welded TWB automotive sheets. Met Mater Int 17:83–98

Lee W, Chung KH, Kim D, Kim J, Kim C, Okamoto K, Wagoner RH, Chung K (2009) Experimental and numerical study on formability of friction-stir welded TWB sheets based on hemispherical dome stretch tests. Int J Plast 25:1626–1654

Lim H, Lee MG, Sung JH, Kim JH, Wagoner RH (2012) Time-dependent springback of advanced high strength steels. Int J Plast 29:42–59

Miles MP, Mahoney MW, Fuller CB (2006) Prediction of bending limits in friction-stir-processed thick plate aluminum. Metall Mater Trans A 37A:399–404

Mishra RS, Ma ZY (2005) Friction-stir welding and processing. Mater Sci Eng R 50:1–78

Moon YH, Kang SS, Cho JR, Kim TG (2003) Effect of tool temperature on the reduction of the springback of aluminum sheets. J Mater Process Technol 132:365–368

Mulder J, Vegter Henk, Jin Jin Ha, van den Boogaard AH (2012) Determination of flow curves under equibiaxial stress conditions. Key Eng Mater 53:504–506

Park S, Lee CG, Kim J, Han HN, Kim SJ, Chung K (2008) Improvement of formability and springBack of AA5052-H32 sheets based on surface friction-stir method. J Eng Mater Technol 130: 041007-1–041007-10

Ramulu PJ, Ganesh Narayanan R, Kailas SV, Reddy J (2013a) Internal defect and process parameter analysis during friction-stir welding of Al 6061 sheets. The Int J Adv Manufact Technol 65: 1515–1528

Ramulu PJ, Kailas Satish V, Ganesh Narayanan R (2013b) Influence of tool rotation speed and feed rate on the forming limit of friction-stir welded AA6061-T6 sheets. Proc Inst Mech Eng Part C: J Mech Eng Sci 227:520–541

Wang JF, Wagoner RH, Carden WD, Matlock DK, Barlat F (2004) Creep and anelasticity in the springback of aluminium. Int J Plast 20:2209–2232

Wang L, Huang G, Zhang H, Wang Y, Yin L (2013) Evolution of springback and neutral layer of AZ31B magnesium alloy V-bending under warm forming conditions. J Mater Process Technol 213:844–850

Woo W, Balogh L, Ungár T, Choo H, Feng Z (2008) Grain structure and dislocation density measurements in a friction-stir welded Aluminum alloy using X-ray peak profile analysis. Mater Sci Eng A 498:308–313

Zadpoor AA, Sinke J, Benedictus R, Pieters R (2008) Mechanical properties and microstructure of friction-stir welded tailor-made blanks. Mater Sci Eng A 494:281–290

Zadpoor AA, Sinke J, Benedictus R (2009) Finite element modeling and failure prediction of friction-stir welded blanks. Mater Des 30:1423–1434

Prediction of Formability of Bi-axial Pre-strained Dual Phase Steel Sheets Using Stress-Based Forming Limit Diagram

Shamik Basak, Kaushik Bandyopadhyay, Sushanta Kumar Panda and Partha Saha

Abstract Dual phase (DP) steel is of great interest for automotive part manufacturers due to its excellent combinations of strength and formability. Complex components involving three-dimensional stampings are usually fabricated through multistage sheet forming operations. The ability of a sheet metal to be deformed into a specific desired shape by distributing strain over arbitrary tool surface depends on complex interaction of material, process and design variables. The strain-based forming limit diagram (ε-FLD) is often used as a measure of formability in the press shop due to convenience of measuring the limiting strain. However, it was reported by previous researchers that the ε-FLD of sheet metal shifts after pre-strain due to the initial forming operations. Hence, this work proposes a mathematical framework for constructing σ-FLD of different pre-strained sheets incorporating Barlat-89 yield criterion with different hardening laws. The formability of biaxially pre-strained DP600 was evaluated experimentally in two stages. The forming behaviour of pre-strained material was predicted by finite element model using the σ-FLD, and the predicted results matched very closely with the experimental data. It was also observed that the σ-FLD was robust and underwent insignificant changes due to the change in the pre-strain path.

Keywords Stress-based forming limit diagram · DP600 · Pre-strain · LS-DYNA

S. Basak · K. Bandyopadhyay · S.K. Panda (✉) · P. Saha
Indian Institute of Technology Kharagpur, Kharagpur 721302, India
e-mail: sushanta.panda@mech.iitkgp.ernet.in

S. Basak
e-mail: shamikmech@gmail.com

K. Bandyopadhyay
e-mail: kaushik.me08@gmail.com

P. Saha
e-mail: psaha@mech.iitkgp.ernet.in

R.G. Narayanan and U.S. Dixit (eds.), *Advances in Material Forming and Joining*,
Topics in Mining, Metallurgy and Materials Engineering,
DOI 10.1007/978-81-322-2355-9_8

1 Introduction and Background

Formability is a measure of the amount of deformation a material can withstand prior to fracture or excessive thinning. Sheet metal forming consists of simple bending, stretching, deep drawing, flanging, etc. to fabricate various complex parts. Therefore, determining the extent to which a material can deform is necessary for designing a reproducible forming operation. In sheet metal forming operations, the amount of useful deformation (formability) is limited by the occurrence of unstable deformation which mainly takes the form of necking or wrinkling. Necking occurs when the stress state leads to cause excessive thinning of the material. Wrinkling occurs when the dominant stresses are compressive, tending to cause thickening of the material. As mechanical properties greatly influence formability, and forming properties may vary from coil to coil, it is essential to test incoming sheet material. However, the outcome of a forming process depends on both material characteristics and process variables such as strain, strain rate and temperature as discussed by Marciniak et al. (2002). In fact, stress and strain fields are so diverse during a forming process so that no single test can reliably predict the formability of materials in all situations. Material properties that have a direct or indirect influence on formability and product quality are ultimate tensile strength, yield strength, Young's modulus, ductility, hardness, the strain hardening exponent and the plastic strain ratio. All of these parameters can be determined by testing a specimen cut from the blank as thoroughly discussed by Mielnik (1991).

There are several methods by which it is possible to determine the formability of sheet metal. These methods consist of simulating tests in laboratory scale viz. limiting dome height (LDH), limiting drawing ratio (LDR), hole expansion ratio (HER), bendability, spring back and forming limit diagram (FLD). Banabic (2000) discussed how these tests influence the formability of incoming sheet metal in detail. The results obtained from these laboratory tests were compared with the measurements obtained from the forming of industrial components. The excellent correlation between both the results popularizes the use of FLD on the press shop floor in real-time forming operations of industrial parts.

1.1 Forming Limit Diagram

FLD is a two-dimensional representation of the maximum principal strains in major (ε_1) and minor (ε_2) strain loci. These maximum surface strains are measured before the onset of necking or splitting in all possible combinations of tension and compression strain paths when a sheet metal undergoes in different forming operations. The development of the FLD started with tests done by Keeler and Backofen (1963) for positive principal strains. At the same time, Keeler (1965) introduced the use of electrochemically etched grid concept to measure the principal strains and strain distributions during forming operations. Besides experimental FLD, Marciniak and

Kuczynski introduced the concept of theoretical FLD which is popularly known as M–K model (Marciniak and Kuczyński 1967). In 1968, Goodwin (1968) used the combination of various mechanical tests and obtained the limiting strain in negative minor principal strain also. Thus, the complete curve is currently known as conventional FLD or strain-based FLD (FLD or ε-FLD). Later in 1973, Marciniak and Kuczynski demonstrated the effect of planar anisotropy on formability of different sheet metals including steel, aluminium and copper (Marciniak et al. 1973). The outcome of these experiments showed that the limiting strains were somewhat greater in the rolling direction compared to that in the transverse direction. In order to plot FLD, out-of-plane stretching tests were performed by Ghosh and Hecker (1974) and they remarked that limiting strains are considerably higher than those resulting from in-plane stretching. In 1975, Hecker (1975) first presented an organized approach to determine FLD in laboratory scale. A complete FLD was drawn by recording the limiting strains of aluminium-killed steel sheets of various widths over a hemispherical punch. Later on, the method was named LDH test. The height of the dome corresponding to plain strain deformation is defined by LDH_0, which is a popular formability index. The FLD is a powerful tool and can be used in industry as following ways:

- It is an assessment tool for comparing the formability of various sheet metals available in the market for fabrication of components.
- FLD is used as a diagnostic tool for predicting failure, and hence, it is used as a damage model in numerical simulation of the stamping processes.
- It gives the graphical representation of material failure tests by observing strain path progression in the FLD curve, and hence, the failure strain state and the cause can be identified.

1.2 Experimental Strain-Based Forming Limit Diagram

To ease of industrial forming operation, FLDs must cover as much strain domain as possible. The maximum values of principal strains can be determined by measuring the principal strains at necking or fracture on sheet metal covered with circular grid marks with particular dimensions in a linear strain path. In order to generate complete FLD, the load path ranging from equi-biaxial tension ($\varepsilon_1 = \varepsilon_2$) to pure shear ($\varepsilon_1 = -\varepsilon_2$) should be generated. In normal practice, the state of strain in simple uniaxial tension ($\varepsilon_1 = -2\varepsilon_2$, for isotropic material) is never exceeded in the blank holder region. In order to get the complete shape of ε-FLD, many researchers came up with different tests such as uniaxial tensile test, hydraulic bulge test (Olsen 1920), Hecker test (1975), Marciniak test (1971), Nakazima test (1968), Hasek test (1973), etc. to get the different regions of ε-FLD. Among them, the Nakazima test (1968) is a popular laboratory scale test by which FLD is generated. The used sheet metal samples usually had the same length but different widths as shown in Fig. 1a.

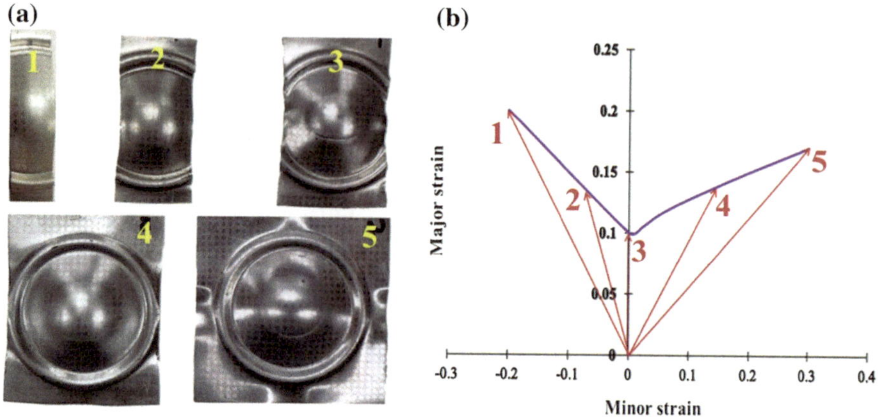

Fig. 1 **a** Formed samples after failure, **b** FLD for various sample dimensions and strain path

A pattern of circular grids of a definite size was applied on the surface of the samples by electrochemical etching method. During the test, the samples were deformed until fracture by a hemispherical punch. Major and minor principal strains at fracture, in the vicinity of crack, were plotted for all specimen dimensions and a forming limit curve was constructed. Different specimen sizes led to different proportional strain paths as shown in Fig. 1b. This forming limit curve distinguishes the range of strains between safe and failed regions in a sheet forming process. The major and minor true strains were calculated from Eq. 1.

$$\varepsilon_1 = \ln\left(1 + \frac{u-d}{u}\right) \quad \text{and} \quad \varepsilon_2 = \ln\left(1 + \frac{v-d}{v}\right) \tag{1}$$

where u = deformed ellipse major axis length, v = deformed ellipse minor axis length and d = initial circular grid diameter.

1.3 Factors Effecting the Forming Limit Diagram

FLD depends on the several crucial factors, and a slight change in these factors may alter the formability limits greatly. The formability increases with the increasing sheet thickness. A survey of a wide range of steels led to the Keeler-Brazier formula (discussed later). The strain hardening exponent, n, is numerically related to the uniform true strain (ε_u) during deformation. The uniform deformation is larger with the increase of n resulting higher limiting strains in FLD. So the formability increases which shift FLD upwards. It is found that higher Lankford anisotropy parameter, r-value, has favourable effect on the limit strain values particularly in the negative range of minor principal strain. It can be easily proved that for simple tension the slope of the strain path trajectory can be calculated by the following expression; $\frac{\varepsilon_2}{\varepsilon_1} = \text{Slope} = \theta = -\frac{r}{1+r}$; where r = Lankford anisotropy coefficient. It follows from the equation that the larger is the anisotropy coefficient, the larger is the slope of the straight strain path. With increase in the strain rate sensitive index (m) value, the ductility of the material increases due to increase in the post-uniform elongation. This increases the formability of the material (i.e. material is more easy for forming) by shifting the FLD upwards. During the experiment, it is mandatory to maintain proportional strain path, i.e. the strain path should be linear and the straight line trajectory must be followed by a point in the principal strain plane until reaching the limiting strain. Due to this reason, FLD is often termed as strain-based FLD or ε-FLD. It is observed that the limiting strain depends on the strain path. The shape and size have great influence on the FLD plotting as discussed in Nakazima et al. (1968) earlier. Specimen with particular dimension follows particular strain path while deforming the specimen. The experimental precision greatly affects the final shape of the FLD. Few such experimental conditions are the precision of the grid measurements, the frictional conditions between different pair of components, temperature of the blank during experiments, accuracy of the press, etc.

Out-of-plane stretch forming and in-plane stretch forming are the two popular methods of determining FLD experimentally. In out-of-plane stretch forming, the deformation surface does not lie in a single plane. As discussed earlier, Hecker (1975) introduced a new out-of-plane stretch forming approach to determine the FLD. Out-of-plane stretch forming have certain important features such as (a) large strain gradients are produced because of the presence of friction and curvature of different surfaces. Hence, forming limits determined using the out-of-plane method show a dependence on tooling geometry variables such as punch radius and die radius; (b) failure occurs in the test specimen at specific locations that are dictated by specimen and tooling geometry. The out-of-plane method is not very sensitive to material defects. Another approach to evaluate forming limit is in-plane stretch forming method. Here, deformation always takes place in uniform and proportional manner within the plane of the sheet sample, without imposing any bending on the specimen and avoiding friction effects. For this cause, in-plane forming limits can be more sensitive to material defects and are not influenced to the same extent by tooling geometry variables. But, it has certain advantages too, as the curvature and friction effects are absent, large strain gradients can be avoided in the in-plane test. Therefore, forming limits can be defined more accurately. Azrin and Backofen (1970) generated in-plane forming limits for proportional strain paths ranging from plane strain to a biaxial strain path. As defined earlier, Marciniak and Kuczynski (1967) developed an ingenious technique to deform sheet specimens' in plane under balanced biaxial conditions. Raghavan (1995) developed modified Marciniak and Kuczynski (MK) tests that allow the sample material to deform freely "in plane" with less effect of tool geometry and friction. This in-plane stretch forming process is very useful to give pre-strain of sheet metal.

1.4 Path Dependency of Strain-Based Forming Limit

Problem arises when same FLD is used to predict pre-strained material behaviour. All the theoretical and experimental works till discussed for constructing FLD, assumed linear strain paths. By conducting an experimental study on aluminium-killed steel, Matsuoka and Sudo (1969) investigated the behaviour of forming limits in a two-stage forming process. They concluded that the forming limits increase with increase in the first-stage strains, but for first-stage balanced biaxial tension, the fracture strains get reduced significantly. This fact was confirmed by the study of Kobayashi et al. (1972) on aluminium-stabilized-steel material. They also added that for plane strain deformation in first stage the limiting strain was slightly reduced. Muschenborn and Sonne (1975) performed their study on two different materials: titanium-stabilized and aluminium-killed steels. They also confirmed that uniaxial pre-staining shifts the FLD in a direction where major strains were increased but biaxial pre-straining shifted the FLD in a direction where major strain reduced. In 1978, Laukonis and Ghosh (1978) carried out an extensive study on cold rolled annealed aluminium-killed steel and 2036-T4 aluminium alloy. Their findings confirmed the downward shift of FLD on both materials for biaxial pre-strain. However, the limiting strain was reasonably unchanged in case of plane strain pre-straining of 2036-T4 aluminium.

Along with the experimental studies, sheet-metal-researchers also tried to simulate the effect of pre-straining on limiting strain theoretically. This effort was pioneered by Rasmussen. He attempted to predict strain-based FLD for two-stage forming process by using M–K theoretical model assuming isotropic hardening models (Rasmussen and Wanheim 1981). He did further research on the forming limit by including curvilinear strain paths (Rasmussen 1982) and concluded that the reduction in limiting strain occurred when curvilinear path approaches to plane strain condition. Another remarkable theoretical study on multiple strain paths was done for isotropic, rate sensitive material by Barata da Rocha and Jalinier (1984). They developed a model incorporating anisotropy of the material with strain path-dependent FLD. A couple of researchers, Lu and Lee (1987), developed their limiting strain model by incorporating both isotropic and kinematic hardening model. They proposed that FLD calculations using isotropic hardening model were comparatively easier.

Recent major experiment of Graf and Hosford (1993, 1994) showed the serious limitation of stain FLD while subjected to various levels of initial pre-strain in different orientations. They concluded that plane strain and uniaxial pre-strain actually increase the limiting strain, whereas the biaxial pre-strain reduces the forming limit and all the FLDs' cusps or low points in the plane strain condition. Hence, no single curve is there in principal strain space which may represent forming limit universally whatever may be the initial pre-strain condition. Therefore, finding a single path-independent curve to characterize the forming limits was of the prime interest of the sheet-metal-researchers.

1.5 Stress-Based Forming Limit Diagram

From the previous discussion, it can be concluded that characterization of forming limits is a significant challenge in complex processes as the conventional strain-based FLD is sensitive to strain path. Hence, it is very difficult to trace dynamic FLD for each element. The most promising solution for dealing with strain path effects in FLD is to use a stress-based FLD curve (σ-FLD). It was observed that stress-based FLD is not affected by the strain path, and it can be applied without modification to analyse of all forming problems. Since stress value cannot be determined directly from the Nakazima et al. (1968) experiments, calculation of plastic deformation is needed. Kleemola and Pelkkikangas (1977) reported that the forming limits of mild steel, copper and brass depend on the magnitude and type of pre-strain. Interestingly, the authors' also reported that the observed shift was not present when strain points were converted to stress point. Another group of researchers Arrieux et al. (1987) reported a similar finding for pre-strained aluminium. However, neither the knowledge of the discovery of a path-independent stress-based forming limit is widespread nor its significance is appreciated. This lack of awareness is partly due to the belief that the conventional FLD method is satisfactory when limited to analysis of the first draw die, which generally receives the most attention. Recently in 1999, Stoughton (2000) reintroduced this concept. He converted the ε-FLD reported by Graf and Hosford (1993) to σ-FLD and concluded that σ-FLD is not affected by changing of the strain path (Stoughton 2001; Stoughton and Zhu 2004; Stoughton and Yoon 2005). More recently, Yoshida et al. (2007) concluded that if the material was described by isotropic hardening model, the σ-FLDs were path independent.

Hence within the framework of constitutive assumptions, there exists a single curve in principal stress plane which represents formability of sheet metal. The prime assumption behind the construction of ε-FLD and σ-FLD is plane stress condition ($\sigma_1, \sigma_2 \neq 0$ and $\sigma_3 = 0$), i.e. the through-thickness normal component of stress, σ_3, is neglected.

1.6 Extended Stress-Based Forming Limit Diagram

Another FLD which was developed recently to take care of normal component of stress, σ_3, is known as extended stress-based FLD (XS-FLD). A study by Koc and Altan (2001) showed that during hydroforming operation necking/splitting of tube takes place under the influence of not only in-plane stresses (σ_1, σ_2) but also through-thickness compressive principal component of stresses (σ_3). They also explained that during hydroforming, due to internal pressure when the tube was initially expanding, the stress state in the thin-walled tube could be considered as plane stress. But, after the tube wall touched the die, a through-thickness compressive stress due to the internal pressure acted along with in-plane stresses. Hence in these cases, prediction of failure by ε-FLD or σ-FLD is erroneous and leads to a wrong prediction of formability. Recently, Simha et al. (2005, 2007) introduced a new measure (XS-FLD) of forming limit to predict the onset of necking in sheet metal underwent non-proportional loading, as well as under three-dimensional stress states. They converted σ-FLD to XS-FLD with suitable assumptions and concluded that XS-FLD could predict necking under a three-dimensional state of stress. They used tubes of two alloys, EN-AW 5018 aluminium alloy and dual phase DP600 steel. In one hand, XS-FLD is as robust as σ-FLD and in other hand it is also taking account of stress in thickness direction.

2 Theoretical Models for Strain-Based FLD

Previous section it is seen that FLD depends on several factors. It is difficult to verify experimentally the influence of each parameter individually since it is virtually impossible to change only one parameter at a time. Moreover, experimental determination of FLD for all materials is not only tedious but also expensive. Therefore, an extensive effort was made to develop various mathematical and analytical models to predict the forming limits of sheet metals theoretically. Different theoretical methods which were proposed to explain the localized necking of steel sheets are discussed in this section. Al 2008 T4 material properties (Stoughton 2000) are used (strain hardening coefficient (n) 0.285,´ strength coefficient (K) 539 MPa and Linkford anisotropy coefficient ($\bar{r}$) 0.58) in this section to formulate all the theoretical FLDs.

In 1952, Swift (1952) developed a theoretical model based on Considere's analysis (1885) in order to determine the limiting strain in biaxial tension region. It predicts the onset of diffuse necking with the assumption that plastic instability occurs at a maximum load (so-called Modified Maximum Force Criterion) for proportional loading along the both principal directions. Critical major strain for diffuse necking can be formulated by Eq. 2.

$$\varepsilon_1 = \frac{2n(1+\rho+\rho^2)}{(\rho+1)(2\rho^2-\rho+2)} \tag{2}$$

where n = strain hardening coefficient and ρ is a strain ratio defined as: $\rho = \mathrm{d}\varepsilon_2/\mathrm{d}\varepsilon_1$.

Hill (1952) assumed that a local neck may forms with an angle to the direction of the major principal stress and the angle of the neck (α) with respect to loading axes may be expressed as: $\alpha = \tan^{-1}(\sqrt{-\rho})$, where ρ is the strain ratio. Clearly, the equation makes sense when the strain ratio is negative, i.e. the minor principal strain should be negative. He also assumed that the necking direction is coincident with the zero-elongation direction. The criterion for occurring instability is given by the following equation: $\frac{\mathrm{d}\sigma}{\mathrm{d}\varepsilon} = \frac{\sigma}{1+r}$, where r is the Lankford anisotropic coefficient. Using

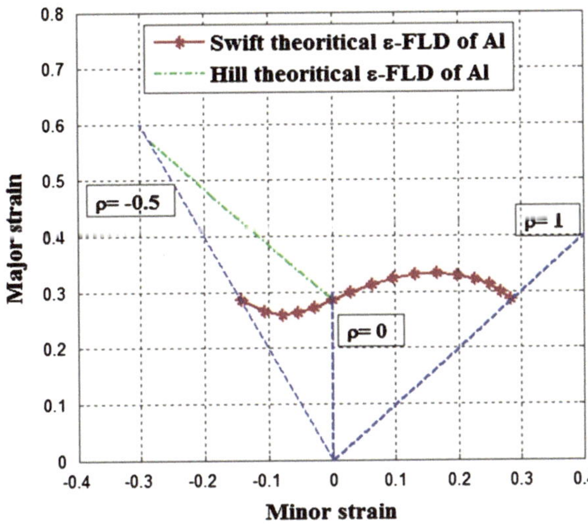

Fig. 2 Forming limit diagram obtained by Swift theory and Hill theory

Hollomon power law of true stress–true strain relationship, the critical condition for negative strain ratio is defined as: $\varepsilon_1 = \frac{n}{1+\rho}$.

Figure 2 is the comparison between predicted forming limit by both of the FLD models. It can be observed from the Swift FLD that diffuse necking is observed in biaxial stretching, plane strain and uniaxial tension, when the major strain reaches a maximum value. Interestingly, this maximum major strain value is equal to the strain hardening exponent (n value) obtained from power hardening law. The limiting strain measured by this method is always less than that of experimental data especially in the second quadrant of the FLD. This disadvantage restricts the use of Swift model for practical purpose. Also it can be seen that for Hill's model, under plain strain condition, localized necking occurs at $\varepsilon_1 = n$, and it steadily increases to $\varepsilon_1 = 2n$, at pure uniaxial region. Thus, Hill's model does not depend upon different yield criteria but it depends on the different values of hardening coefficients.

Stören and Rice (1975) proposed a FLD model based on the J2 deformation theory of plasticity which says that localized necking is caused by the vertex developed on the subsequent yield surface (Stören and Rice 1975). This method can predict the localized necking over the entire range of FLD. However, it underestimates the limit for localized necking at the left-hand side of FLD. This drawback was eliminated by Zhu et al. (1987) by incorporating the moment equilibrium in addition to the force equilibrium condition adopted by Stören and Rice. They found that the discontinuity of shear stress inside and outside of the localized band is actually zero. The bifurcation analysis with vertex theory, proposed by Stören and Rice, was later revised by Zhu et al. (2001). The modified vertex theory, which involves a generalized material constitutive relation, is achieved by applying Hosford's higher-order yield criterion (Hosford 1979) and the concepts of plasticity theory. Later, analytical expression of the theoretical FLD was developed by Chow et al. (2003) with the adoption of Hosford higher-order yield criterion. Final theoretical form developed by Chow et al. (2003) is presented in Eqs. (3) and (4).

$$\varepsilon_{\text{major}}^{\text{LHS}} = \frac{(1+r_0)[\alpha^{\gamma-2} + r_{90}(1+\rho)(1-\alpha)^{\gamma-2}]f^{\gamma}(\alpha)}{(\gamma-1)(1+\rho)(1+\rho\alpha)[\alpha^{\gamma-2} + (r_{90} + r_0\alpha^{\gamma-2})(1-\alpha)^{\gamma-2}]} + \frac{(\gamma-1)n-1}{(\gamma-1)(1+\rho)} \quad (3)$$

$$\varepsilon_{\text{major}}^{\text{RHS}} = \frac{(1+r_0)[\alpha^{\gamma-2} + r_{90}(1-\alpha)^{\gamma-2}]f^{\gamma}(\alpha)}{(\gamma-1)(1+\rho\alpha)[\alpha^{\gamma-2} + (r_{90} + r_0\alpha^{\gamma-2})(1-\alpha)^{\gamma-2}]} + \frac{(\gamma-1)n-1}{(\gamma-1)(1+\rho\alpha)} \quad (4)$$

where $f(\alpha) = \frac{1}{[r_{90}(1+r_0)]^{\frac{1}{\gamma}}}[r_{90} + r_0|\alpha|^{\gamma} + r_0 r_{90}|1-\alpha|^{\gamma}]^{\frac{1}{\gamma}}$.

α = stress ratio, ρ = strain ratio and γ = Hosford's exponent, r_0 and r_{90} are the anisotropy coefficients in rolling direction and in transverse direction, respectively. By using the above-mentioned equation, the theoretical FLD (shown in Fig. 3) is computed with varying orders of yield function, i.e. γ = 2, 4 and 6.

Perturbation analysis is another method to predict plastic instability. In this method, the sheet material is assumed to be homogeneous at the beginning of deformation. However, after every increment of plastic deformation, a perturbation is considered to affect the homogeneous flow. This concept was initially developed to study the dynamics of flow in fluids, but it was adapted to the plastic flow in solids by researchers such as Zbib and Aifantis (1989) in order to study shear bands and localized necking of sheet samples deformed in uniaxial tension. In 1991, Dudziniski and Molinari (1991) first suggested the concept of the effective instability and then successfully adopted the instability method to formulate the FLD.

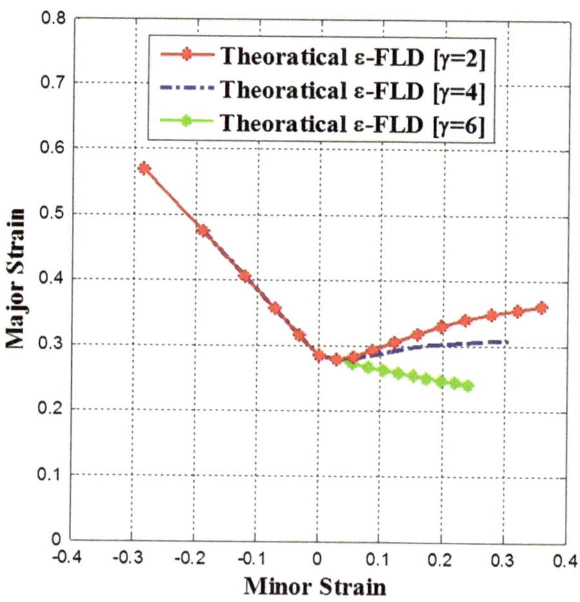

Fig. 3 Forming limit diagram obtained from Stören-Rice model

The effective instability analysis gives the basic trend of M–K model analysis (discussed later in this chapter). The dependence of the limit strains on the value of the instability intensity parameter is equivalent to the dependence of the limit stains on the value of the initial defect in the M–K model analysis. Advantage of this method is that an explicit expression for the critical strain can be obtained unlike M–K model. However, linear perturbation theory can determine the forming limits for some strain rate sensitive materials only. For steels, this theory under predicts the forming limits.

In 1984, Jones and Gillis (1984) proposed a new theory (known as J–G theory) for prediction of localized necking. He assumed that the material is homogeneous till the maximum load and a strain concentration under constant load. After that the localized neck formed due to rapid load decrease. Initially, this method was suitable to predict right-hand side of FLD but Choi et al. (1989) extended this theory for negative minor strain also. But due to computational difficulty, this method is not as user friendly as M–K model. This sheet non-homogeneity model (M–K model) is based on the hypothesis of the existence of imperfections in sheet metal. Marciniak and Kuczynski (1967) developed this model, and it is well known as M–K damage model. Initially sheet metal has geometrical imperfections (thickness variation shown in Fig. 4) and/or structural imperfections (inclusions and gaps). In the forming process, these imperfections progressively evolve and the plastic forming of the sheet metal is almost completely localized in them, leading to the necking of the sheet metal. The theoretical model proposed by Marciniak assumes that the specimen has two regions: region "P" with a uniform thickness t_{0P}, and region "Q" with the thickness t_{0Q}. The initial geometrical non-homogeneity of the specimen is described by the so-called "coefficient of geometrical non-homogeneity", *f*, expressed as the ratio of thickness in the "Q" region to "P" region.

In the M-K model, the strain and stress states in the two regions are analysed and the principal strain ε_{1Q} in region "Q" in relation to the principal strain ε_{1P} in region "P" is monitored. When the ratio of these strains $\varepsilon_{1Q}/\varepsilon_{1P}$ becomes too large (infinitely large in theory, but greater than 10 in practice), one may consider that the entire straining of the specimen is localized in region "Q".

Fig. 4 Geometrical model of Marciniak–Kuczynski theory

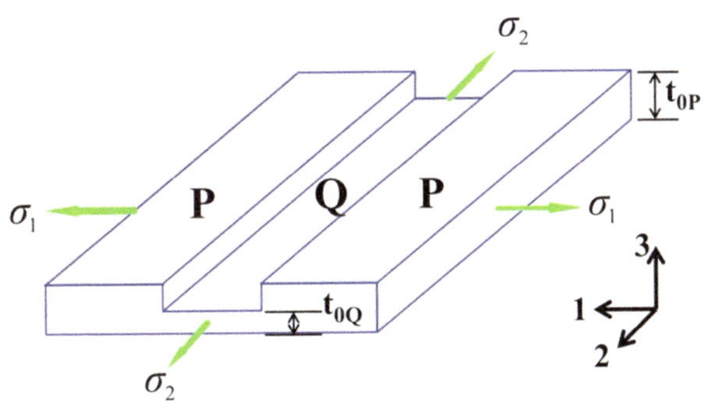

This model was intensely used and developed by researchers due to the advantages it offers: it has an intuitive physical background; it correctly predicts the influence of different process or material parameters on the limit strains; the predictions are precise enough; the model can be easily coupled with finite element simulation software for sheet metal forming processes. But in this model, the prediction results are very sensitive to the constitutive equations used, as well as to the values of the non-homogeneity parameter (*f*). Also, the shape and position of the FLD depend on the value of the *f*-coefficient. If $f = 1$ (geometrically homogeneous sheet), the curve becomes coincident with the first bisector. Thus, this theory cannot model the strain localization for geometrically homogeneous sheets. M–K method tends to give very high limit strain when the strain ratios near balanced biaxial tension. In other words, the M–K method under predicts the limit strains near plane strain deformation, but over predicts them in the equi-biaxial stretching regime.

Keeler and Brazier (1977) proposed an empirical relationship (Eq. 5) for calculating the limit true strains corresponding to plane true strain, FLD_0 (located in the FLD where it crosses the major strain axis) using thickness and n-value of the sheet material.

$$\begin{aligned} \varepsilon_1 &= \mathrm{FLD}_0^{\mathrm{True}} - \varepsilon_2, \quad \text{for } \varepsilon_2 < 0 \quad \text{and} \\ \varepsilon_1 &= \ln[0.6 \times (\exp(\varepsilon_2) - 1) + \exp(FLD_0^{True})], \quad \text{for } \varepsilon_2 > 0 \end{aligned} \tag{5}$$

where $\mathrm{FLD}_0^{\mathrm{True}} = \ln[1 + (0.233 + 0.413 \times t) \cdot \frac{n}{t}]$.

Here, n = strain hardening coefficient obtained from hardening laws and t = sheet metal thickness. The empirical Keeler-Brazier curve is shown in Fig. 5.

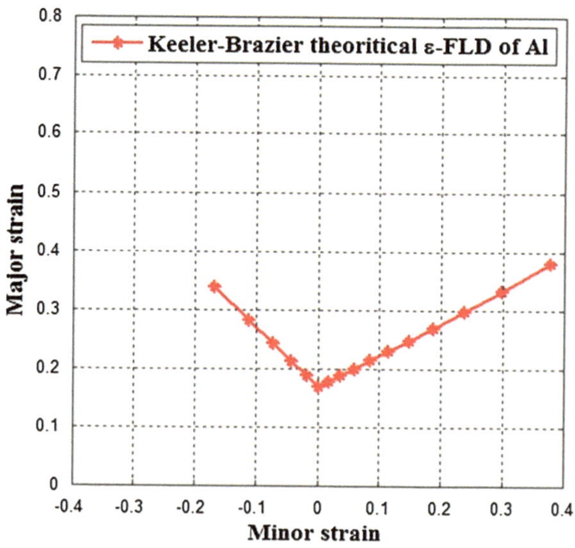

Fig. 5 Forming limit diagram obtained by Keeler-Brazier theory

2.1 Theoretical Models for Estimation of Stress-Based FLD

2.1.1 Phenomenological Approach

The important elements which are required to depict the plastic behaviour of a material in a general stress state are as follows:

- A *yield criterion* expressing a relationship between the stress components at the moment when plastic yielding occurs.
- An *associated flow rule* describes the manner in which plastic strain depends upon a given combination of stress.
- A *hardening rule* describing the evolution of the initial yield stress during forming process.

When stress reaches the yield point of the material, the transition from the elastic to plastic state occurs. The yield point in uniaxial tension is established using the stress–strain curve of the material. Yield surface generally defined as $f(\bar{\sigma}, Y) = \bar{\sigma} - Y = 0$, where $\bar{\sigma}$ is the equivalent stress which is a scalar function of deviatoric stress tensor and Y is the yield stress obtained from a simple test (tension, compression or shearing). In this chapter Barlat-89 yield criterion is used in combination with Hollomon, Swift and modified Voce hardening laws to convert ε-FLD into σ-FLD.

2.1.2 Barlat-89 Yield Criterion

In 1989, Barlat and Lian (1989) proposed a yield criterion in the form of

$$f = a|k_1 + k_2|^M + a|k_1 - k_2|^M + c|2k_2|^M = 2\bar{\sigma}^M \tag{6}$$

where a, c, h and p are material anisotropy constants related to r_0 and r_{90}. M is chosen according to the crystallographic structure of the material. The best approximation for the exponent was given by: $M = 6$ (for BCC materials) and $M = 8$ (for FCC materials).

$$k_1 = \frac{\sigma_x + h\sigma_y}{2}; \quad k_2 = \sqrt{\left[\left(\frac{\sigma_x - h\sigma_y}{2}\right)^2 + p^2\sigma_{xy}^2\right]}$$

Using another identification procedure based on the coefficients r_0 and r_{90} it is obtained:

$$a = 2 - c = 2 - 2\sqrt{\frac{r_0}{1 + r_0}\frac{r_{90}}{1 + r_{90}}}; \quad h = \sqrt{\frac{r_0}{1 + r_0}\frac{1 + r_{90}}{r_{90}}}.$$

As in the present sheet forming experiments, the anisotropic thin rolled sheets were considered, and the shear stress was assumed to be neglected so as to reduce the complexity in the analytical formulation. Hence the parameters boil down to:

$$k_1 + k_2 = \sigma_x \quad \text{and} \quad k_1 - k_2 = h \cdot \sigma_y$$

For sheet metals, x is parallel to the rolling direction (RD), y in transverse direction (TD) and z in the normal direction. If the anisotropic orientation of the sheet metal coincides with the principal direction of stress tensor (Fig. 6), then for plane stress condition $\sigma_3 = 0$ (in sheet metal forming) the Eq. (6) reduces to:

$$f = a|\sigma_1|^M + a|h\sigma_2|^M + c|\sigma_1 - h\sigma_2|^M = 2\bar{\sigma}^M \tag{7}$$

Applying flow rule, $\mathrm{d}\varepsilon_{ij} = d\lambda \frac{\partial \bar{\sigma}(\sigma_{ij})}{\partial \sigma_{ij}}$ (where $\mathrm{d}\lambda$ is a positive scalar that depends on the stress state) in Eq. (7), and upon simplification the equation can be written in the form of stress ratio ξ, defined as the ratio of effect stress ($\bar{\sigma}$) to major stress (σ_1) as:

$$\xi = \frac{\bar{\sigma}}{\sigma_1} = \left[\frac{1}{2}\{a + a|h\alpha|^M + c|1 - h\alpha|^M\}\right]^{\frac{1}{M}} \tag{8}$$

where $\alpha = \frac{\sigma_2}{\sigma_1}$.

Again strain ratio $\rho = \frac{\mathrm{d}_{\varepsilon_2}}{\mathrm{d}_{\varepsilon_1}}$ can be written in the form of α as,

$$\rho = \frac{a|h\alpha|^{M-1} - ch|1 - h\alpha|^{M-1}}{a + c|1 - h\alpha|^{M-1}} \tag{9}$$

From the definition of plastic work ($\bar{\sigma} \cdot \mathrm{d}\bar{\varepsilon} = \sigma_1 \cdot \mathrm{d}\varepsilon_1 + \sigma_2 \cdot \mathrm{d}\varepsilon_2 + \sigma_3 \cdot \mathrm{d}\varepsilon_3$) and further imposing the plane stress condition ($\sigma_3 = 0$), the effective strain can be expressed in the form of Eq. (10):

$$\bar{\varepsilon} = \frac{\varepsilon_1}{\xi}(1 + \rho\alpha). \tag{10}$$

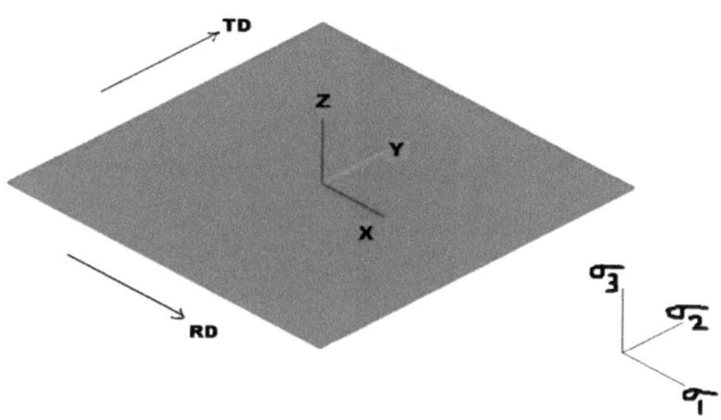

Fig. 6 Axis orientation in sheet metal forming

2.1.3 Hardening Laws

Strain hardening is the strengthening of a metal by plastic deformation. Three hardening laws are considered in this work, namely Hollomon, Swift and modified Voce model.

The Hollomon equation is given by

$$\sigma = K \cdot \bar{\varepsilon}_P^n \tag{11}$$

where $\bar{\sigma}$ and $\bar{\varepsilon}_P$ are, respectively, the effective stress and plastic strain. K (known as strength coefficient) and n (known as stain hardening exponent) are material constants.

The Swift equation is given by

$$\bar{\sigma} = K \cdot (\varepsilon_0 + \bar{\varepsilon}_P)^n \tag{12}$$

The parameters K, n and ε_0 are material constants.

Modified Voce equation is given by

$$\bar{\sigma} = A - B_0 \cdot e^{(-C \cdot \bar{\varepsilon}_P)} + B_1 \cdot \bar{\varepsilon}_P \tag{13}$$

where A, B_0, B_1 and C are the material constants. All these material constants can be calculated by fitting the experimentally determined true stress–true strain curve. B_1 is approximately identified in this work as a half of yield stress value for each of the investigated steels as defined by Panich et al. (2013).

2.1.4 Steps for Constructing σ-FLD from ε-FLD

The major and minor strain values after pre-strain are defined by ε_{1i} and ε_{2i}, respectively, and that after step-II is defined by ε_{1f} and ε_{2f}, respectively. Amount of effective plastic strain induced in the specimen during pre-strain was calculated from the Eq. (10) by calculating the ratios such as ρ, α and ξ as formulated by the Eqs. (8) and (9). Similarly, effective plastic strain increment was estimated during step-II deformation by estimating different derived parameters such as ρ, ξ and α. After calculating total effective plastic strain induced during forming, total effective stress at the end of the step-II was calculated by hardening laws given by following Eq. (14).

$$\bar{\sigma} = f\{\bar{\varepsilon}(\varepsilon_{1i}, \varepsilon_{2i}) + \bar{\varepsilon}(\varepsilon_{1f} - \varepsilon_{1i}, \varepsilon_{2f} - \varepsilon_{2i})\} \tag{14}$$

Finally, principal stresses were calculated with the ratios ξ and α. These major and minor principal stress values were mapped in stress locus to obtain σ-FLD.

2.2 Theoretical Models of Extended Stress-Based FLD

Simha et al. (2007) introduced two assumptions to get XS-FLD from ε-FLD.

- The materials are assumed to follow isotropic hardening with the J2 deformation theory of plasticity. Strain path proportionality is the main assumption behind XS-FLD.
- They introduced two stress invariants, namely equivalent stress (plotted along y axis) and mean stress (plotted along x axis). These two stress invariants characterize the formability limit under plane stress loading and they are also representative of the forming limit under three-dimensional stress states.

The principal stress pair (σ_1, σ_2) can be calculated from corresponding principal strain pair $(\varepsilon_1, \varepsilon_2)$ as shown by steps of the previous section using suitable hardening laws. After that the two stress invariants can be calculated by the following formula:

$$\text{Equivalent stress} = \sum \sqrt{(\sigma_1^2 + \sigma_2^2 - \sigma_1 \cdot \sigma_2)} \text{ and Mean stress} = \sum_{\text{hyd}} = \frac{\sigma_1 + \sigma_2}{3}$$

It should be noted that the through-thickness component of stress is not present in case of sheet metal forming operation. Otherwise, mean stress and effective stress should be calculated accordingly to take care of that component.

For validation and comparison of three damage models, data of as-received and different pre-strain condition were collected from literature (Graf and Hosford 1993) for material Al 2008 T4 (Fig. 7a). All the ε-FLDs were converted to σ-FLD incorporating Barlat-89 anisotropic yield criterion coupled with Hollomon hardening law. The final σ-FLDs are shown for all pre-strain condition in Fig. 7b, and it can be seen that all the scattered pre-strain ε-FLD curves merged into a single line. From final principal stress value, XS-FLD was also computed. It is shown in Fig. 7c and found to be equally effective as σ-FLD. Hence, the insensitivity of σ-FLD and XS-FLD on strain path was verified.

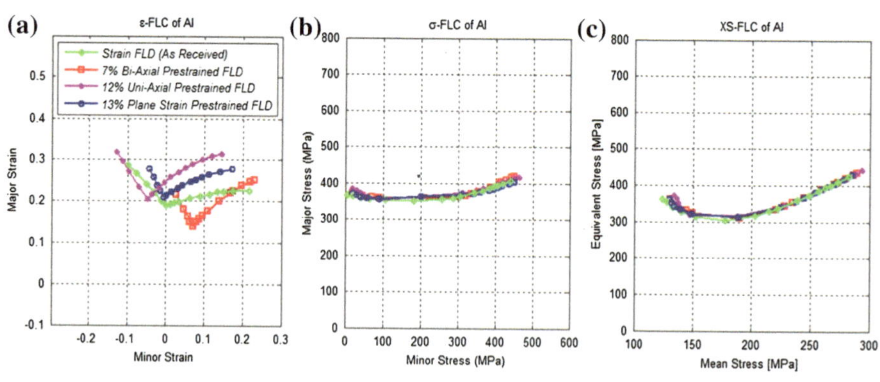

Fig. 7 **a** ε-FLD (Graf and Hosford 1993) **b** σ-FLD and **c** XS-FLD for different pre-strain conditions

3 Motivation and Objective

Recently, automotive industries are interested in the use of various advanced high strength steels (AHSSs) to fabricate light weight components with improved vehicle performances. Among the AHSS family of steels, the DP steels are the most favourite to the manufacturing industries due to their inherent combinations of high strength and ductility.

However, sheet metal forming process requires several passes in different forming operations to give the final shape of the product. The FLD shifts after each of the forming operations in a total process. So, it is very difficult to predict the nature of fracture or stress state into the material by strain-based FLD. For these cases, σ-FLD is the convenient option as σ-FLD is robust and no significant changes in it by changing the strain path. Clearly the change in formability in multistage forming process of DP steel is very much critical and should be studied thoroughly. Hence, the main goal of the present work was to experimentally design a two-stage stretch forming set up to evaluate the forming behaviour of different biaxial pre-strained DP steels. The σ-FLD was evaluated theoretically incorporating Barlat-89 anisotropic yield criterion with different hardening laws. Finally, the developed σ-FLD was implemented in the FEM to predict the forming behaviour of pre-strained materials, and the results were validated with experimental results. In this research work, the main interest was the capability of estimated σ-FLD to predict the failure in the material deformed under complex strain path.

4 Experimental Procedure

4.1 Material Characterization

Commercial grade DP steel of a particular type, namely DP600 with a thickness of 1.2 mm, was investigated in this study. The material characterization was performed by standard uniaxial tensile tests as per ASTM E8 M standard (ASTM 2003). The stress–strain response of the material was evaluated along three directions viz. 0, 45 and 90° with respect to rolling direction of the sheet. Various properties, such as ultimate tensile strength, 0.2 % offset yield strength, total elongation, *r*-value (plastic strain ratio) were evaluated and reported in Table 1.

Table 1 Uniaxial tensile properties of DP600 steel used in this study

Grades and gauges	Yield strength (MPa)	Ultimate tensile strength (MPa)	% Total elongation	Anisotropic properties		
				r_0	r_{45}	r_{90}
DP600 (1.2)	407	665	19.12	0.96	0.93	1.05

Table 2 Material constants evaluated incorporating different hardening models

Hollomon law			Swift law				Modified Voce law				
K (MPa)	n	R^{2a}	ε_0	K (MPa)	n	R^{2a}	A	B_0	B_1	C	R^{2a}
1125	0.21	0.9836	0.005	1140	0.20	0.9980	839	361	202	10	0.9843

[a]Here, R^2 is defined as the coefficient of determination which indicates how well data points fit a line or curve

The true stress–true strain data from uniaxial tensile test of DP600 were fitted and extrapolated with the Hollomon, Swift and modified Voce hardening laws and the different coefficient values for different hardening laws are determined by best fitting curves. The different coefficient values were tabulated in Table 2.

4.2 Multistage Forming Experiment

Complex strain paths were induced in the sheet metal using combinations of two linear strain paths. The biaxial pre-strain was induced to the sheet metal in the step-I, and out-of-plane stretch forming operation was carried out to determine formability in step-II. The details are mentioned below.

STEP-I: To induce pre-strain into the specimen

To conduct this test, an in-plane stretch forming set up (Fig. 8a) was conceptualized, designed and fabricated. The sheet metal together with brass washer (Fig. 8a) was together kept over the die. Thick film of lubrication was provided between the punch and the washer before the flat punch deformed the test specimen. As the punch moved down, the hole in the washer expands in radial direction inducing radial friction forces in the contact region between the test piece and washer. This friction prevents the test piece from getting fractured near the punch corner.

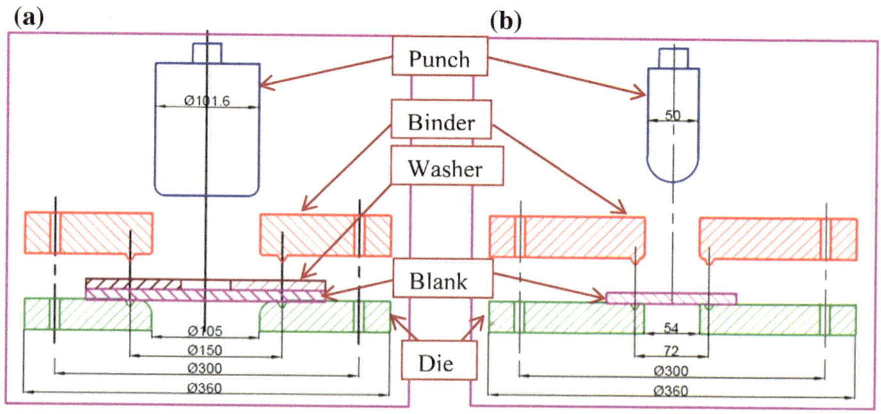

Fig. 8 Set-up for **a** biaxial pre-strain (Step-I), **b** out-of-plane stretch forming (Step-II)

Circular grids of 2.5 mm diameter were marked on the blanks by electrochemical etching technique to measure major and minor strains in the specimen after deformation. Due to radial friction forces at the sheet-washer interface, it was found that nearly equi-biaxial pre-strain was induced in the specimen. It is important to note that the amount of pre-strain induced in the material was predicted by FE simulations, and accordingly, the punch travel was planned in the experiment. Two different levels of pre-strain, i.e. 6.3 and 10.9 % were planned in the first stage of deformation.

STEP-II: Stretch forming operation on pre-strained specimen

After measuring the pre-strain, the bottom portion was trimmed by wire EDM. This resultant blank was treated as input test piece for Step-II. The formability of the pre-strained blank was evaluated in terms of LDH by stretch forming using a hemi-spherical die punch set up as shown in Fig. 8b. The experiments were stopped when a visible neck or initiation of fracture was observed on the specimens.

5 Finite Element Analysis

The success of the sheet forming numerical simulation lies in its ability to reproduce the experimental data with proper selection of material model, boundary conditions and forming limit (damage model). In this work, FE simulations of the both the forming steps were carried out using LS-DYNA 971. All tooling surfaces such as the die, punch and blank holder (or binder) were modelled as rigid bodies, and these were meshed as 1-mm-thick shell elements taking proper care of the warp age angle (20°) for the punch and die corner. The deformable blank and the washer were modelled as four-nodded quadrilaterals Belytschko–Tsay elements (Hallquist 2003). The quarter symmetry model was developed to reduce the computational time. Barlat-89 yield criterion was used as the material model for the blank incorporating the planar and normal anisotropy described in the Sect. 2.1.2. The die was fixed, and both the punch and binder were allowed to move in the z direction (along the punch axis). The coefficient of friction between the punch and blank was 0.01 in first step as thick film lubrication was applied to prevalent biaxial straining in Step-I. However, the friction value of 0.15 was assigned in all the dry conditions of the forming operations. The modelling and simulations were done for two different pre-straining cases: (i) 6.3 % pre-strain and (ii) 10.9 % pre-strain, and these simulations helped in predicting the required punch displacement for experimental trial. The predicted major and minor strain data points at each step of the deformation in the FE simulation were interposed into the ε-FLD. In this work, the experimental ε-FLD for DP600 with 1.2 mm sheet thickness was taken from the literature (Kilfoil 2007). Calculated stress values in both principal directions were plotted on already estimated σ-FLD obtained from experimental ε-FLD. The time step in which the strain or stress value reached to the limiting value was referred to predict failure.

6 Results and Discussion

6.1 Limiting Dome Height (LDH) Prediction

The final deformed dome obtained after the LDH testing of circular pre-strained DP600 is shown in Fig. 9. It was observed in the case of both the materials that the formability index-LDH decreased as the amount of pre-strain increased. The pre-strain during the first stage of deformation exhausted some ductility of the material reducing the limiting strains, hence the reduction in LDH during the second stage of deformation as compared to a fresh as-received material. It is also observed that the LDHs predicted by ε-FLD and σ-FLD are equal, and comparable with that of experimental results for as-received material. Whereas in case of pre-strained conditions, the result predicted by σ-FLD is closer to the experimental result compared to the results predicted by ε-FLD as shown in Table 3. This fact can be well explained through the nature of shifting of ε-FLD curve for pre-strained condition. In case of DP600, nearly bi-axial strain path was observed during pre-straining. It was already reported that the ε-FLD shifted downwards by lowering the limiting strain, and this resulted smaller dome height. In order to get insight into the formability changes in LDH tests of pre-strained materials, the deformation path and the thickness distribution were analysed in the subsequent section.

6.2 Strain and Stress Path During Different Steps

The strain and stress path of the failed region obtained from FE results of both the stages of forming of DP600 steel are shown in Fig. 10a with reference to its ε-FLD

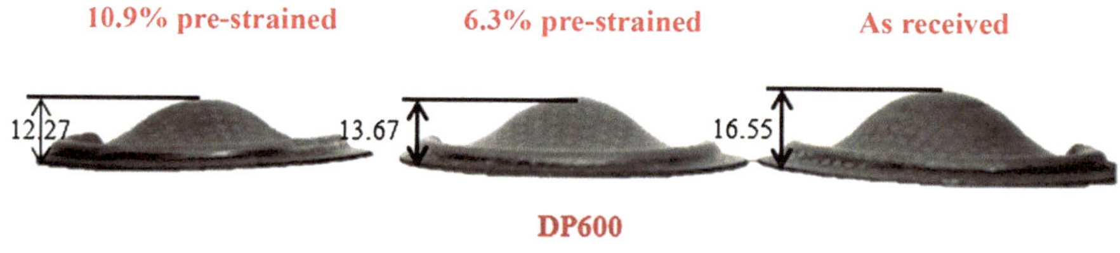

Fig. 9 Comparison of LDH of DP600 at different pre-strain conditions

Table 3 Comparison of LDH of DP600 at different pre-strain conditions

Material	Experimental data		Predicted data (by LS-DYNA)		
	% of pre-strain	Final dome height (mm)	% of pre-strain	Final dome height (mm)	
				Using ε-FLD	Using σ-FLD
DP600	As-received	16.55	As-received	17.18	17.18
	6.3 %	13.67	6.2 %	14.82	13.35
	10.9 %	12.27	11.0 %	14.02	11.82

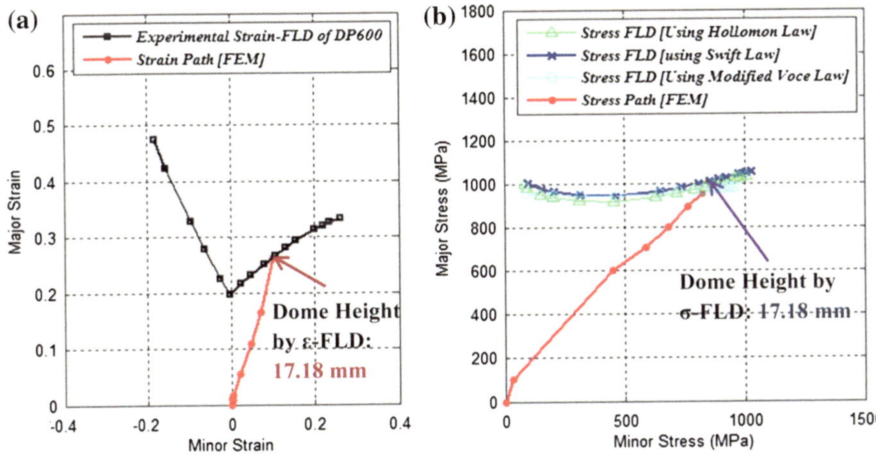

Fig. 10 **a** Strain path and **b** stress path during forming of DP600 as-received material

and σ-FLD. Further it can be observed from Fig. 10b that Swift hardening law is predicting better forming limit in comparison with that of Hollomon and modified Voce hardening laws incorporating Barlat-89 yield criteria. Strain data are taken at different time steps from LS-DYNA simulation. However, the final strain value was measured from experiment and it found a good correlation with simulation data.

In Fig. 11, stain and stress paths have shown for 6.3 % pre-strain condition for DP600. Also from the Fig. 11a, it can be noticed that the slope of the strain path changes from first step to second step in ε-FLD. In the first step, the strain path is

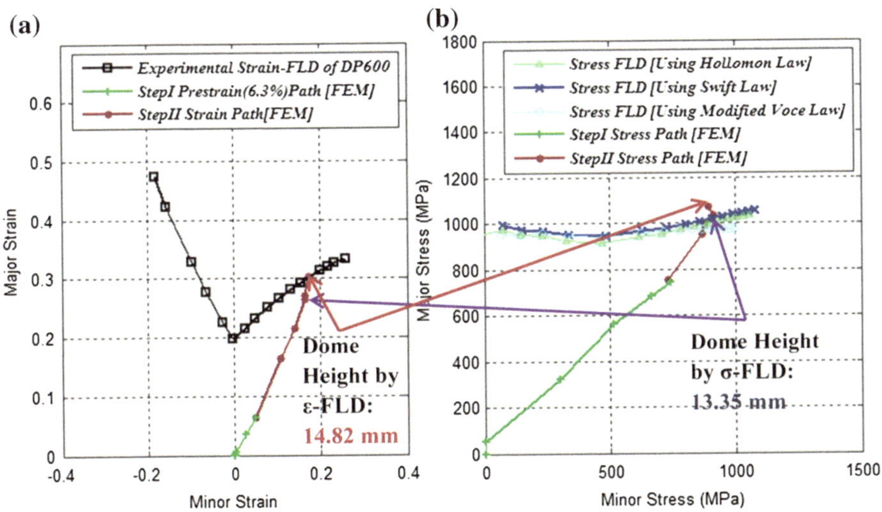

Fig. 11 **a** Strain path and **b** stress path during forming of 6.3 % pre-strained DP600 material

more towards biaxial side compared to second step. This is because a thick film of lubrication was applied which reduced the friction in the first step and helped to achieve nearly bi-axial pre-strain. On contrary, the second step stretch forming operation was performed in completely dry condition. It can be observed that the actual failure takes place far before the time step when ε-FLD predicts failure. The σ-FLD is able to predict the correct time step when the failure occurs. It is clear from these plots that ε-FLD is unable to predict formability in the present multistage forming process. However, the σ-FLD is insensitive to the deformation history, and hence it is suitable for complex forming process.

6.3 Validation of Load Progression and Thickness Distribution

The load–displacement curves of the blank were obtained through data acquisition system during stretch forming operation. For DP600, in case of biaxial stretch forming operation, the load requirement for deformation increases with the increase in punch travel distance. As punch travels more, the contact area between punch and the blank increases which increase resisting force for deforming the sheet. Load–displacement data are plotted for as-received material as well and pre-strained material. For pre-strained material, the peak load requirement is less compared to as-received condition. Similar trend was also predicted from FE simulation as shown in Fig. 12.

The deformed cups of DP600 were cut along the rolling direction, and the thickness variation was measured at different distances from the reference pole (pole is defined as cup centre or the topmost position of the dome) shown in Fig. 13a. The observed thinning pattern is symmetrical with respect to pole. The thickness variation was measured and plotted as shown in Fig. 13a. Negligible thinning was observed at the pole. As the tools (either punch or dies) were in direct contact with blank, the frictional force between them did not allow thinning to

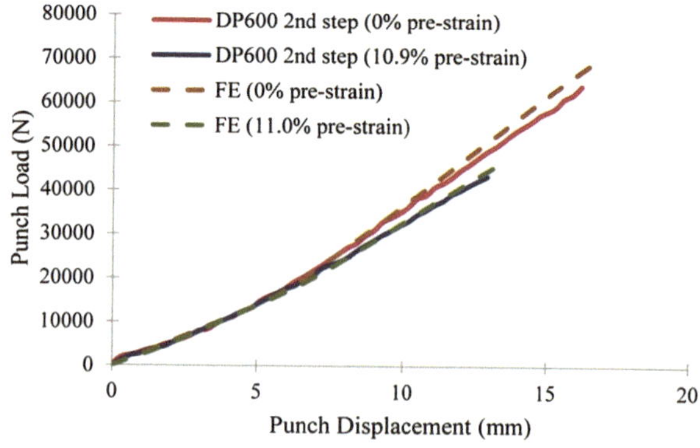

Fig. 12 Comparison of load–displacement data of as-received and 10.9 % pre-strained DP600

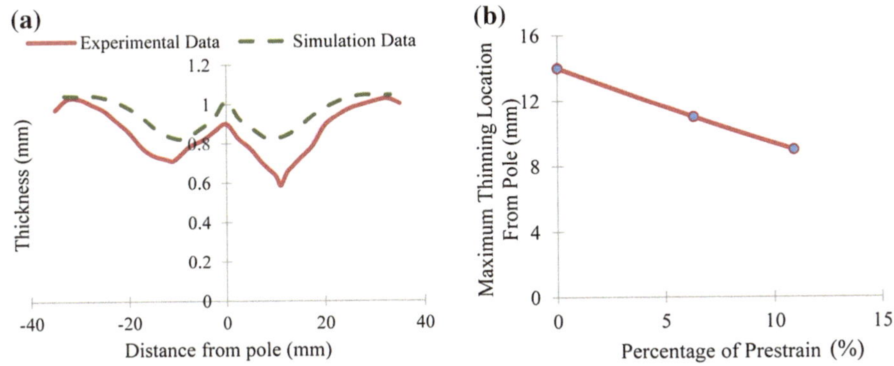

Fig. 13 **a** Thickness distribution in a 6.3 % pre-strained stretch formed cup **b** maximum thinning location for different pre-strain DP600

occur at the pole and flange. However, maximum thinning or failure has developed at a certain distance away from the pole corresponding to the location where the cup was starting to be out-of-contact with the punch. The location of the maximum thinning development was measured in the case of all the pre-strain deformed cups of both the materials, and the result was shown in Fig. 13b. It can be observed that the location of failure depends on the level of pre-strain and types of material. The area of contact between the punch and blank increases with the increase in LDH, hence the location of maximum thinning development.

7 Outlook

In one word, "formability" is an outcome of complex interaction between constitutive properties of sheet metal as well as external influencing factors associated with a forming operation, being limited by the onset of localized necking. This chapter deals with theoretical prediction of FLD which is affected by many complex parameters such as strain hardening coefficient, modes of stretching, strain rate sensitivity index, hardening laws and most importantly the strain path. Among the theoretical models, M–K model is the most efficient and useful method to predict the formability limit precisely. However, in the current study, the basic mechanical and forming behaviour of DP600 is investigated experimentally and numerically. Nonlinear strain path was introduced in the form of pre-strain to know the change in formability limit at different pre-strain condition. Experimentally two-stage forming processes were fabricated to analysis the formability of DP600 steel under different levels of pre-strain conditions. Stress-based FLD and extended stress-based FLD were formulated analytically to examine the effectiveness of them as a damage model in numerical simulation. Finally, the forming behaviour in terms of LDH, load displacement and thickness distribution was studied experimentally and numerically. The followings are the major conclusions.

I. A serious restriction of ε-FLD is that it applies only for proportional loading condition. For complex strain path operation, it gives a false prediction of formability. This apparent path-dependent problem vanishes when forming limit is calculated in stress co-ordinate both in σ-FLD and XS-FLD.
II. Stress-based FLDs were evaluated for different pre-strain conditions using ε-FLD by Barlat-89 yield criteria and different hardening models, and it was observed that it is insensitive to the deformation history and depends only on the stress state.
III. LDH predicted by FE simulation incorporating ε-FLD is more as compared to σ-FLD for bi-axially pre-strained DP600 specimens.
IV. Barlat-89 yield criterion coupled with Swift hardening law predicts formability of pre-strain DP600 closely to the experimental LDH, thickness distribution and load progression results

Acknowledgments The authors are greatly thankful to Prof. Y.N. Zhou of Department of Mechanical and Mechatronics Engineering, University of Waterloo, Canada, for providing the sheet metals used in the present study.

References

Arrieux R, Boivin M, Le Maître F (1987) Determination of the forming limit stress curve for anisotropic sheets. CIRP Ann Manuf Technol 36(1):195–198
ASTM E8-01 (2003) Standard test methods for tension testing of metallic materials. Ann Book ASTM Stand, ASTM, Philadelphia
Azrin M, Backofen WA (1970) The deformation and failure of a biaxially stretched sheet. Metall Trans 1(10):2857–2865
Banabic D (ed) (2000) Formability of metallic materials: plastic anisotropy, formability testing, forming limits. Springer, Berlin
Barata da Rocha A, Jalinier JM (1984) Plastic instability of sheet metals under simple and complex strain path. Trans Iron Steel Inst Jpn 24:133–140
Barlat F, Lian K (1989) Plastic behavior and stretchability of sheet metals. Part I: a yield function for orthotropic sheets under plane stress conditions. Int J Plast 5(1):51–66
Choi W, Gillis PP, Jones SE (1989) Calculation of the forming limit diagram. Metall Trans A 20 (10):1975–1987
Chow CL, Jie M, Hu SJ (2003) Forming limit analysis of sheet metals based on a generalized deformation theory. J Eng Mater Technol 125(3):260–265
Considere A (1885) Use of the iron and steel in buildings. Ann Des Ponts Chaussees 9:574–575
Dudzinski D, Molinari A (1991) Perturbation analysis of thermoviscoplastic instabilities in biaxial loading. Int J Solids Struct 27(5):601–628
Ghosh AK, Hecker SS (1974) Stretching limits in sheet metals: in-plane versus out-of-plane deformation. Metall Trans 5(10):2161–2164
Goodwin GM (1968) Application of strain analysis to sheet metal forming problems in the press shop (No. 680093). SAE Technical paper
Graf AF, Hosford WF (1993) Calculations of forming limit. Metall Trans A 24(11):2497–2501

Graf A, Hosford W (1994) The influence of strain-path changes on forming limit diagrams of A1 6111 T4. Int J Mech Sci 36(10):897–910
Hallquist JO (2003) LS-DYNA keyword manual. Livermore Software Technology Corporation, Livermore, CA
Hasek V (1973) On the strain and stress states in drawing of large un-regular sheet metal components. Berichte aus dem Institut für Umformtechnik, Universität Stuttgart, Nr. 25, Girardet, Essen (in German)
Hecker SS (1975) Simple technique for determining forming limit curves. Sheet Metal Indus 52 (11):671–676
Hill RT (1952) On discontinuous plastic states, with special reference to localized necking in thin sheets. J Mech Phys Solids 1(1):19–30
Hosford WF (1979) On yield loci of anisotropic cubic metals. In proceedings of the 7th North American metalworking conference, SME, Dearborn, MI (pp. 191–197)
Jones SE, Gillis PP (1984) An analysis of biaxial stretching of a flat sheet. Metall Trans A 15 (1):133–138
Keeler SP (1965) Determination of forming limits in automotive stampings (No. 650535). SAE Technical paper
Keeler SP, Backofen WA (1963) Plastic instability and fracture in sheets stretched over rigid punches. ASM trans Q 56(1):25–48
Keeler SP, Brazier WG (1977) Relationship between laboratory material characterization and press-shop formability. In Proc. Conf. on Microalloy 75(1977):517–528
Kilfoil LJ (2007) In-plane plane strain testing to evaluate formability of sheet steels used in tubular products. MS Thesis, Queen's University, Ontario, Canada, Chapter 2, p. 33
Kleemola HJ, Pelkkikangas MT (1977) Effect of predeformation and strain path on the forming limits of steel, copper and brass. Sheet Met Ind 64(6):591–592
Kobayashi,T, Ishigaki H, and Abe T (1972) Effect of strain ratios on the deforming limit of sheet steel and its application to actual press forming. In 12th biennial congress–international deep drawing research group, (pp. 8.1–8.4)
Koc M, Altan T (2001) An overall review of the tube hydroforming (THF) technology. J Mater Process Technol 108(3):384–393
Laukonis JV, Ghosh AK (1978) Effects of strain path changes on the formability of sheet metals. Metall Trans A 9(12):1849–1856
Lu ZH, Lee D (1987) Prediction of history-dependent forming limits by applying different hardening models. Int J Mech Sci 29(2):123–137
Marciniak Z (1971) Limits of sheet metal formability. WNT, Warsaw in Polish
Marciniak Z, Kuczyński K (1967) Limit strains in the processes of stretch-forming sheet metal. Int J Mech Sci 9(9):609–620
Marciniak Z, Kuczyński K, Pokora T (1973) Influence of the plastic properties of a material on the forming limit diagram for sheet metal in tension. Int J Mech Sci 15(10):789–800
Marciniak Z, Duncan, JL, Hu SJ (2002) Mechanics of sheet metal forming. Chapter-1, 2. Butterworth-Heinemann
Matsuoka T, Sudo C (1969) Effect of strain path on the fracture strain of steel sheet. sumitomo search,1: 71–80
Mielnik EM (1991) Metalworking science and engineering. McGraw-Hill, Inc, USA, p 976
Muschenborn W, Sonne HM (1975) Effect of strain path on the forming limits of sheet metal. Arch Eisenhuttenwes 46(9):597–602
Nakazima K, Kikuma T, Hasuka K (1968) Study on the formability of steel sheets. yawata Tech Rep, 264: 8517–8530
Olsen TY (1920) Machines for ductility testing. Proc Am Soc Mat 20:398–403
Panich S, Barlat F, Uthaisangsuk V, Suranuntchai S, Jirathearanat S (2013) Experimental and theoretical formability analysis using strain and stress based forming limit diagram for advanced high strength steels. Mater Des 51:756–766
Raghavan KS (1995) A simple technique to generate in-plane forming limit curves and selected applications. Metall Mat Trans A 26(8):2075–2084

Rasmussen SN (1982) Assessing the influence of strain path on sheet metal forming limits. In: 12th biennial congress–international deep drawing research group, (No. 1, pp. 83–93)

Rasmussen SN, Wanheim T (1981) Theoretical prediction of strain path dependence of limit strains in sheet materials. CIRP Ann Manuf Technol 30(1):179–184

Simha CHM, Gholipour J, Bardelcik A, and Worswick MJ (2005, August) Application of an extended stress-based flow limit curve to predict necking in tubular hydroforming. In: AIP conference proceedings (Vol. 778, No. A, p. 511). IOP Institute of Physics Publishing ltd

Simha C, Grantab R, Worswick MJ (2007a) Computational analysis of stress-based forming limit curves. Int J Solids Struct 44(25):8663–8684

Simha CHM, Gholipour J, Bardelcik A, Worswick MJ (2007b) Prediction of necking in tubular hydroforming using an extended stress-based forming limit curve. J Eng Mater Technol 129 (1):36–47

Stören S, Rice JR (1975) Localized necking in thin sheets. J Mech Phys Solids 23(6):421–441

Stoughton TB (2000) A general forming limit criterion for sheet metal forming. Int J Mech Sci 42 (1):1–27

Stoughton TB (2001) Stress-based forming limits in sheet-metal forming. J Eng Mater Technol 123(4):417–422

Stoughton TB, Yoon JW (2005) Sheet metal formability analysis for anisotropic materials under non-proportional loading. Int J Mech Sci 47(12):1972–2002

Stoughton TB, Zhu X (2004) Review of theoretical models of the strain-based FLD and their relevance to the stress-based FLD. Int J Plast 20(8):1463–1486

Swift H (1952) Plastic instability under plane stress. J Mech Phys Solids 1(1):1–18

Yoshida K, Kuwabara T, Kuroda M (2007) Path-dependence of the forming limit stresses in a sheet metal. Int J Plast 23(3):361–384

Zbib HM, Aifantis EC (1989) A gradient-dependent flow theory of plasticity: application to metal and soil instabilities. Appl Mech Rev 42(11S):S295–S304

Zhu Y, Dodd B, Caddell RM, Hosford WF (1987) Convexity restrictions on non-quadratic anisotropic yield criteria. Int J Mech Sci 29(10):733–741

Zhu X, Weinmann K, Chandra A (2001) A unified bifurcation analysis of sheet metal forming limits. J Eng Mater Technol 123(3):329–333

Size Effect on Mechanical Behaviour of SS304

J. Sahu and S. Mishra

Abstract Of the available micro-manufacturing technology, micro-forming draws the attention of manufacturers and researchers due to its high precision, low energy consumption and low cost products with better mechanical properties. However, due to the size effect, the traditional macro-forming techniques cannot be applied to micro-forming directly. Therefore, manufacturing of micro-components is a challenging field. When the parts are miniaturized, adhesive force and surface tension play an important role. The microstructure, grain orientation, grain boundary and surface integrity significantly influence the micro-forming process. Three FCC materials (SS304) of 30, 50 and 90 μm thicknesses were selected as working material to study the size effect. The effects of thickness, surface grain, cross-sectional grain size on mechanical behaviour were investigated. The grain distribution through surface and cross section was quantified. The effect of thickness on YS and UTS at nearly same grain size was studied, and it was observed that the strength decreases with increasing thickness from 30 to 90 μm.

Keywords Micro-forming · Size effect · Thickness · Mechanical behaviour

1 Introduction

There is currently a growing demand for micro-forming in the field of electronic products, electro-mechanical systems (MEMS), medical equipment and sensor technology. Examples of some typical products are micro-screw, IC carriers, fasteners, connecting elements, vehicle aircraft, packaging for consumable goods, cans

J. Sahu · S. Mishra (✉)
Department of Mechanical Engineering,
Indian Institute of Technology Bombay, Mumbai, India
e-mail: sushil.mishra@iitb.ac.in

J. Sahu
e-mail: jambeswar@gmail.com

R.G. Narayanan and U.S. Dixit (eds.), *Advances in Material Forming and Joining*,
Topics in Mining, Metallurgy and Materials Engineering,
DOI 10.1007/978-81-322-2355-9_9

for drinks and frames for TV/computer screens/monitors/displays, etc. (Engel and Eckstein 2002; Razali and Qin 2013). Earlier to this lithographic technologies, especially Lithographie, Galvanoformung and Abformung (LIGA) processes have dominated in the manufacturing of micro-components. The LIGA process has good accuracy, but the process is slow and costly. Therefore, this process is not suitable for mass production (Wang et al. 2007). But micro-forming produces precise, low energy consumption and low cost products with better mechanical properties. The material surface topography and microstructure is invariant with the dimension of the material. Hence, the ratio of dimension of part to the microstructure parameter or surface topography parameter differs with miniaturization. These factors lead to size effect (Saotome and Inoue 2000; Geiger et al. 1994). The size effect concerns with the friction and material behaviour, which are generally relevant for almost all forming processes. The flow stress variation due to the miniaturization is investigated by geometrically similar tensile and upsetting tests. It is observed that when the size is scaled down, flow stress decreases significantly (Kals 1999; Kocan and Prejs 2000; Raulea et al. 1999; Picard and Michel 1999). Upsetting test for copper, CuZn13 and CuSn6 specimen shows that flow stress decreases with scaling down the size of material (Geiger et al. 1997; Miyazaki et al. 1979). This test is carried out keeping grain size constant. Huang and Chang (2012) investigated the size effect of pure iron steel of three different grain sizes and four different thicknesses. It is observed from tensile test that for each thickness the flow stress decreases with increasing grain size. Due to these effects, the conventional macro-forming phenomena cannot be applied to micro-forming directly. When the parts are minimized, the orientation, grain size, number of grain in cross section, grain distribution, etc. play an important role. Therefore, manufacturing a micro-component by micro-forming process is a challenging field and the grain and specimen size effect study is needed for micro-forming.

2 Experimental Details

Commercially available materials (SS304) of three sizes 30, 50 and 90 μm are considered for the present work. The sample for tensile test is prepared as dimension gauge length 22 mm, width 5 mm and total length 50 mm, and the experimental work is carried out that includes annealing, tensile test, microstructure analysis and tensile test. Initially, the specimen was annealed in Muffle furnace at 1050 °C and it was observed that the thin sheet was burnt out. Then, it was annealed at 800 °C for 15, 20, 25 and 30 min in order to get different grain sizes and the material was taken out for air cooling. Instron 3345 single vertical column type tensile test machine of capacity 5 kN is used for tensile testing with a strain rate of $10^{-4}s^{-1}$. The microstructure image was taken by TSL OIM™ orientation imaging microscopy on a FEI Quanta TM 200 Hv scanning electron microscopy (EBSD) and was analysed by TSL OIM Analysis 7 software. All the video conditions were kept same for all scans, and analysis is carried out by keeping CI 0.1 as CI is the important parameter. The EBSD sample was prepared by mechanical polishing followed by diamond polishing and electro-polishing. For mechanical polishing, 2500 grade paper was used followed by diamond polishing by cotton cloth and 1 μm grade diamond paste. The electrolyte used for electro-polishing is 80 % methanol and 20 % perchloric acid.

3 Result and Discussion

Tensile test was carried out to study the size effect on mechanical behaviour of received material. Stress strain curve for all the thicknesses is shown in Fig. 1a–c. For repeatability, three samples were tested for each condition. It is observed that the total elongation of received SS304 is 2–3 % and flow stress is the range of 1300–1400 MPa. Due to very low elongation and very high load requirement, the material cannot be used for further deformation. To find the cause of very low elongation, the ESBD microstructure analysis is carried out. The EBSD results are shown in Fig. 2, which shows that the grain size is very small and heavily cold worked which cannot be indexed. Size effect is also observed for as-received material as shown in Fig. 3. To study the effect of miniaturization, the indicative curve is plotted between flow stress and thickness (Fig. 3). It is observed that flow stress attains maximum at 50 μm thickness and decreases towards 30 and 90 μm thicknesses. But the flow stress of 90 μm thick is more than 30-μm-thick materials. From this analysis, it is conformed that there is a mutual effect of grain size and thickness on mechanical properties. Therefore, SS304 is annealed at 800 °C for 15, 20, 25 and 30 min to generate different grain sizes. The annealed

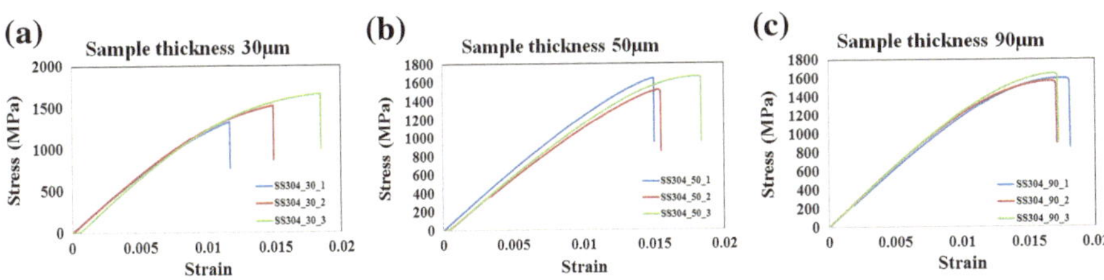

Fig. 1 **a–c** Tensile curve of received material

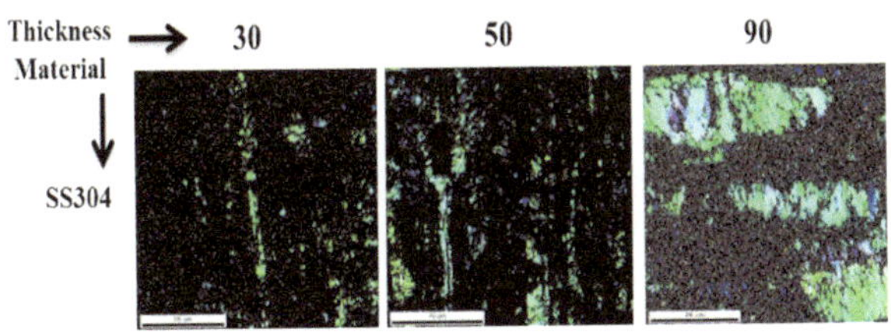

Fig. 2 Microstructure of SS304 (as received)

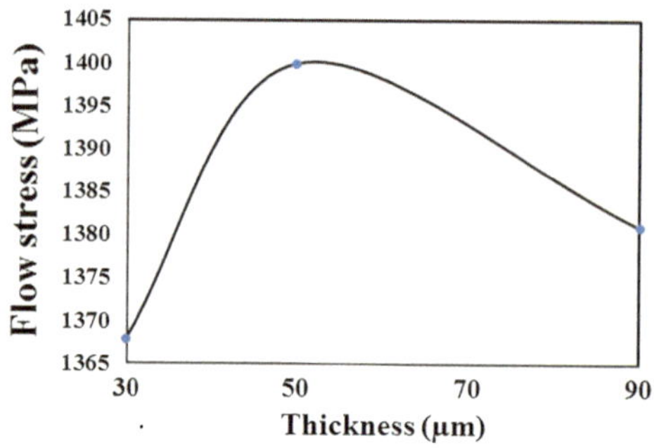

Fig. 3 Effect of thickness on flow stress

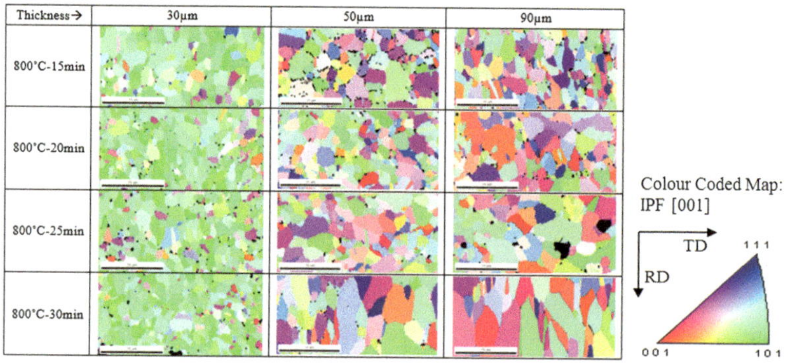

Fig. 4 Surface grain modification by heat treatment of SS304 (15, 20, 25 and 30 min)

Fig. 5 Effect of annealing time on surface grain

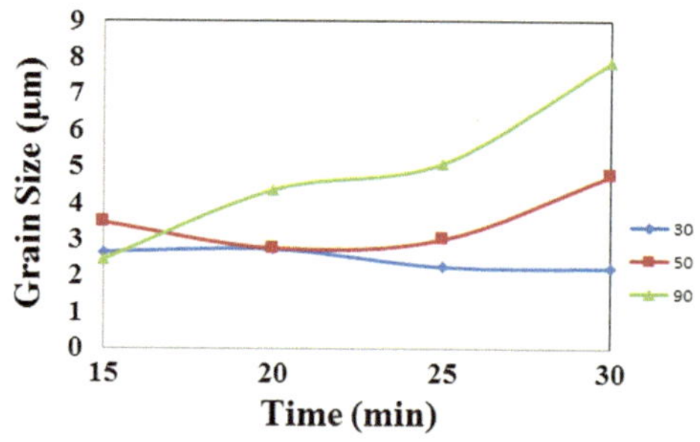

sample of different thicknesses for various time is prepared for EBSD scan. The EBSD scan is shown in Fig. 4.

The effect of annealing time on surface grain size of SS304 foils at 800 °C for 15, 20, 25 and 30 min is described in Fig. 5. It is observed that the surface grain is increasing significantly with respect to time in case of 90 μm SS304, but for 50 μm grain size, it decreases from 15 to 20 min and then increases. In 30 μm SS304, grain size remains same for 15–20 min and then decreases for 25–30 min. Figure 5 shows that cross-sectional grains are decreasing with respect to annealing time in case of 90 μm SS304, but for 50 μm grain size, it decreases from 15 to 20 min and then increases 25 min and remains same for 30 min. In 30 μm SS304, grain size remains same for 15–20 min and then slightly increases for 25–30 min. It is clearly observed from Figs. 5 and 6 that the surface grains are larger than inside grains. Therefore, the 3D structures of grains are elliptical in shape.

3.1 Effect of Annealing Time on Elongation

The percentage elongation with respect to annealing time is plotted in Fig. 7. The percentage elongation decreases with increasing annealing time in 30 and 50 μm, but in 90 μm SS304, the elongation decreases from 15 to 20 min, increases to 25 min and then decreases up to 30 min. The maximum elongations are 13, 25 and 30 % for 30, 50 and 90 μm and attain at 15, 15 and 25 min, respectively. Annealed samples have very high elongation when compared to the received material elongation as shown in Table 1.

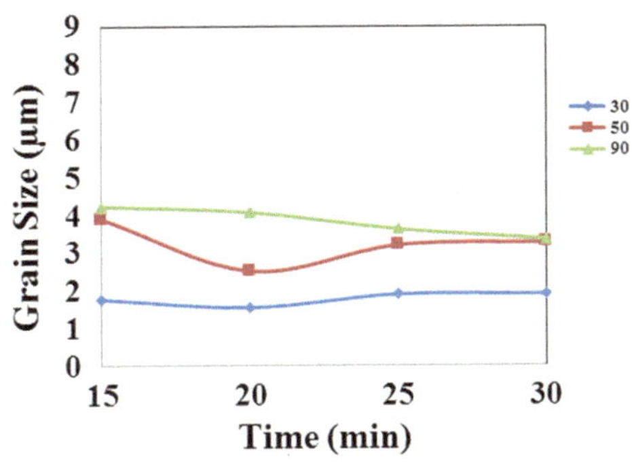

Fig. 6 Effect of annealing time on cross-sectional grain

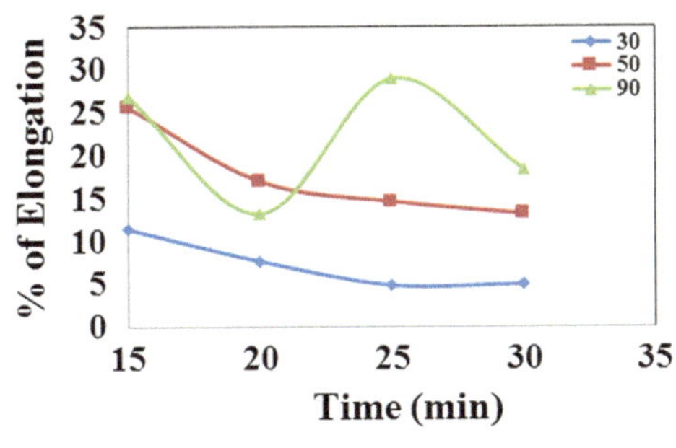

Fig. 7 Percentage of elongation with annealing time

Table 1 Percentage of elongation of as-received material

Thickness	% of Elongation
90	2.28
50	2.08
30	2.55

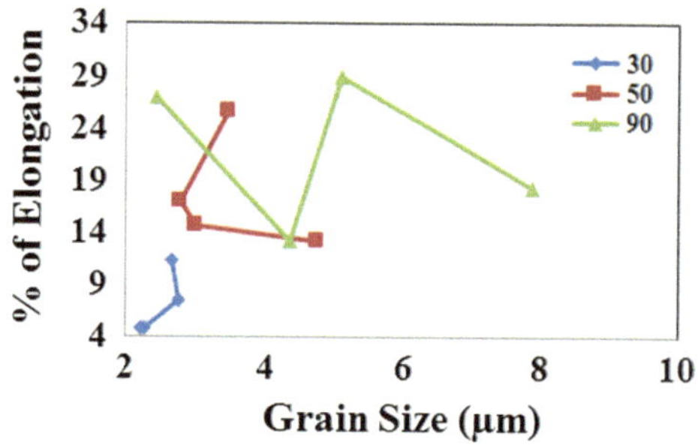

Fig. 8 Effect of grain size on total elongation

3.2 *Effect of Grain Size on Elongation*

The total elongation of different thicknesses of specimen at different grain sizes is shown in Fig. 8. There is no clear trend found between grain size and elongation, but it is observed that the percentage elongation varies significantly for different thicknesses. The maximum elongation is achieved at 2.7, 3.5 and 5 μm grain for 30, 50 and 90 μm, respectively.

3.3 *Effect of Annealing Time on Flow Stress*

The flow stress (MPa) of as-received and annealed sample for 15, 20, 25 and 30 min at 800 °C is shown in Table 2 and plotted in Fig. 9. It is observed that the flow stress decreases for 30 μm and increases for 90 μm SS304 with increasing

Table 2 Flow stress of annealed and un-annealed SS304

Thickness (μm)	Un-annealed	15 min	20 min	25 min	30 min
30	1368.97	761.31	606.25	539.63	561.25
50	1400.44	778.80	372.82	691.75	356.69
90	1381.66	639.82	678.66	846.37	766.36

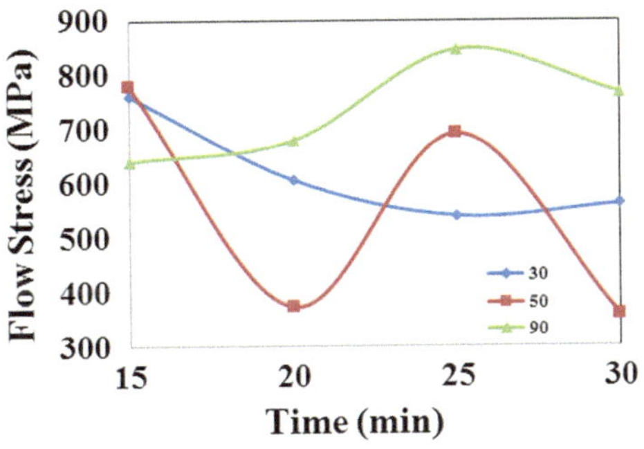

Fig. 9 Effect of annealing time on flow stress

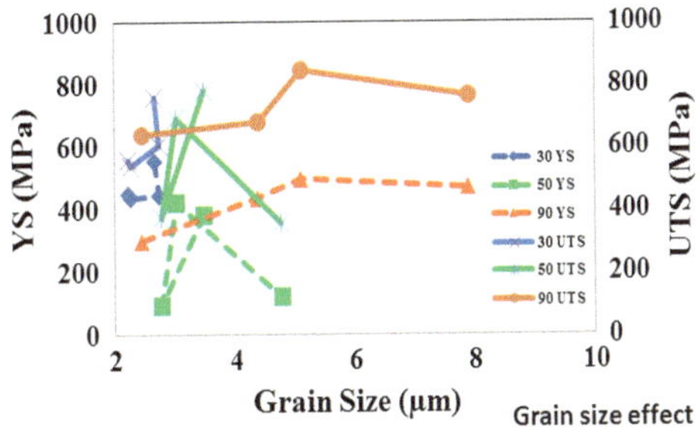

Fig. 10 Effect of grain size on YS and UTS

annealing time. For 50 μm, no clear trend was observed on flow stress. Sometimes it is seen that the flow stress increases with increasing grain size. The behaviour may be due to twining effect, grain orientation effect and thickness effect.

3.4 Grain Size Effect (Grain Size—YS and UTS)

The grain size effect on yield stress and UTS is shown in Fig. 10. The yield stress and UTS are increasing with grain size for 90 μm SS304 and attain maximum at 5 μm grain size. In 50 μm SS304, maximum strength is achieved at 3–3.5 μm grain size and beyond that strength decreases. The strength is maximum at 3 μm grain size in 30 μm thickness SS304.

3.5 Thickness Effect (YS, UTS and GAM at Same Grain Size)

Three nearly same grain size are chosen for three different thicknesses as shown in Table 3; the yield strength, UTS and GAM values are plotted against thickness for SS304 (Fig. 11). It is observed that, strength is decreasing with increasing thickness from 30 to 90 μm. EBSD is carried out to study microstructure of the three conditions. Figure 12 depicts various microstructures. It is observed that the gauss

Table 3 Yield strength and UTS at nearly same grain size for different thicknesses

Thickness (μm)	Grain size (μm)	YS (MPa)	UTS (MPa)	GAM
30	2.659	553.76	761.31	0.530309
50	3.477	384.49	778.80	0.938186
90	2.445	301.02	639.82	1.03154

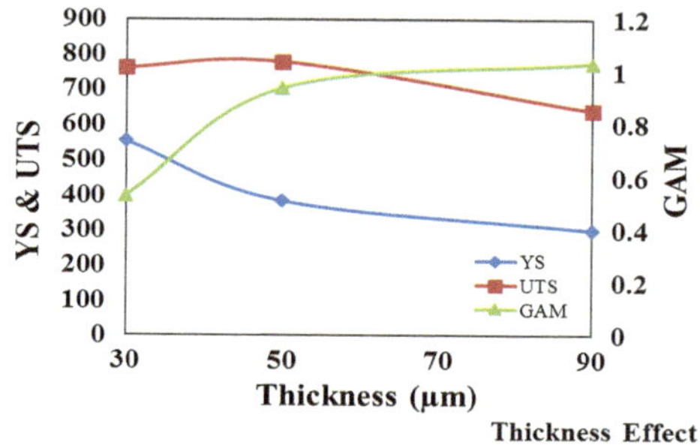

Fig. 11 Effect of thickness on YS, UTS and GAM

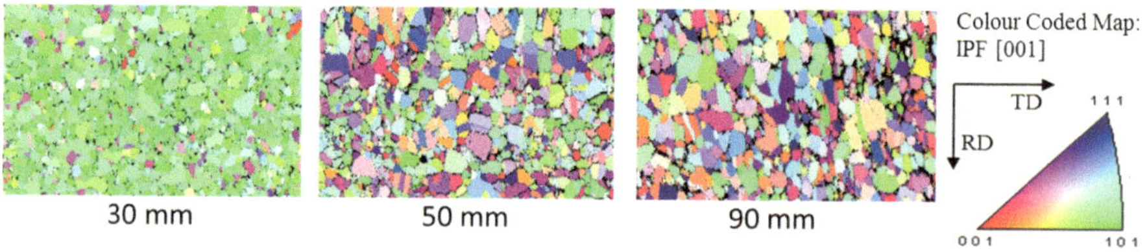

Fig. 12 Microstructure of 30-, 50- and 90-μm-thick SS304 at grain size 2.659, 3.477 and 2.445 μm, respectively

average misorientation (GAM) is increasing from 30- to 90-μm-thick SS304 and 30 μm sheet has single orientation grain ([101] orientation). The [101] orientation is diluting in 50- and 90-μm-thick sheet.

4 Conclusions

In present research, a significant grain growth was encountered at 800 °C with annealing time of 15, 20, 25 and 30 min. The surface grains are found to be larger than cross-sectional grains. The total elongation and the flow stress have significant variation with respect to grain size and sample size. The percentage elongation is maximum for 2.5, 3.5 and 5 μm grain size at 30, 50 and 90 μm thickness of SS304, respectively. The grain size effect on yield strength and UTS is observed, and it is found that in 90 μm SS304 both strengths increase with increasing grain size. In 30 and 50 μm, no clear trend is found. The effect of thickness on YS and UTS at nearly same grain size is studied. It is observed that the strength decreases with increasing thickness from 30 to 90 μm as the texture diluting from 30 to 90 μm thickness.

References

Engel U, Eckstein R (2002) Microforming-from basic research to its realization. J Mater Process Technol 125–126:35–44

Geiger M, Engel U, Vollertsen F, Kals R, Mebner A (1994) Metal forming of microparts for electronics. Prod Eng 2(1):13–18

Geiger M, Mebner A, Engel U (1997) Production of microparts-size effects in bulk metal forming. Similarity Theor Prod Eng 4(1):33–38

Huang Y, Chang Y (2012) A study of the micro-strengthening process of pure iron sheet. J Mater Process Tech 212(3):713–717

Kals RTA (1999) Fundamentals on the miniaturization of sheet metal working processes. In: Geiger M, Feldmann K (eds) ReiheFerti-gungstechnik Erlangen, Band 87, Meisenbach, Bamberg

Kocanda A, Prejs T (2000) The effect of miniaturisation on the final geometry of the bent products. In: Proceedings of the 8th international conference on metal forming, pp 373–378

Miyazaki S, Shibata K, Fujita H (1979) Effect of specimen thickness on mechanical properties of polycrystalline aggregates with various grain sizes. Acta Metall 27(5):855–862

Picard P, Michel JF (1999) Characterization of the constitutive behaviour for very small components in sheet metal forming. In: Proceedings of the 2nd ESAFORM conference on material forming. Guimaraes, pp 169–170

Raulea LV, Govaert LE, Baaijens FPT (1999) Grain and specimen size effects in processing metal sheets. In: Geiger M (ed) Advanced technology of plasticity. Proceedings of the 6th international conference on technology of plasticity Nuremberg, vol 2. Springer, Berlin, pp 939–944

Razali AR, Qin Y (2013) A review on micro-manufacturing, micro-forming and their key issues. Procedia Eng 53:665–672

Saotome Y, Inoue A (2000) New amorphous alloys as micro materials and the processing technology. Proceedings of the IEEE 13th annual international conference on micro electro mechanical systems, pp 288–292

Wang C, Shan D, Guo B, Zhou J, Sun L (2007) Key problems in microforming processes of microparts. J Mater Sci Technol 23(2):283–288

Prediction of Weld-Induced Angular Distortion of Single-Sided and Double-Sided Fillet Joints by SAW Process

Arpan Kumar Mondal, Pankaj Biswas, Swarup Bag and Manas M. Mohapatra

Abstract In the present work, a numerical elasto-plastic thermo-mechanical model has been developed to predict the weld-induced angular distortion of single-sided and double-sided fillet joints by submerged arc welding (SAW) process. The welding was carried out by using recyclable flux-filled backing strip in single pass. The angular deformations for both the cases have been measured experimentally. It has been found that the maximum magnitude of angular deformation is lower in case of double-sided fillet joint. A detail comparative study of the angular deformation between single- and double-sided fillet joints has been presented in this study. It has been observed that the developed elasto-plastic thermo-mechanical model is well comparable with experimental results.

Keywords Submerged arc welding · Elasto-plastic · Thermo-mechanical model · Single- and double-sided fillet weld joints · Single pass · Angular deformation

1 Introduction

Submerged arc welding (SAW) is well known for the high efficiency and high metal deposition, and therefore, it is extensively used for fabricating large structural parts, very thick sections, etc. Being a fusion welding process, SAW deposits large amount of heat to melt the electrode as well as the work piece, due to which a locally high thermal gradient generates and which is inevitable. The thermal gradient causes residual stresses followed by the distortion of the welded part upon

A.K. Mondal (✉) · P. Biswas · S. Bag
Department of Mechanical Engineering, Indian Institute of Technology Guwahati, Guwahati 781039, Assam, India
e-mail: m.arpan@iitg.ernet.in

M.M. Mohapatra
Department of Mechanical and Industrial Engineering, Indian Institute of Technology Roorkee, Roorkee 247667, Uttarakhand, India

R.G. Narayanan and U.S. Dixit (eds.), *Advances in Material Forming and Joining*, Topics in Mining, Metallurgy and Materials Engineering, DOI 10.1007/978-81-322-2355-9_10

cooling and which also decides the final grain structure in different zones of the welded plate, i.e. heat-affected zone and fusion zone. The distortion in welded joints may be categorised as longitudinal, transverse and angular, while assembling several parts, welding deformation is a serious concern. Referring to Deng et al. (2011), angular deformation affects the structural integrity, dimensional accuracy as well as the quality and performance of the final part and thus leads to an unstable structure. Submerged arc welded stiffened panels are the part of any large structure and can be seen in bridges, shipbuilding, etc. (Deng et al. 2011; Biswas et al. 2009).

The angular deformation created by localised thermal gradient during the welding process reduces the proper functioning of the welded structure. Therefore, it is essential to estimate the amount of angular deformation induced in a welded structure to study the mechanism of weld-induced distortion. As several parameters involve in this whole process, it becomes a complex and difficult phenomena to be analysed by the theoretical and experimental processes only. The angular deformation can be determined experimentally, or it can be predicted by using numerical techniques (Biswas et al. 2009). Finite element (FE) analysis of welding is highly effective in predicting thermo-mechanical behaviour. It has been more than four decades ever since Ueda and Yamakawa (1971) proposed a thermal elastic plastic FE method to analyse the thermal transient stresses induced in a butt and fillet weld under a moving electrode. Following the pioneering work of Ueda and Yamakawa (1971) and many researchers, various researchers (Deng et al. 2011; Biswas et al. 2009); Radaj 1992; Hong et al. 1998) have successfully developed a large number of numerical models based on the FE method to predict temperature distribution, analysing the welding residual stress and distortion both in 2D and 3D modes.

Teng et al. (2001) have investigated the residual stresses and distortion of T-joint fillet welds using two-dimensional FE analyses. Tsai and Jung (2004) modelled the angular distortion of T-joints using plasticity-based distortion analysis. Michaleris and DeBiccari (1997) successfully predicted the distortions during fusion welding using temperature-dependent material properties of the steel in the FE modelling. Mahapatra et al. (2007) analysed the effect of tack position on angular deformation in single-sided submerged arc welded fillet joints. Tsai and Cheng (1999) investigated distortion mechanism and the effects of welding sequence on thin panel distortion using FE analysis.

The works of the above researchers indicate that angular deformation in a welded joint is a matter of real concern which strongly affects the performance of the structure. However, it has been observed that adequate importance has not been provided to the work related to comparative study of angular deformation between the single- and double-sided fillet joints. The present study aims to develop a 3D FE-based thermo-mechanical model to predict the residual stress and angular deformation in submerged arc welded mild steel thick plates. The basic welding mechanism of SAW is shown in Fig. 1. The numerical model considers elasto-plastic response of material for predicting residual stresses and the angular deformation, temperature-dependent material properties and latent heat of melting and solidification. The numerical results have been validated with experimentally

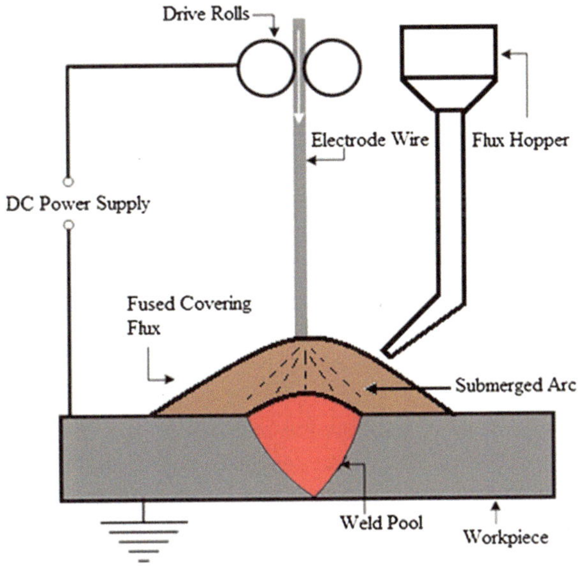

Fig. 1 Schematic view of submerged arc welding system

measured angular deformation to address the reliability of developed model. Finally, the developed model has been utilised to do a comparative study between single- and double-sided submerged arc welded fillet joints.

2 Mechanism of Welding Distortion

During the welding process, locally rapid melting and solidification occurs in specific zones of the base plate (FZ and HAZ), and the existence of a high temperature gradient in the weld zone creates a very high stress concentration in the weld zone and nearby HAZ (Asim et al. 2011; Kong and Kovacevic 2012), which generally exceeds the yield strength of the material. If the stresses generated from thermal expansion/contraction exceed the yield strength of the parent metal, localised plastic deformation of the metal occurs. Plastic deformation causes a permanent reduction in the component dimensions and distorts the structure (Lucas et al. 2014).

Initially, when the weld pool is formed upon the base metal, heat is transferred rapidly and heat-affected zone is formed, but due to this extreme heating, the metal at HAZ gets expanded and compressive stresses are formed in the surrounding cold base metal to counter this expansion. But, tensile stresses arise upon cooling when the contraction of the weld metal as well as the HAZ takes place and which is resisted by the adjacent cold base metal. Presence of large amount of residual stresses in the welded structure can evidently reduce the fatigue strength, can cause crack generation and shorten the lifetime of the welded part (Arai et al. 1995;

Kong and Kovacevic 2012). In a nutshell, the principle of the distortion can be assumed as the material shrinkage during heating and cooling cycle of the welding process (Weisman 1976). The change in temperature and stresses during welding are explained by many researchers (Chen 2011). Let us consider a typical butt joint (Fig. 2) where welding is being done by SAW, the welding plate can be divided in three zones i.e. welded metal zone (A), weld poll zone (B) and the metal to be welded (C). At the cooled region where the welding is yet to be done, the temperature still has not raised therefore the temperature gradient will be almost zero and the welding thermal stresses are zero.

The weld pool (zone B) will have the maximum temperature therefore the temperature gradient at the center of the weld pool will be very high and the distribution from center of the weld line to the away from the weld line will be very steep in nature. As the metal is in molten state in this weld pool region it cannot support any load, thus the stresses at the center of the weld pool (section 2–2) is almost zero. But the stresses are compressive in the heat affected zone (HAZ) away from the center of the molten weld pool region because the expansion of these areas is restrained by surrounding colder metal.

If we consider a section (3–3) in the weld pool section slightly away from the welding arc where welding is just completed we can see change in temperature is less steep and the distribution is more even. In this region the weld metal and HAZ is cooled enough, the resulting stress is tensile in nature in the weld line as they try to shrink. The stresses are compressive in nature further away from the weld line.

Finally the last section (welded metal zone) where welding is already finished long ago, which is located far away from the welding arc and due to this the metal already cools down and again the temperature gradient due to welding is zero. At this cooled region higher tensile stresses are present in the HAZ and compressive stresses exist away from the weld line.

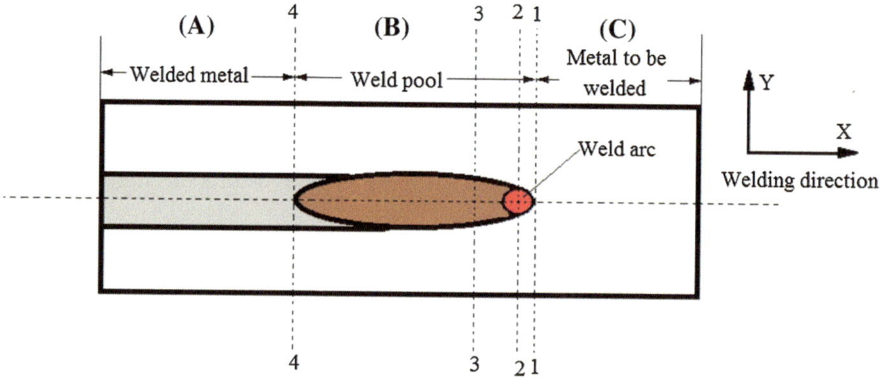

Fig. 2 Schematic diagram of the different weld zones during welding

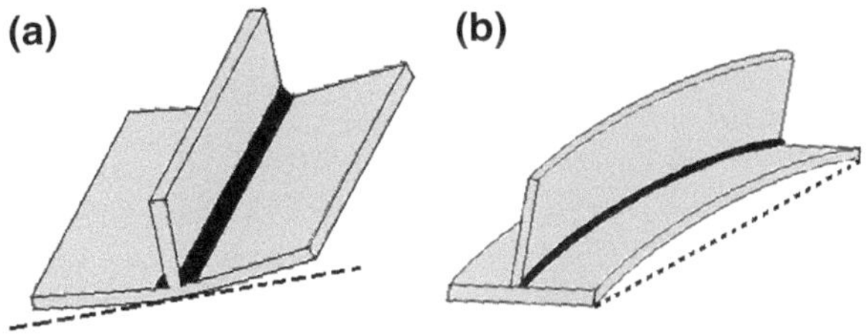

Fig. 3 Various types of welding distortions in fillet joints **a** angular deformation, **b** longitudinal distortion (Awang 2002)

2.1 Types of Welding Deformations

The residual stresses resulting from the thermal gradients produce internal force leading to the shrinkage of the material. Depending upon the shape and shrinkage pattern of the final welded structure, the final deformations may be of different types, e.g. bending, buckling and rotation. Generally, the welding deformations in fillet joins (Awang 2002) can be found as shown in Fig. 3.

3 Finite Element Modelling

In submerged arc welding, the heat is supplied from the arc formed between the electrode and the base plate. The heat source model in welding has an important effect on the heat distribution pattern in the vicinity of the weld zone where fusion zone and the HAZ are formed. As the plate thickness number for SAW is 0.88, the heat flow can be considered as two-dimensional (Sorensen 1999). For the process of submerged arc welding, the heat distribution of heat source may be characterised as a distribution of heat flux on the weldment surface. In this study, the heat from the welding arc is applied at any given instant of time as a normally distributed heat flux which is expressed as

$$q_{\text{sup}}(r) = \frac{3Q}{\pi \bar{r}^2} \exp\left[-3\left(\frac{r}{\bar{r}}\right)^2\right] \quad (1)$$

where Q is heat input from arc and is equal to the multiplication of arc voltage, current and efficiency; 'r' is the radial distance from the centre of heat source on the plate surface; and $\bar{r}$ is the effective arc radius that defines the region in which 95 % of heat flux is deposited (Mahapatra et al. 2007; Friedman 1975; Pathak and Datta 2004). The temperature-dependent material properties of mild steel (MS) used for the transient heat transfer analysis and elasto-plastic analysis are given in Table 1. Temperature-dependent enthalpy and yield stress for MS are given in Tables 2 and 3, respectively. The melting and transformation temperatures of MS are 1495 and 723 °C, respectively.

Table 1 Temperature-dependent material properties of mild steel (Adak and Mandal 2003; Brown and Song 1992)

Temperature (°C)	Thermal conductivity (W/m K)	Specific heat (J/kg K)	Thermal expansion coefficient (10^{-6}/°C)	Young's modulus (GPa)	Poisson ratio
0	51.9	450	10	200	0.2786
100	51.1	499.2	11	200	0.3095
300	46.1	565.5	12	200	0.331
450	41.05	630.5	13	150	0.338
550	37.5	705.5	14	110	0.3575
600	35.6	773.3	14	88	0.3738
720	30.64	1080.4	14	20	0.3738
800	26	931	14	20	0.4238
1450	29.45	437.93	15	2	0.4738
1510	29.7	400	15	0.2	0.499
1580	29.7	735.25	15	0.00002	0.499
3500	42.1	400	15.5	0.00002	0.499

Table 2 Temperature-dependent enthalpy for mild steel (Adak and Mandal 2003)

Temperature (°C)	0	100	200	300	400	500	600	700	800	900	1000	>2500
Enthalpy (MJ/m^3)	0	360	720	1100	1500	1980	2500	3000	3700	4500	5000	9000

Table 3 Temperature-dependent yield stress for mild steel (Adak and Mandal 2003)

Temperature (K)	293	373	573	773	973	1073	1273	1473	1673
Yield stress (σ_y) (MPa)	398	379	305	192	41	36	28	20	12

Temperature-dependent material properties used for the transient heat transfer and elasto-plastic analysis can be found elsewhere (Adak and Mandal 2003; Brown and Song 1992). However, temperature-dependent convection coefficients and yield stress for steel are considered from published literature (Adak and Mandal 2003).

3.1 Thermal Analysis of Submerged Arc Welding

Figure 4 shows the fillet joint arrangement with different axis. Distributed heat source was applied to the weld zone, and convective heat loss was considered in the remaining surfaces. The general governing equation for heat conduction for a

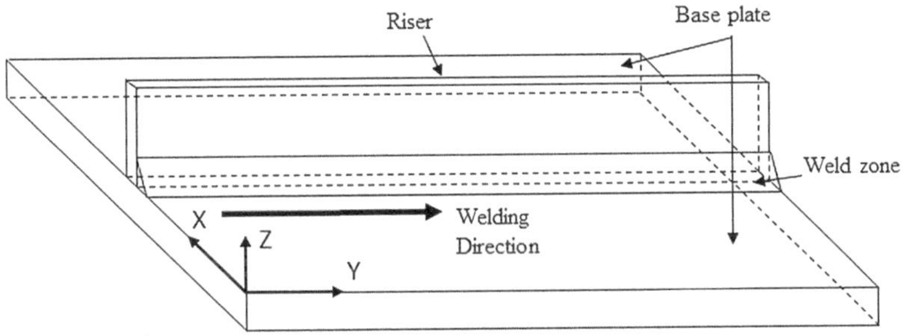

Fig. 4 Schematic of the fillet joint with rectangular coordinate system

homogenous, isotropic solid without heat generation in rectangular coordinate system (x, y, z) is shown below.

$$\frac{\partial}{\partial x}\left[k\frac{\partial T}{\partial x}\right]+\frac{\partial}{\partial y}\left[k\frac{\partial T}{\partial y}\right]+\frac{\partial}{\partial z}\left[k\frac{\partial T}{\partial z}\right]=\rho c\frac{\partial T}{\partial t} \tag{2}$$

Equation (2) can be written as

$$\rho c\frac{\partial T}{\partial t}=-\{L\}^T\{q\} \tag{3}$$

where $\{L\}^T T=\vec{\nabla}T$ and $\{L\}^T\{q\}=\vec{\nabla}.\{q\}$, $\{L\}=\begin{Bmatrix}\frac{\partial}{\partial x}\\ \frac{\partial}{\partial y}\\ \frac{\partial}{\partial z}\end{Bmatrix}$ = vector operator.

Fourier's law is used to relate the heat flux vector to the thermal gradient

$$\{q\}=-[D]\{L\}T$$

where $[D]=\begin{bmatrix}k & 0 & 0\\ 0 & k & 0\\ 0 & 0 & k\end{bmatrix}$ = conductivity matrix.

Equation (3) can be written as,

$$\rho c\frac{\partial T}{\partial t}=\{L\}^T([D]\{L\}T) \tag{4}$$

Boundary conditions

The governing equation (Eq. 2) of the thermal behaviour of submerged arc welded plate is subjected to the following initial and boundary conditions.

Initial condition 1
A specified initial temperature is considered for the entire elements of the specimen as

$$T = T_{\infty} \quad \text{for} \quad t = 0 \tag{5}$$

where T_{∞} is ambient temperature. q_{sup} represents the heat flux supplied to the work surface in $^{\text{W}}/_{\text{m}^2}$ from an external welding arc. q_{conv} represents heat loss from the work surface at temperature, T by convection with a heat transfer coefficient and h_f into an external ambient temperature. q_{conv} can be represented as

$$q_{\text{conv}} = h_f(T - T_{\infty}) \tag{6}$$

It should be noted that heat flux q_{sup} and convection loss q_{conv} do not occur over the same boundary segment. If heat loss by convection exists, then heat gain by q_{sup} does not or vice versa.

Let $\vec{n}$ is the unit outward normal vector represented as $\vec{n} = \vec{i}l_x + \vec{j}l_y + \vec{k}l_z$. The quantity q_n represents the component of the conduction heat flux vector normal to the work surface and it is given as

$$q_n = -k\vec{\nabla}T \cdot \vec{n} \tag{7}$$

where $\vec{\nabla}T = \vec{i}\frac{\partial T}{\partial x} + \vec{j}\frac{\partial T}{\partial y} + \vec{k}\frac{\partial T}{\partial z}$. The normal component of the heat flux vector at the surface becomes

$$q_n = -k\left(l_x\frac{\partial T}{\partial x} + l_y\frac{\partial T}{\partial y} + l_z\frac{\partial T}{\partial z}\right) = -k\frac{\partial T}{\partial n} \tag{8}$$

To develop the second- and third-boundary conditions, the energy balance considered at the work surface as heat supply = heat loss.

Boundary condition 1
A specific heat flow acting over weld zone surface as

$$q_n = -q_{\text{sup}} \Rightarrow -k\frac{\partial T}{\partial n} = -q_{\text{sup}} \tag{9}$$

Boundary condition 2
Considering heat loss due to convection over the surface except the weld zone (Newton's law of cooling)

$$q_n = q_{\text{conv}} \Rightarrow -k\frac{\partial T}{\partial n} = h_f(T - T_{\infty}) \Rightarrow -k\frac{\partial T}{\partial n} = -h_f(T_{\infty} - T) \tag{10}$$

3.2 Mechanical Analysis

Thermal-induced expansion and shrinkage and phase transformation-induced volume change are the main causes of the residual stress and distortion of welded parts (Kong and Kovacevic 2012). By considering the elastic and plastic material properties, the stress and strain relationships in submerged arc welded metal parts are given as follows (Biswas et al. 2009). The stress–strain relationship can be represented as

$$\{\sigma\} = [D]\{\varepsilon_e\} \tag{11}$$

where

$\{\sigma\}$ = stress vector = $[\sigma_X \sigma_Y \sigma_Z \sigma_{XY} \sigma_{YZ} \sigma_{XZ}]^{\mathrm{T}}$,
$[D]$ = stress–strain correlation matrix (stiffness matrix) and
$\{\varepsilon_e\}$ = elastic strain vector.

The elastic strain is given as:

$$\varepsilon_e = \varepsilon - \varepsilon_{\mathrm{th}} - \varepsilon_{\mathrm{pl}} - \varepsilon_{\Delta v} - \varepsilon_{\mathrm{Trp}} \tag{12}$$

where ε is total strain vector, $\varepsilon_{\mathrm{th}}$ is thermal strain vector, $\varepsilon_{\mathrm{pl}}$ is plastic strain vector, $\varepsilon_{\mathrm{Tr}\Delta v}$ is phase transformation-induced volume change strain vector, and $\varepsilon_{\mathrm{Trpl}}$ is phase transformation-induced plastic strain vector. As phase transformation effect was ignored in this study, the effect of $\varepsilon_{\mathrm{Tr}\Delta v}$ and $\varepsilon_{\mathrm{Trpl}}$ was ignored.

If we consider the plastic and thermal strain part, the elastic strain vector becomes

$$\{\varepsilon_e\} = \{\varepsilon\} - \{\varepsilon_{\mathrm{th}}\} - \{\varepsilon_{\mathrm{pl}}\} \tag{13}$$

$$\varepsilon = \text{total strain vector} = [\varepsilon_X\ \varepsilon_Y\ \varepsilon_Z\ \varepsilon_{XY}\ \varepsilon_{YZ}\ \varepsilon_{XZ}]^{\mathrm{T}} \tag{14}$$

$$\varepsilon_{\mathrm{th}} = \text{thermal strain vector} = \Delta T[\alpha_X\ \alpha_Y\ \alpha_Z\ 0\ 0\ 0]^{\mathrm{T}} \tag{15}$$

$$\Delta T = T_n - T_\infty \tag{16}$$

Here, T_n is the instant temperature at a particular time; ε_X ε_y and ε_z are direct strain in the X, Y and Z directions, respectively; ε_{XY}, ε_{YZ} and ε_{ZX} are shear strain in the X–Y, Y–Z and Z–X planes, respectively; and α_X, α_Y and α_Z are coefficient of thermal expansion in X, Y and Z directions, respectively.

The structural analysis involved large displacements (strain) and rate-independent thermo-elasto-plastic material model along with temperature-dependent material properties. Kinematic work hardening was assumed in the analysis. This is a common material model used in the welding analysis together with von-Mises yield criterion and associated flow rules (Biswas and Mandal 2010; Lindgren 2001; Wang et al. 2011; Fanous et al. 2003).

4 Implementation of Numerical Procedure by Using APDL

In the present work, a numerical elasto-plastic thermo-mechanical model has been developed for predicting thermal history and angular distortions of submerged arc welded plates with fillet joint configuration. Moving distributed heat source is used in the 3D FE modelling to make it more realistic. First of all, the heat transfer analysis is carried out to find out the nodal transient temperatures over the entire welded plate. Then, the nodal temperatures were applied as load for structural analysis where residual stress and corresponding angular deformation have been predicted.

The weld bead region during SAW is filled up by the filler material due to the melting of the electrode. The effect of filler metal deposition has been taken into account by implementing element death and birth technique to deactivate or reactivate the selected elements (Biswas et al. 2009). Thus, the elements of the weld bead will be non-existent before the welding and become existent after the welding. During the death period, an elements' stiffness value is multiplied by a large reduction factor. The mass and energy of deactivated elements are also excluded from the model. The strain value of an element is set to zero as soon as that element is killed. Similarly, during the birth period, the elements are activated by returning the stiffness, mass, element loads, etc., to their full original values (Biswas et al. 2009). von-Mises yield criterion and associated flow rules were considered as stated in the literatures of Cheng (2005) and Alberg (2005).

The von-Mises yield criterion is as follows:

$$\sigma_{eq} = \sqrt{\frac{1}{2}\left[(\sigma_1 - \sigma_2)^2 + (\sigma_2 - \sigma_3)^2 + (\sigma_3 - \sigma_1)^2\right]} \tag{17}$$

where $\sigma_1, \sigma_2, \sigma_3$ are principal stresses and σ_{eq} is the equivalent one-dimensional stress.

The boundary conditions which prevented rigid body motions are also incorporated in the modelling. Eight nodded brick elements are used for the thermal analysis and similar eight nodded elements are used for structural analysis. The plasticity is assumed as rate independent and is modelled by assuming bi-linear isotropic hardening behaviour along with associated flow rule. Figure 4 shows the schematic diagram of the single sided fillet weld joint considered for the 3D FE modelling. Moving distributed heat flux of the welding arc is applied on the weld zone. The remaining part of the butt joint surface is exposed to the atmosphere where heat loss takes place due to convection. Figure 5 shows the meshed view of the single-sided fillet joint.

The steps involved in the coupled thermo-mechanical analysis that carried out in the present work are represented by a block diagram as shown in Fig. 6. The solution was obtained using the ANSYS 14 package.

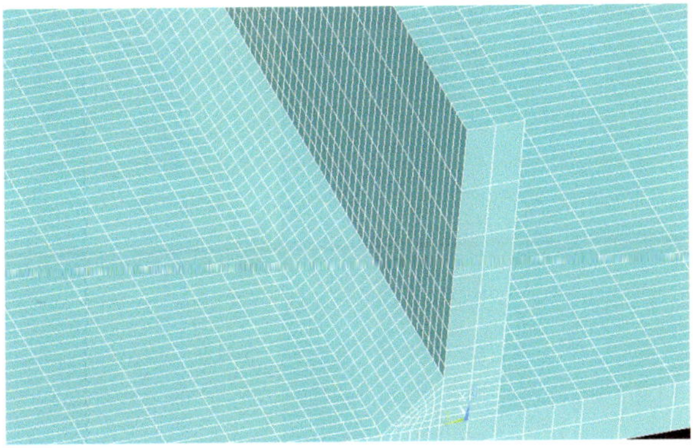

Fig. 5 Meshed view of the single-sided fillet joint

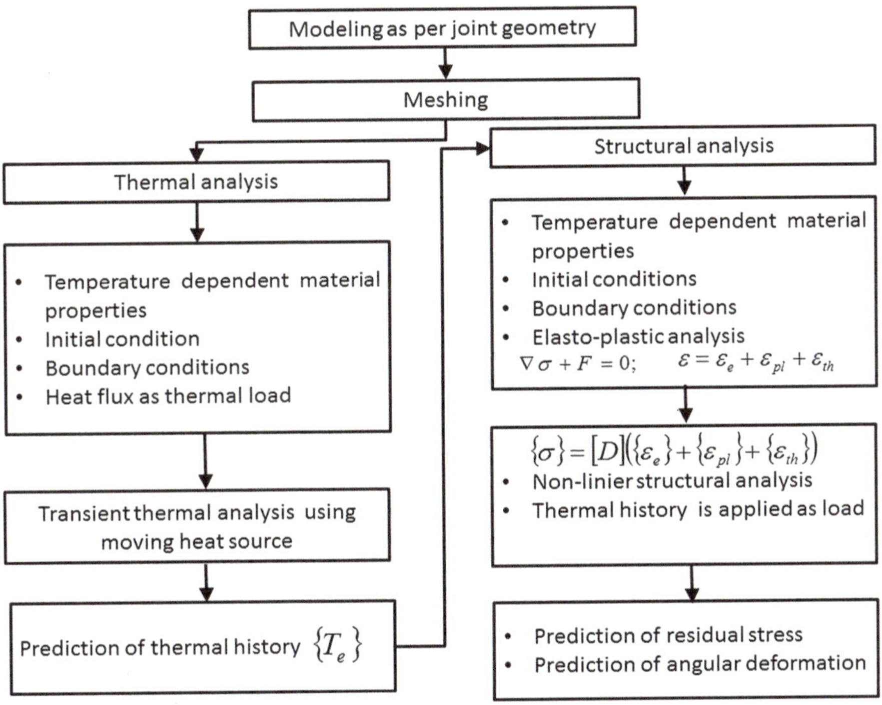

Fig. 6 Steps involved in the coupled thermo-mechanical analysis

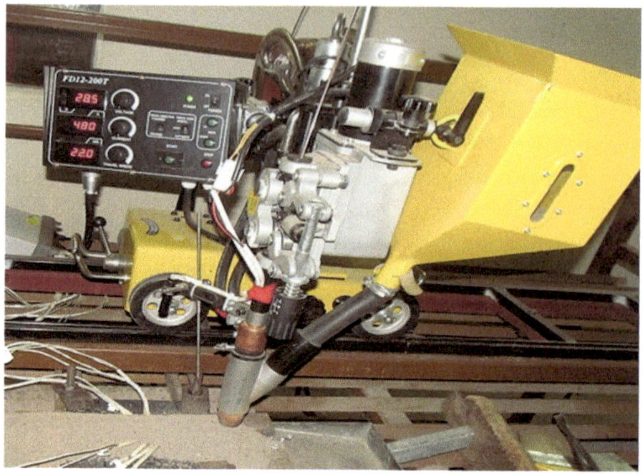

Fig. 7 Submerged arc welding set-up

5 Experimental Investigations

In the present investigation, a constant current SAW set-up, copper-coated MS electrode of 3.1 mm diameter and granular flux have been used for carrying out the experiments.

The MS plates used in this study were of dimension 8 mm thick 100 mm × 200 mm with web height of 50 mm for each half of the fillet joint. For double-sided fillet welding, two sides were welded one after another and transient thermal history was predicted accordingly. Figure 7 shows the ADORE submerged arc welding set-up. Figure 8a, b shows the submerged arc welded specimens of single- and double-sided fillet joints. The experiments conducted to study the angular deformation for single-sided fillet joint and double-sided fillet joint. Before welding the tack, welded samples were marked to measure the coordinates before and after the welding to estimate magnitude of angular distortions. The angular distortions are measured by a co-ordinate measuring machine (CMM) (Fig. 8c).

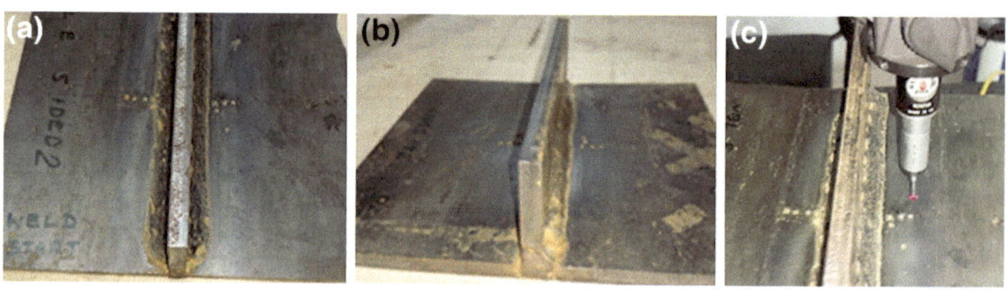

Fig. 8 Submerged arc welded fillet joints, **a** double and **b** single sided, **c** angular deformation measurement

6 Results and Discussions

A comparison of the predicted angular distortion from the FE analysis with the angular distortion from the experimental analysis is shown in Table 4. The maximum magnitudes of angular deformation are considered both for fillet side and opposite side. For both cases, the percentage deviation of the predicted results from the experimental results lie in between 9.07 and 17.57 % and thus shows the accuracy of the developed FE model. The developed numerical model is used to simulate the residual stress and distortions both for single- and double-sided fillet joints. This section provides the detailed analysis of obtained results and possible conclusion on the comparison between single- and double-sided fillet joints. The residual stress distribution pattern for fillet welding was validated with published literature of Ma et al. (1995). Figure 9 illustrates the comparison between predicted results and published data for longitudinal (S_x) and transverse (S_y) residual stress. A good agreement can be observed in Fig. 9, and this indicates that the FE model for present parameters set can be used in a reliable manner. After validating the present study, the same have been used to study the welding residual stress distribution for both single- and double-sided fillet welding.

The comparison of simulation results of single- and double-sided fillet welding is shown in Figs. 10, 11, 12, 13 and 14. Figures 10 and 11 represent the longitudinal and transverse residual stress distribution pattern, respectively.

Table 4 Comparison of angular deformation between simulation experiments

	Angular deformation opposite side			Angular deformation fillet side		
	Exp. (mm)	Pred. (mm)	Error (%)	Exp. (mm)	Pred. (mm)	Error (%)
Single sided	0.082	0.071	15.49	0.745	0.683	9.07
Double sided	0.991	0.84	17.97	0.863	0.734	17.57

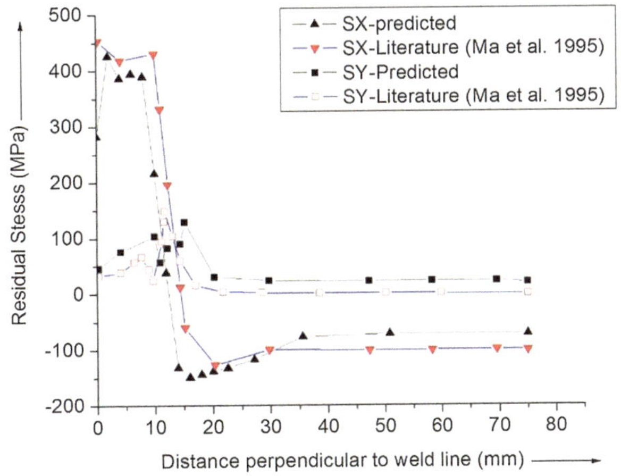

Fig. 9 Comparison of residual stresses with published literature (Ma et al. 1995)

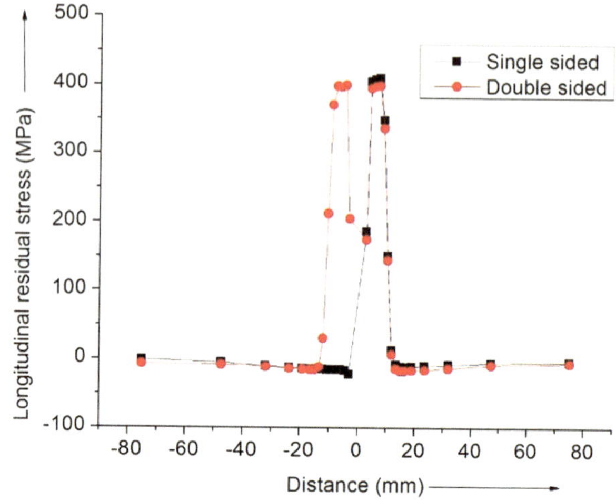

Fig. 10 Comparison between longitudinal residual stresses

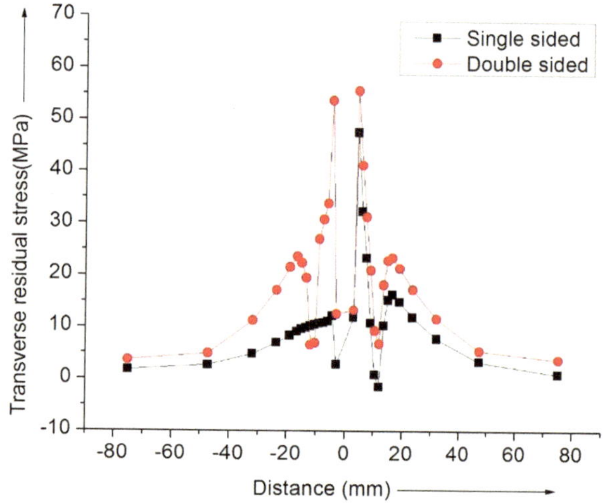

Fig. 11 Comparison between transverse residual stresses

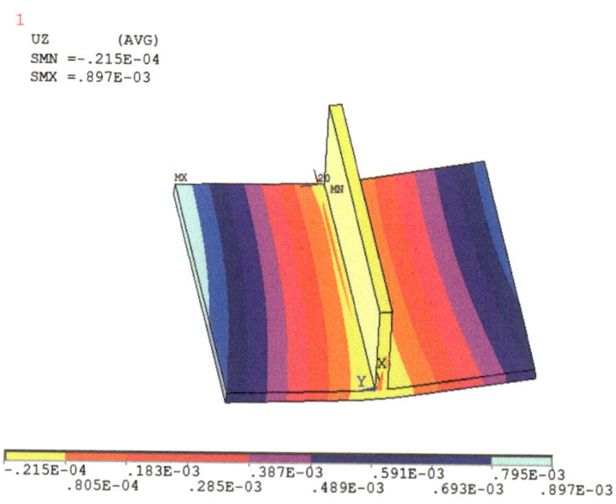

Fig. 12 Contour plot of angular deformation of double-sided fillet weld

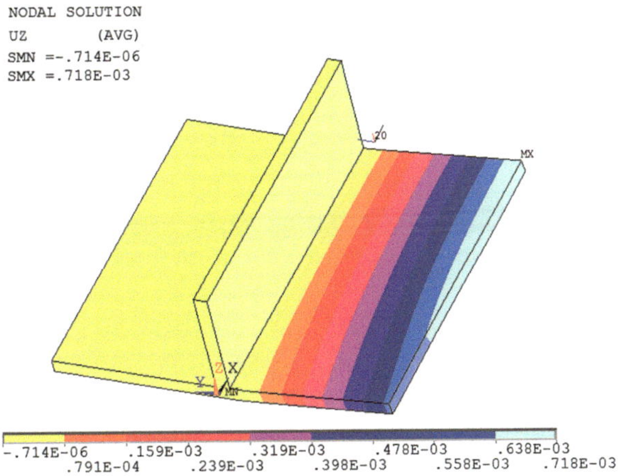

Fig. 13 Contour plot of angular deformation of single-sided fillet weld

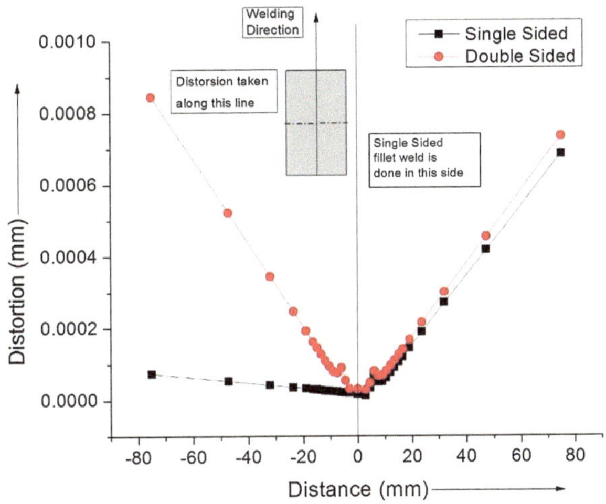

Fig. 14 Comparison between the angular deformations

From Figs. 10 and 11, it was observed that within and closer to the welding region, the residual stress is tensile in nature and away from the weld line, it is compressive in nature.

The maximum value of longitudinal residual stress is 409 MPa (tensile) for single-sided fillet weld but for double-sided fillet weld is 399 MPa. In case of single-sided fillet joint, the residual stress is unbalanced in both sides. The maximum magnitude of transverse residual (Fig. 11) stress is very less than compared to longitudinal residual stress. For single-sided fillet joint, it is 47 MPa, and for double-sided fillet, it is 555 MPa.

Figures 12 and 13 show contour plots of double- and single-sided fillet joints, respectively. It can be seen from Fig. 12 that within and closer to the welding region, the amount of deformation increases with the distance perpendicular to the welding line. In Fig. 13, the amount of angular deformation in the single-sided fillet weld is very less in the opposite side of the fillet. The maximum value of angular deformation in case of double-sided fillet weld is 0.215 mm, and in case of single-sided fillet weld, it is 0.718 mm.

Figure 14 illustrates the distortions of the single- and double-sided fillet welded MS plate. The distortions are taken perpendicular to the weld line from centre of weld to both sides of the weld at the middle length of the weld line. The angular deformation in weld is the product of the weld shrinkage force during cooling and residual stress after complete solidification. It is clear from Fig. 14 that the maximum magnitude of the angular distortion both in single- and double-sided fillet joints is almost same. However, in case of single-sided fillet joint, the angular deformation is completely unsymmetrical in both sides of the fillet. As in case of double-sided fillet joint, the welding is done in both sides of the fillet that is why the angular deformation developed is well balanced in both the sides, but cumulative value of the angular deformation is more in case of double-sided fillet joint.

7 Conclusions

A comparative study was done between the welding-induced residual stress and angular deformation in single- and double-sided fillet joints' configuration by means of FE modelling and simulation. The following conclusions can be derived from the present investigation:

- A three-dimensional FE model for predicting residual stresses for SAW has been developed utilising the nonlinear transient thermo-mechanical analysis assuming elasto-plastic behaviour of MS.
- The maximum value of angular deformation in case of double-sided fillet weld is 0.215 mm, and in case of single-sided fillet weld, it is 0.718 mm.
- The angular deformation in single-sided fillet joint was unbalanced in both sides of the fillet. The cumulative angular deformation was more in double-sided fillet joint.
- The maximum value of residual stress is more in case of single-sided fillet joint compared to that of double-sided fillet joint.
- The angular deformation of single-sided fillet joint was less than the double-sided fillet joint, which implies the single-sided fillet joint may be preferred than double-sided fillet joints, thus saving time and resources.
- The results obtained through analysis and those obtained from published literatures compared fairly well with a variation of about 9–17 % only.

References

Adak M, Mandal NR (2003) Thermo-mechanical analysis through a pseudo-linear equivalent constant stiffness system. Proc Inst Mech Eng Part M J Eng Marit Environ 217:1–9

Alberg H (2005) Simulation of welding and heat treatment modelling and validation. PhD thesis, Lulea University of Technology, Sweden

Arai Y, Kikuchi M, Watanabe T, Nakagaki M (1995) Residual stress due to welding and its effect on the assessment of cracks near the weld interface. Int J Press Vessels Piping 63(3):237–248

Asim K, Lee J, Pan J (2011) Failure mode of laser welds in lap-shear specimens of high strength low alloy (HSLA) steel sheets. Fatigue Fract Eng Mater Struct 35:219–236

Awang M (2002) The effects of process parameters on steel welding response in curved plates. MS thesis, Department of Mechanical Aerospace Engineering, College of Engineering and Mineral Resources, West Virginia University, Morgantown, West Virginia, USA

Biswas P, Mandal NR (2010) Thermomechanical finite element analysis and experimental investigation of single-pass single-sided submerged arc welding of C-Mn steel plates. Proc Inst Mech Eng Part B J Eng Manuf 224:627–639

Biswas P, Mahapatra MM, Mandal NR (2009) Numerical and experimental study on prediction of thermal history and residual deformation of double-sided fillet welding. Proc Inst Mech Eng Part B J Eng Manuf 224:125–134

Brown S, Song H (1992) Implication of three-dimensional numerical simulation of welding of large structures. Weld J 71:55s–62s

Chen B-Q (2011) Prediction of heating induced temperature fields and distortions in steel plates. Masters dissertation, Centre for Marine Technology and Engineering (CENTEC), Instituto Superior Técnico, Technical University of Lisbon, Portugal

Cheng W (2005) In-plane shrinkage strains and their effects on welding distortion in thin-wall structures. PhD thesis, The Ohio State University, USA

Deng D, Ma N, Murakawa H (2011) Finite element analysis of welding distortion in a large thin-plate panel structure. Trans Jpn Weld Soc 40:89–100

Fanous FZI, Maher YA, Wifi SA (2003) 3-D Finite element modeling of the welding process using element birth and element movement techniques. Trans ASME J Press Vessel Technol 125:144–150

Friedman E (1975) Thermo mechanical analysis of the welding process using the finite element method. Trans ASME 97:206–213

Hong JK, Tsai CL, Dong P (1998) Assessment of numerical procedures for residual stress analysis of multipass weld. Weld J 77:372s–382s

Kong F, Kovacevic R (2012) Development of a comprehensive process model for hybrid laser-arc welding. In: Kovacevic R (ed) Welding processes. Intech, New York, pp 165–190. ISBN 978-953-51-0854-2

Lindgren LE (2001) Finite element modelling and simulation of welding part 2: improved material modelling. J Therm Stresses 24:195–231

Lucas B, Verhaeghe G, Leggatt R (2014) http://www.twi-global.com/technical-knowledge/job-knowledge/distortion-types-and-causes-033/

Ma NX, Ueda Y, Murakawa H (1995) FEM analysis of 3D welding residual stresses and angular distortion in T-type fillet welds. Trans Jpn Weld Res Inst 24:115–122

Mahapatra MM, Datta GL, Pradhan B, Mandal NR (2007) Modelling the effects of constraints and single axis welding process parameters on angular distortions in one-sided fillet welds. Proc Inst Mech Eng Part B J Eng Manuf 221(B3):397–407

Michaleris P, DeBiccari A (1997) Prediction of welding distortion. Weld J 76:172s–180s

Pathak AK, Datta GL (2004) Three-dimensional finite element analysis to predict the different zones of microstructures in submerged arc welding. Proc Inst Mech Eng Part B J Eng Manuf 218:269–280

Radaj D (1992) Heat effects of welding: temperature field. Residual stress, distortion. Springer, Berlin

Sorensen MB (1999) Simulation of welding distortions in ship section. PhD thesis, Department of Naval Architecture and Offshore Engineering, Technical University of Denmark. ISBN 87-89502-13-2

Teng TL, Fung CP, Chang PH, Yang WC (2001) Analysis of residual stresses and distortions in T-joint fillet welds. Int J Press Vessels Piping 78:523–538

Tsai CL, Cheng WT (1999) Welding distortion of thin-plate panel structures. Weld J 78:156s–165s

Tsai CL, Jung GH (2004) Plasticity-based distortion analysis for fillet welded thin- plate T-joints. Weld J 83:177s–187s

Ueda Y, Yamakawa T (1971) Analysis of thermal elastic–plastic stress and strain during welding by finite element method. Trans Jpn Weld Soc 2:90–100

Wang J, Ma N, Murakawa H, Teng B, Yuan S (2011) Prediction and measurement of welding distortion of a spherical structure assembled from multi thin plates. Mater Des 32:4728–4737

Weisman C (1976) Welding handbook: fundamentals of welding, vol 1, 7th edn. American Welding Society

Optimization of Process Parameters in Submerged Arc Welding Using Multi-objectives Taguchi Method

Abhijit Saha and Subhas Chandra Mondal

Abstract Submerged arc welding (SAW) is one of the oldest automatic welding processes to provide high quality of weld. The quality of weld in SAW is mainly influenced by independent variables such as welding current, arc voltage, welding speed, and electrode stick out. The prediction of process parameters involved in SAW is very complex process. Researchers attempted to predict the process parameters of SAW to get smooth quality of weld. This paper presents an alternative method to optimize process parameters of SAW of IS: 2062, Gr B mild steel with multi-response characteristics using Taguchi's robust design approach. Experimentation was planned as per Taguchi's L8 orthogonal array. In this paper, experiments have been conducted using welding current, arc voltage, welding speed, and electrode stick out as input process parameters for evaluating multiple responses namely weld bead width and bead hardness. The optimum values were analyzed by means of multi-objective Taguchi's method for the determination of total normalized quality loss (TNQL) and multi-response signal-to-noise ratio (MRSN). The optimum parameters for smaller bead width and higher bead hardness are weld current at low level (12.186 A), arc voltage at low level (12.51 V), welding speed at low level (12.25 mm/min), and electrode stick out at low level (12.29 mm). Finally, confirmation experiment was carried out to check the accuracy of the optimized results.

A. Saha
Haldia Institute of Technology, Haldia, West Bengal 721657, India
e-mail: alfa.nita2010@gmail.com

S.C. Mondal (✉)
Indian Institute of Engineering Science and Technology, Shibpur, Howrah 711103, India
e-mail: scmondall@gmail.com

R.G. Narayanan and U.S. Dixit (eds.), *Advances in Material Forming and Joining*,
Topics in Mining, Metallurgy and Materials Engineering,
DOI 10.1007/978-81-322-2355-9_11

1 Introduction

Welding is a process of joining different materials. It is more economical and is a much faster process compared to both casting and riveting. In industries and research organizations, most widely used welding methods are shield metal arc welding (SMAW), gas metal arc welding (GMAW), gas tungsten arc welding (GTAW), and submerged arc welding (SAW). SAW process is one of the oldest automatic welding processes introduced in 1930s to provide high quality of weld. This process gets its name from the fact that the welding arc, between the ends of a copper-coated bare wire electrode and the work, is actually submerged under a layer of granular fusible flux, which blankets the molten weld metal and the base metal near the joint, and protects the molten weld metal from atmospheric contamination as shown in Fig. 1.

The SAW process is widely used in heavy fabrication industries because it offers high production rate, high melting efficiency, easy control of process parameters and process variables, smooth bead geometry, ease of automation, and low operator skill requirement (Apps et al. 1963; Kanjilal et al. 2006). Quality of welded joint is judged by the weld bead geometry and also the mechanical properties, i.e., weld quality is strongly characterized by the weld bead geometry, which plays an important role in influencing the mechanical properties of welded joints (Aesh 2001; Bapat et al. 2009). The quality of weld in SAW is mainly influenced by

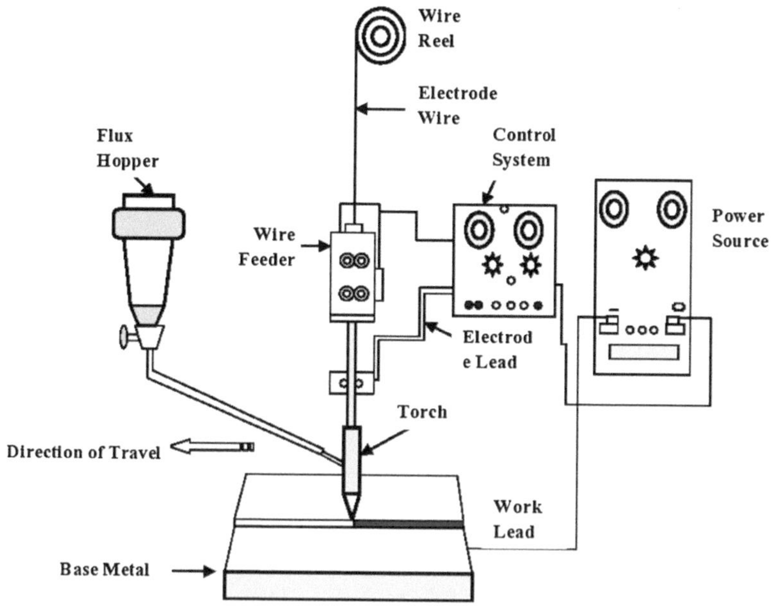

Fig. 1 Submerged arc welding

independent variables such as welding current, arc voltage, welding speed, and electrode stick out. The prediction of process parameters involved in SAW is very complex process.

Researchers and practioners have several attempted to predict the process parameters of SAW to get smooth quality of weld. Tarng and Yang (1998) applied Taguchi method to the optimization of the SAW process. Taguchi method was used to formulate the experimental layout, to analyze the effect of each welding parameter on welding performance, and to predict the optimal setting for each welding parameter. For the prediction and optimization of the weld bead volume for SAW, a mathematical model was developed and also studied using the weld bead geometry and shape relationship in SAW of pipes (Gunaraj and Murugan 1999a, b). Chang et al. (2002) applied gray-based Taguchi methods for optimization of SAW process parameters in hardfacing. They considered multiple weld qualities and determined optimal process parameters based on gray relational grade from gray relational analysis proposed by Taguchi. Sabbaghian and Roostaazad (2005) successfully applied Taguchi method to optimize the process performance in the production of lipase.

Dhas et al. (2007) elaborate the study of welding procedures generation for the SAW process. Several research works have already been carried out in the field of SAW for parametric optimization. Biswas et al. (2007) studied the effect of process parameters on output features of submerged arc weld by using Taguchi method. Bandyopadhyay et al. (2008) analyzed the gray-based Taguchi method for optimization of bead geometry in submerged arc bead on plate welding. Taguchi's orthogonal array design and signal-to-noise ratio have been used to derive objective functions which are optimized within experimental domain. The objective functions have been selected in relation to parameters of bead geometry viz bead width, bead reinforcement, depth of penetration, and depth of HAZ. Taguchi' approach followed by gray relational analysis has been applied to solve the multi-response optimization problem (Mondal et al. 2013). The significance of the factors on overall output features of weldment has also been evaluated quantitatively by ANOVA. An optional result has been verified through additional experiment. This indicates application feasibility of the gray-based Taguchi's technique for continues improvement in product quality in manufacturing industry. Benyounis and Olabi (2008) while performing the study on optimization of different welding processes developed mathematical relationship between welding input process parameters and weld joint output parameters using statistical and numerical approaches and suggested that these approaches are useful for determination of welding input parameters which lead to desired weld quality.

Bhattacharya et al. (2012) has done experimental investigation for multi-response optimization on plain carbon steel 200 × 75 × 12 mm as test material. The multi-response from the observation was converted to gray analysis using MATLAB code. Finally, they concluded that welding current is most significant for maximizing depth of penetration and minimizing bead height and bead width. Shen et al. (2012) performed a series of measurements on specimens of submerged arc-welded plates of ASTM A709 Grade 50 steel. The bead reinforcement, bead width, penetration depth, HAZ size, deposition area, and penetration area increased with increasing heat input but the bead contact angle decreased with it. The electrode melting efficiency increased initially and then decreased with increasing heat input, but the plate melting efficiency and percentage dilution changed only slightly with it. Cooling time exhibited a very good linear relationship with the total nugget area, heat transfer boundary length, and nugget parameter. Hari and Pandey (2013) have shown from their work that HAZ width rises more effectively with wire feed rate. With negative polarity, dilution rate decreases more than 20 %. HAZ area varies linearly with heat input. Singh et al. (2013) have done a review study on the effect of flux composition on its behavior and bead geometry. With the help of their detailed review, they showed that flux constituents have a major effect on flux behavior and bead shape geometry. The load carrying capacity of the welded joint does not only depend on microstructure, but it is also affected by the physical behavior of the flux and bead geometry. The main characteristics which are affected by flux constituents are arc stability, slag detachability, capillarity, viscosity, and basicity index.

In the present paper, multi-objective Taguchi method was applied to optimize the welding parameters under simultaneous consideration of multiple weld quality characteristic such as higher weld bead hardness and smaller bead width size. The optimization approach starts with the calculation of total normalized quality loss (TNQL) under simultaneous consideration of response weighting factor and followed by the observation of multiple signal-to-noise ratio (MSNR). Finally, experimental confirmation test was also conducted to validate the optimized parameters.

2 Taguchi and Multi-objective Taguchi Method

Taguchi's robust design is a simple, systematic, and more efficient technique for optimizing the process parameters. In this method, main parameters which are assumed to have influence on process results are located at different rows in a designed orthogonal array (OA). With such arrangement, completely randomized experiments can be conducted. An advantage of the Taguchi method is that it emphasizes a mean performance characteristic value close to the target value rather than a value within certain specification limits, thus improving the product quality. In this present work, optimization of welding operations using Taguchi's robust design methodology with multiple performance characteristics is proposed. In order to optimize the multiple performance characteristics, Taguchi parametric design approach was not applied directly. Since each performance characteristic may not have the same measurement unit and of the same category in the S/N ratio analysis. Therefore, to solve problems of this kind, steps 1–3 followed first then the traditional Taguchi technique for single response optimization has performed.

Step 1 Normalize the loss function corresponding to each performance characteristics as follows:

$$y_{ij} = \frac{L_{ij}}{L_i^*} \tag{1}$$

where y_{ij} is the normalized quality loss associated with the ith quality characteristic at the jth trial condition, and it varies from a minimum of zero to a maximum of 1. L_{ij} is the quality loss or MSD for the ith quality characteristic at the jth trial, and L_i* is the maximum quality loss for the ith quality characteristic among all the experimental runs.

Step 2. Apply a weighting method to determine the importance of each normalized loss function. The TNQL function Y_j in the jth experiment as given under:

$$Y_j = \sum_{i=1}^{k} W_i y_{ij} \tag{2}$$

where W_i represents the weighting factor for the ith quality characteristic and k is the total number of quality characteristics.

Step 3. In multi-objective optimization, a single overall S/N ratio for all quality characteristics is computed in place of separate S/N ratios for each of the quality characteristic. This overall S/N ratio is known as multiple S/N ratio (MSNR). Transform the total loss function into a MSNR as follows:

$$\text{MSNR} = -10\log_{10}(Y_j) \tag{3}$$

Based on the MSNR, the optimal factors or level combination is determined like the traditional Taguchi technique. Finally, the optimal process parameters are verified through the confirmation experiment.

Usually, there are three categories of quality characteristic in the analysis of S/N ratio: nominal the better, the lower the better (LB), and the higher the better (HB). In this study, the LB and HB principles are considered to minimize the bead width and maximize the bead hardness, respectively.

The summary statistics η(dB) of the HB performance characteristics is expressed as follows:

$$\eta = -10\log_{10}\left[\frac{1}{n}\sum_{i=1}^{n}\frac{1}{y_i^2}\right] \tag{4}$$

where $i = 1, 2, \ldots, n$.

The summary statistics η(dB) of the LB performance characteristics is expressed as follows:

$$\eta = -10\log_{10}\left[\frac{1}{n}\sum_{i=1}^{n}\frac{1}{y_i^2}\right] \tag{5}$$

where i = 1, 2, …, n.

This relation is used to determine the S/N ratio η(dB) for surface roughness.

3 Experimental Process SetUp and Procedures

In this study, experiment was conducted on "TECHNOCRATS PLASMA SYSYTEMS PVT LTD," MODEL-1000, automatic SAW equipment with a constant voltage, rectifier-type power source with a 1000 A capacity was used to join the two mild steel plates of size 200 mm length × 50 mm width × 12 mm thickness. Copper-coated electrode Automelt EH-14, wire size: 3.20 mm diameter, of coil form and basic fluoride-type granular flux were used. The welded sample, weld bead measurement method, and hardness measurement method are shown in Figs. 2, 3, and 4. The chemical composition of the work piece and flux is listed in Tables 1 and 2. Three welding parameters such as welding current, arc voltage, welding speed, and electrode stick out were selected for experimentation with two levels of each factor. The value of the welding process parameter at the different levels is tabulated in Table 3. Experimental process was conducted using L8 orthogonal array in Taguchi method which has nine rows corresponding to the

Fig. 2 Welded portion of the workpiece

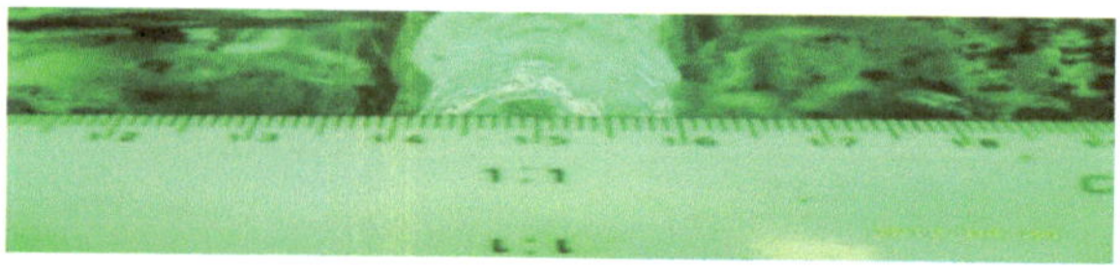

Fig. 3 Measurement of the weld bead width

Fig. 4 Measurement of hardness by rockwell hardness tester

Table 1 Chemical composition of the base metal IS: 2062, Gr.B

Element	Carbon	Manganese	Silicon	Sulfur	Phosphorous
Percentage	0.16	0.76	0.24	0.022	0.028

Table 2 Chemical composition of the flux: Automelt, B 31

Compositions	$SiO_2 + TiO_2$	$CaO + MgO$	$Al_2O_3 + MnO$	$Ca + F_2$
Percentage	25	20	30	35

Table 3 Welding parameters with different level

Symbol	Welding parameters	Level 1	Level 2
A	Welding current, A	300	360
B	Arc voltage, V	25	28
C	Welding speed, mm/min	900	1000
D	Electrode stick out, mm	19	25

number of experiments, and the experimental results for the weld bead geometry using the L8 orthogonal array are shown in Table 4.

4 Multi-objective Optimization

From Table 5, quality loss values for different quality characteristics (higher is best for weld bead hardness and smaller is better for bead width) in each experimental run are calculated using Eqs. (4) and (5). These quality loss values are shown in Table 6. The normalized quality loss values for both quality characteristics in each

Table 4 Measured values from the test specimen as per L8 orthogonal array

Trial no.	Welding current (A)	Arc voltage (V)	Welding speed (mm/min)	Electrode stick out (mm)	Bead width measured (mm)	Measured weld bead hardness (kgf)
1	300	25	900	19	15	40.50
2	300	25	1000	25	14.7	51.00
3	300	28	900	25	16	50.00
4	300	28	1000	19	14.6	49.50
5	360	25	900	25	14.5	39.00
6	360	25	1000	19	14	43.50
7	360	28	900	19	19	41.00
8	360	28	1000	25	20	46.00

Table 5 Quality loss values for weld bead width and weld bead hardness

Trial no.	A	B	C	D	Quality loss values (dB)	
					Bead width	Weld bead hardness
1	1	1	1	1	−23.5218	32.1491
2	1	1	2	2	−23.3463	34.1514
3	1	2	1	2	−24.0824	33.9794
4	1	2	2	1	−23.2871	33.8921
5	2	1	1	2	−23.2274	31.82129
6	2	1	2	1	−22.9226	32.76979
7	2	2	1	1	−25.5751	32.25568
8	2	2	2	2	−26.0206	33.25516

Table 6 Normalized quality loss values

Trial no.	A	B	C	D	Normalized quality loss values	
					Bead width	Weld bead hardness
1	1	1	1	1	1.0000011	0.94137
2	1	1	2	2	0.9925408	1
3	1	2	1	2	1.0238332	0.993944
4	1	2	2	1	0.9900202	0.992407
5	2	1	1	2	0.9874823	0.931771
6	2	1	2	1	0.9745241	0.959544
7	2	2	1	1	1.0872923	0.943247
8	2	2	2	2	1.1062334	0.972649

experimental run have been calculated using Eq. (3) that is shown in Table 6. The TNQL values and MSNR for multiple quality characteristics for weld bead hardness and bead width have been calculated using Eqs. (2) and (1), respectively.

Table 7 Total normalized quality loss values (TNQL) and multiple S/N ratios (MSNR)

Trial no.	A	B	C	D	TNQL	MSNR (dB)
1	1	1	1	1	0.485343	3.139514
2	1	1	2	2	0.498135	3.026528
3	1	2	1	2	0.504444	2.971868
4	1	2	2	1	0.495607	3.048627
5	2	1	1	2	0.479813	3.189277
6	2	1	2	1	0.483517	3.155882
7	2	2	1	1	0.507635	2.944486
8	2	2	2	2	0.519721	2.842301

Table 8 Multiple S/N response (average factor effect at different level)

Symbols	Factors	Mean of multiple S/N ratio (dB)	
		Level 1	Level 2
A	Welding current	12.18654[a]	12.13195
B	Arc voltage	12.5112[a]	11.80728
C	Welding speed	12.24514[a]	12.07334
D	Electrode stick out	12.28851[a]	12.02997

[a]Optimum level

These results are shown in Table 7. In calculating TNQL values, two equal weights for weld bead hardness and bead width were used. The effect of different control factors on MSNR is shown in Table 8. The optimum levels of different control factors for higher weld bead hardness and lower bead width obtained are weld current at level 1 (12.18654 A), arc voltage at level 1 (12.5112 V), welding speed at level 1(12.24514 mm/min), and electrode stick out at level 1 (12.28851 mm).

5 Confirmation Tests

After obtaining the optimal level of the welding parameters, the next step is to verify the improvement of the performance characteristics using this optimal combination. After determining the optimum conditions, a new experiment was conducted with the optimum levels of the welding parameters (A1B1C1D1). Then, the predicted value of MSNR ($\hat{\gamma}$) at the optimum parameter levels was calculated by using the following equation:

$$\hat{\gamma} = \gamma_m + \sum_{i=0}^{p} \left(\bar{\gamma}_j - \gamma_m\right) \quad (6)$$

Table 9 Results of confirmation experiment

	Initial machining parameters	Optimal machining parameters	
		Prediction	Experiment
Setting level	A1B2C1D2	A1B1C1D1	A1B1C1D1
Bead width	16		15
Weld bead hardness	39		40.5
Multiple S/N ratio (dB)	2.122	2.981	3.139

Improvement in multiple S/N ratio = 1.0176

where γ_m is the mean MSNR, $\bar{\gamma}_j$ is the average MSNR at the optimum level, and p is the number of main welding parameters that significantly affects the performance. Table 9 shows the results of the confirmation experiment using the optimal welding parameters. The improvement in MSNR from initial parameter which is A1B2C1D2 setting to the optimal parameters (A1B2C1D1) is found to be 1.017 dB. The results show considerable improvement in both the quality characteristics (bead width and bead hardness) with the multi-response optimization used, as compared the initial value of bead width and bead hardness.

6 Conclusions

Taguchi is interested in finding a "robust" solution to the experimental problem. It seeks an answer that is insensitive to factor variations and noise. It does not predict the best combination of factors to achieve desired goals. Taguchi analysis can provide definitive information if there is only one response, but it does not deal with situations where a number of responses are to be optimized. The time required for conducting experiments using Taguchi method is almost half that needed for the RSM. The technique adapted here can efficiently be applied in any manufacturing/production processes for continuous quality improvement and off-line quality control. In this work, an attempt was made to determine important welding process parameters for multiple performance characteristics weld bead width and bead hardness in the SAW process using Taguchi method. Careful attention is necessary to select the welding parameters to obtain a desirable weld quality, and the major key process parameters affecting the bead geometry are arc voltage, welding current, welding speed, and electrode stick out. Based on the results of the present study, the following conclusions are drawn:

(i) The optimum parameters for smaller bead width and higher bead hardness are as follows: weld current at level 1 (12.18654 A), arc voltage at level 1 (12.5112 V), welding speed at level 1(12.24514 mm/min), and electrode stick out at level 1 (12.28851 mm).

(ii) The confirmation test validated the use of multi-objective Taguchi method for enhancing the welding performance and optimizing the welding parameters in SAW process. The results confirm the efficiency of the approach employed for optimization of welding parameters in this study.

This research definitely indicates some directions for future work. In future, this study can be extended by developing and analyzing response surface methodology and also various techniques such as ANN, genetic algorithm, and influence of flux density on weld bead geometry and HAZ. In the study of HAZ, microstructure can be applied for optimization of process parameter.

References

Aesh MA (2001) Optimization of weld bead dimensions in GTAW of aluminum–magnesium alloy. Mater Manuf Process 16(5):725–736

Apps RL, Gourd LM, Lelson KA (1963) Effect of welding variables upon bead shape and size in submerged arc welding. Weld Met Fabr 31:457–463

Bandyopadhyay A, Datta S, Pal PK (2008) Grey based Taguchi method for optimization of bead geometry in submerged arc bead-on-plate welding. Int J Adv Manuf Technol 39:1136 114

Bapat V, Dey V, Datta GL, Jha MN, Pratihar DK, Saha TK (2009) Optimization of bead geometry in electron beam welding using a genetic algorithm. J Mater Process Technol 209(3):1151–1157

Benyounis KY, Olabi AG (2008) Optimization of different welding processes using statistical and numerical approaches. Adv Eng Softw 39:483–496

Bhattacharya A, Batish A, Kumar P (2012) Experimental investigation for multi response optimization of bead geometry in submerged arc welding using grey analysis. J Inst Eng India Ser C 93(2):123–132

Biswas S, Mahapatra SS, Patnaik A (2007) An evolutionary approach to parameter optimization of submerged arc welding in the hardfacing process. Int J Manuf Res 2(4):462–483

Chang CH, Juang SC, Tarng YS (2002) The use of grey-based Taguchi methods to determine submerged arc welding process parameters in hardfacing. J Mater Process Technol 128:1–6

Dhas R, Edwin J, Gothman K, Kumanan S (2007) Determination of submerged arc welding process parameters using Taguchi method and regression analysis. Indian J Eng Mater Sci 14:177–183

Gunaraj V, Murugan N (1999a) Prediction and comparison of the area of the heat affected zone for the Bead-no-plate and Bead-on-joint in submerged arc welding of pipes. J Mater Process Technol 95(1–3):246–261

Gunaraj V, Murugan N (1999b) Application of response surface methodology for predicting weld bead quality in submerged arc welding of pipes. J Mater Process Technol 88:266–275

Hari OM, Pandey S (2013) Effect of heat input on dilution and heat affected zone in submerged arc welding process. Indian Acad Sci 38:1369–1391

Kanjilal P, Majumder SK, Pal TK (2006) Combined effect of the flux and welding parameters on chemical composition and mechanical properties of submerged arc weld metal. J Mater Process Technol 171:223–231

Mondal SC, Ray PK, Maiti J (2013) Modelling robustness for manufacturing processes: a critical review. Int J Prod Res 52(2):521–538

Sabbaghian ES, Roostaazad R (2005) Application of the Taguchi method to optimize the process conditions in the production of lipase by pseudomonas Aeroginosa B-3556. Iran J Sci Technol Trans B, Eng 29:475–482

Shen S, Oguocha INA, Yannacopoulos S (2012) Effect of heat input on weld bead geometry of submerged arc welded ASTM A709 grade 50 steel joints. J Mater Process Technol 212(1):286–294

Singh B, Khan Z A, Siddiquee A N (2013) Review on effect of flux composition on its behavior and bead geometry in submerged arc welding (SAW). J Mech Eng Res 5(7):123–127

Tarng YS, Yang WH (1998) Application of the Taguchi method to the optimization of the submerged arc welding process. Mater Manuf Process 13(3):455–467

Monitoring of Weld Quality in Friction Stir Welding Based on Spindle Speed and Motor Current Signals

Bipul Das, Sukhomay Pal and Swarup Bag

Abstract The process of friction stir welding passed more than two decades since its invention in the year 1991 in TWI, UK. It involves complex physics and has not been explored fully to understand its physical behavior. Due to the lack of precise mathematical modeling and too many influencing factors that govern the welding process, difficulty arises in direct monitoring of the process based on the process parameters only. Moreover, the influencing parameters are so correlated that the effect of one on the weld quality cannot be isolated from the others for effective monitoring of the process. Thus, a need is realized to develop different methods for the efficacious monitoring of the process with the acquired signals during welding for better control over the outcome of the process. In this study, effectiveness of spindle speed and main motor current signals is investigated for the development of tools which will lead to real-time weld quality prediction.

Keywords Monitoring · Friction stir welding · Signal · Neural network · Weld quality · Prediction

1 Introduction

Friction stir welding (FSW) is a solid-state joining process which results in joining of two metals either similar or dissimilar without melting any one of them. FSW as a welding process is quite new in the community, but still reached a diversified use

B. Das · S. Pal · S. Bag (✉)
Department of Mechanical Engineering, Indian Institute of Technology Guwahati, Guwahati, Assam 781039, India
e-mail: swarupbag@iitg.ernet.in

B. Das
e-mail: bipul.das@iitg.ernet.in

S. Pal
e-mail: spal@iitg.ernet.in

R.G. Narayanan and U.S. Dixit (eds.), *Advances in Material Forming and Joining*, Topics in Mining, Metallurgy and Materials Engineering,
DOI 10.1007/978-81-322-2355-9_12

in the field of automobile, railway, shipbuilding, and aerospace industries to name a few. The abilities of the process which help this to penetrate in so many industrial applications are successful welding of difficult-to-weld materials such as aluminum and magnesium, lightweight applications (as no filler materials are required to achieve the welding), environment friendly (as it does not emit any fume), more flexibility in welding of workpieces of different thicknesses, less power consumption as compared to other conventional, and advance welding processes. Different weld configurations are possible with FSW process as in other welding processes such as butt, lap, T-joint, and fillet joint.

As mentioned, the joining in case of FSW is achieved without melting of the parent materials. The process of FSW is schematically shown in Fig. 1. Here, a rotating tool is first plunged in the line of joining between the two workpiece materials which are held together tightly by fixtures. During plunging, first the tool pin is plunged completely into the material and once the pin plunging is over; shoulder of the tool is plunged into the workpiece. The depth through which the shoulder penetrates the material is called as the plunging depth. The tool is kept in this position for a certain time, and this period in referred as the dwell time. After this, the tool is traversed along the line of joining and retracted back once the welding is over. Tool used in FSW process consists of a pin, shoulder, and shank for holding. Different pin profiles can be used in FSW process such as straight cylindrical, cylindrical threaded, and taper threaded. Schematic of a typical straight cylindrical tool is shown in Fig. 2a. Tool in FSW performs two primary functions: (a) localized heating and (b) material flow. In the initial stage of pin plunge, the heating is a result from the friction between the pin and workpiece. Some additional heating results from deformation of materials. The friction between workpiece and shoulder results in the highest component of heating for the plasticization of the material. The shoulder also provides the confinement for the heated volume to be retained in the welding zone beneath the tool shoulder. The second function of the tool is to stir and move the plasticized material around the tool. Pin facilitates the stirring of the material around the tool and moves the material for proper mixing.

Fig. 1 Schematic representation of FSW process

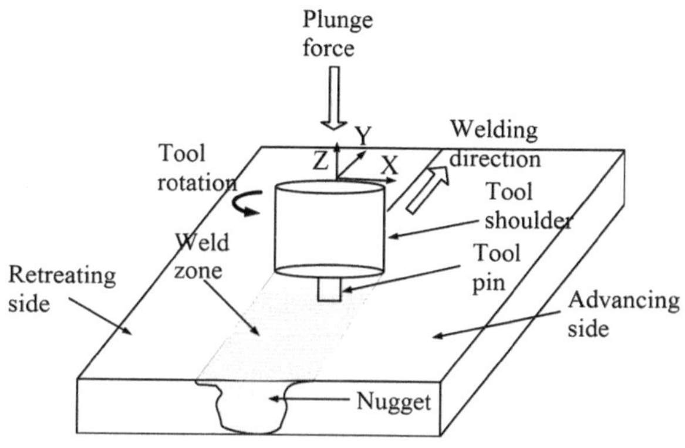

FSW process involves many process parameters which influences the weld quality. The two most influencing process parameters are tool rotational speed (TRS) and welding speed (WS) (Mishra and Ma 2005). The key idea of joining in FSW process lies in severe plasticization of the workpiece material. The major component of heating for the plasticization of material is achieved from the frictional heat generated due to the friction between the rotating tool and workpiece material. This heat is again aided by the adiabatic heat generated due to plastic deformation of the material because of the stirring action of the rotating tool (Mishra and Ma 2005). The rotating tool mixes the plasticized material around the tool. Traversing of the tool distributed the plasticized material behind the tool and carried the heat for the plasticization of the material ahead the tool pin as the tool progresses along the joining line. The material gets solidified during cooling leading to the bond formation in FSW. Thus, it is evident that TRS and WS are most influencing process parameters. Apart from these two parameters, shoulder diameter of the tool, pin length, pin diameter, tool tilt angle, and plunging depth are few process parameters which also play an important role in the joining process of FSW, and all the mentioned parameters can be seen in Fig. 2a, b.

Advancing side and retreating side are the two most frequently used terms associated with FSW process. The side of the weld where the tool rotation direction is in the similar sense with the tool traversing direction is called the advancing side. The side of the weld where the tool rotation direction is in the opposite sense of the tool traverse direction is called the retreating side. These two can be seen in Fig. 1. Threadgill (2007) had reported on standardization of language used to describe FSW process. He proposed a few terms for the microstructural classification of friction stir welded sample.

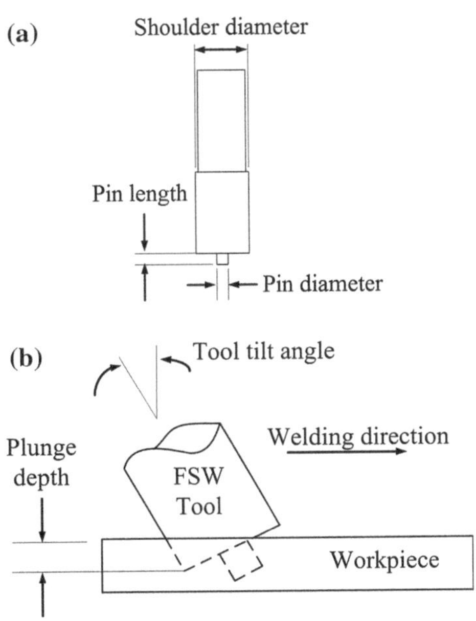

Fig. 2 **a** Typical straight cylindrical FSW tool, **b** process parameters associated with tool

Nugget zone: Frictional heating and intense plastic deformation during the FSW results in generation of a recrystallized fine-grained microstructure within the stirred zone. This zone is usually referred to as nugget zone or dynamically recrystallized zone. In this zone, recrystallization occurs as the temperature reaches a critical value above which the material crosses critical dislocation density for recrystallization. Depending on process parameters, tool geometry, temperature, and thermal conductivity of the workpiece, material shape and the nugget size can vary. In the literature, two types of nugget shapes were reported. Basin-shaped nugget that widens near the upper surface and the other is the elliptical nugget. The interface between the recrystallized nugget zone and the parent material is relatively diffused in the retreating side of the tool, but quite sharp on advancing side of the tool (James et al. 1999).

Thermo-mechanically affected zone (*TMAZ*): This zone is the transition zone between the nugget zone and the base metal. TMAZ is characterized by highly deformed structure. Although TMAZ underwent plastic deformation, recrystallization does not occur in this zone due to insufficient deformation strain. It was revealed that the grains in the TMAZ usually contain a high density of subboundaries (Sato et al. 1999).

Heat-affected zone (*HAZ*): Beyond TMAZ, there is a HAZ. This zone experiences a thermal cycle but does not undergo any plastic deformation, and the grain structure is same as the base metal.

The process of FSW is governed by too many process parameters. The choices of combination of different process parameters need to be explored through proper investigation to find the effects on the weld quality. Researchers have reported different combinations of process parameter settings over different materials for better yielding of joint quality in terms of ultimate tensile strength (UTS), yield strength (YS), percentage elongation (E), and microhardness values of the welded joints. The variation of UTS of 7075-T6 aluminum alloy-welded joints with TRS was found in the work by Rezaei et al. (2011). With the increase in the TRS, UTS and YS were found to increase up to a certain speed and then decreases as the TRS increases. Liu et al. (2013) varied the WS keeping the TRS fixed and defined a parameter named the revolutionary pitch (WS/TRS). UTS, YS, and E were reported to increase with the increase in the revolutionary pitch and then decrease after a certain value of the same. Similar reporting on the variation of weld quality in FSW process can be found in the related works by different researchers (Li and Liu 2013; Lim et al. 2004; D'Urso et al. 2009; Kamp et al. 2009; Buffa et al. 2013; Hamilton et al. 2012; Miyazawa et al. 2012). Researchers have presented the effect of various process parameters over the quality of the weld, but none has reported on the inconsistency of the weld quality over the same range of process parameters. So a complex process like FSW may not be effectively monitored using the process parameters. This leads to the exploration of different indirect methods for the monitoring of the FSW process, and mostly preference is based on the signals acquired during the process.

In indirect monitoring methods, the process is monitored through the observation of different influencing process parameters to predict their effects on the process.

Signals related to different process parameters or the overall process could be capable of yielding more relevant information for the monitoring of the process. The use of force signals for monitoring FSW process was reported by different researchers (Yang et al. 2008; Boldsaikhan et al. 2009; Fleming et al. 2008; Gibson et al. 2014; Ramulu et al. 2013). In all the cases, force dynamometers were used for the acquisition of the force signals to correlate the behavior of these signals with the process parameters. In these research works, it is observed that Fourier transformation (FT) was the signal processing technique used for extracting the signal features. Acoustic emission signals were acquired by a few researchers (Chen et al. 2003; Huang and Kovacevic 2011; Zeng et al. 2006; Yen et al. 2013) using acoustic emission sensors where the signal features were used for the monitoring of the process. Wavelet transformation (WT) was the signal processing technique used by Chen et al. (2003) to characterize the acquired acoustic emission signal during FSW process.

The question might arise at this point if acquisition of signal be sufficient enough for developing an efficient monitoring system. The answer to this is obvious not because signal as a whole cannot be used for monitoring of the system. Relevant information from the signal needs to be presented to systems which may be capable of predicting some output in an understandable way for realistic justification and to develop decision models for the process. Modeling of FSW process for the prediction of different weld characteristics was attempted by many researchers. The most widely used tool is the artificial neural network (ANN) modeling. Okuyucu et al. (2007) developed a single-hidden layer ANN for the prediction of hardness, E, YS and UTS with TRS and WS as the two input parameters. Prediction capabilities of ANN were compared with the response surface method by Lakshminarayanan and Balasubramanian (2009) for the prediction of tensile strength using three process parameters and signal features acquired from the force signals. Mechanical and microstructural properties of friction stir welded joints were predicted by Buffa et al. (2012) using ANN.

The above survey reveals that mostly force signals and/or acoustic emission signals were acquired in FSW process monitoring and ANN was the most popular method among researchers for the modeling of FSW process. Apart from these signals, other signals may also be used for effective monitoring of FSW process. The principle of FSW operation lies in the plunging of a rotating tool into the workpiece material, and once the suitable plasticization and heat requirement is achieved, the tool is traversed with a suitable speed for the welding. If the rotation as well as traversing phenomena of the tool can be captured through signals, the signals will give more direct information as compared to other signals which need some advanced signal processing techniques before implementing it to any prediction mechanism.

2 Sensors and Signals During Friction Stir Welding

The necessity of incorporating signals for the efficient monitoring of the process is realized from the above paragraphs. The need of selection of suitable sensors for acquiring signals comes into picture. The question that needs to be answered is how many sensors will be sufficient for the development of efficient monitoring system. This can be seen in the context of a number of influencing parameters that need to be monitored during the process. The more is the number of sensors, the more will be the sensor data, i.e., signals, but all the acquired signals may not be useful for monitoring the process. Signals should contain some worthy information regarding the process and then only they will be useful for further analysis. The discussion regarding signals and sensors will be restricted to those signals which will be presented herewith. Different possible signals and related sensors were outlined in the earlier section of this article. Apart from those signals, a few other signals can also be acquired during the FSW process; the features of which can be effectively correlated with the physical outcome of the process.

The FSW process is governed by mainly the rotational speed of the tool. If this parameter can be monitored, it is expected to provide more straightforward information for monitoring the overall process output. The current signal from the main spindle motor which is responsible for the rotation of the tool is one of the signals which were not considered by researchers. For the acquisition of this signal, Hall effect current transducers are mostly used as these kind of sensors are not affected by the change in the ambient conditions and produce output independent of rate of detected field, which leads to less noisy signals. The other signal that can be acquired is the current signal from the motor which is responsible for the traversing of the tool. As WS also plays an important role in obtaining a competent welding, monitoring this parameter is also expected to enhance the monitoring of the overall process. TRS can also be monitored in terms of signals through rotational speed-detecting sensors which can be an effective addition to the monitoring of the TRS along with the main spindle motor current. Voltage signal is another solution for the monitoring of both TRS and WS from the respective motors.

The data from the sensors need to be stored for further processing, and for this purpose, data acquisition system is required. The prime objective of any data acquisition system is to convert the analog input signals to digital input signals and to store them for further processing. That is why sometimes data acquisition systems are also referred as analog-to-digital converter. The selection of data acquisition system depends on the output signal type from the sensors. In signal acquisition, sampling rate or sampling frequency is a crucial parameter. Analog-to-digital converter converts an analog signal into stream of digital numbers, each representing the analog signal's amplitude, at a moment in time. Each number is called a 'sample.' The number of sample per second is called as the 'sampling rate,' measured in samples per second. This sampling rate or sampling frequency has to be selected based on the frequency of the signals to be acquired. According to Nyquist–Shannon theorem, the sampling frequency of the signals should be at least

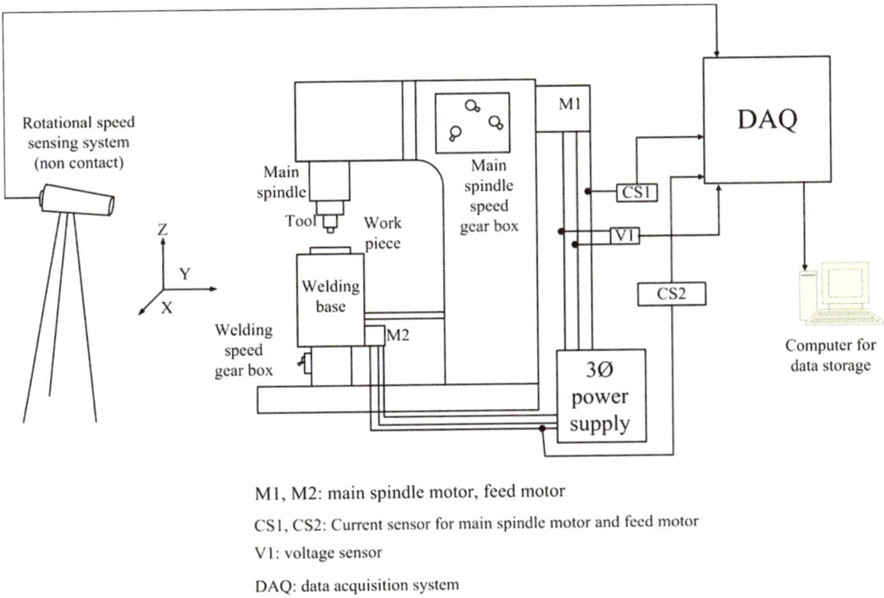

Fig. 3 Arrangement of different sensors in FSW process

two times of the frequency of the signals to be acquired (Shannon 1998). Mathematically, if a signal is having a frequency of f_s, then sampling frequency should be $F_s \geq 2 f_s$, where F_s is the sampling frequency as per the theorem to avoid aliasing error. A typical arrangement of different sensors with the FSW system is shown as line diagram in Fig. 3. Figure 4a–d shows typical main spindle motor current signal, TRS signal, feed motor current signal, and main spindle motor voltage signal acquired during FSW process under same operating condition.

3 Analysis of the Signals

In the process of FSW, there are four significant stages which make this process completely different from others. First is the plunging of the rotating tool in which initially the tool pin is plunged completely, and then, the shoulder is plunged up to a distance called as the plunge depth. Then, after a defined period called as the dwell time, the tool is traversed along the joining line to achieve the welding. Once the welding is over, the rotating tool is retracted back. All these stages carry notable information for successful and competent welding in FSW process. So, if these stages can be captured through some signals, then it would be of more importance to analyze the process as a whole for desired output as anomaly in any of these stages will surely affect the joining process and the quality of the weld will deteriorate.

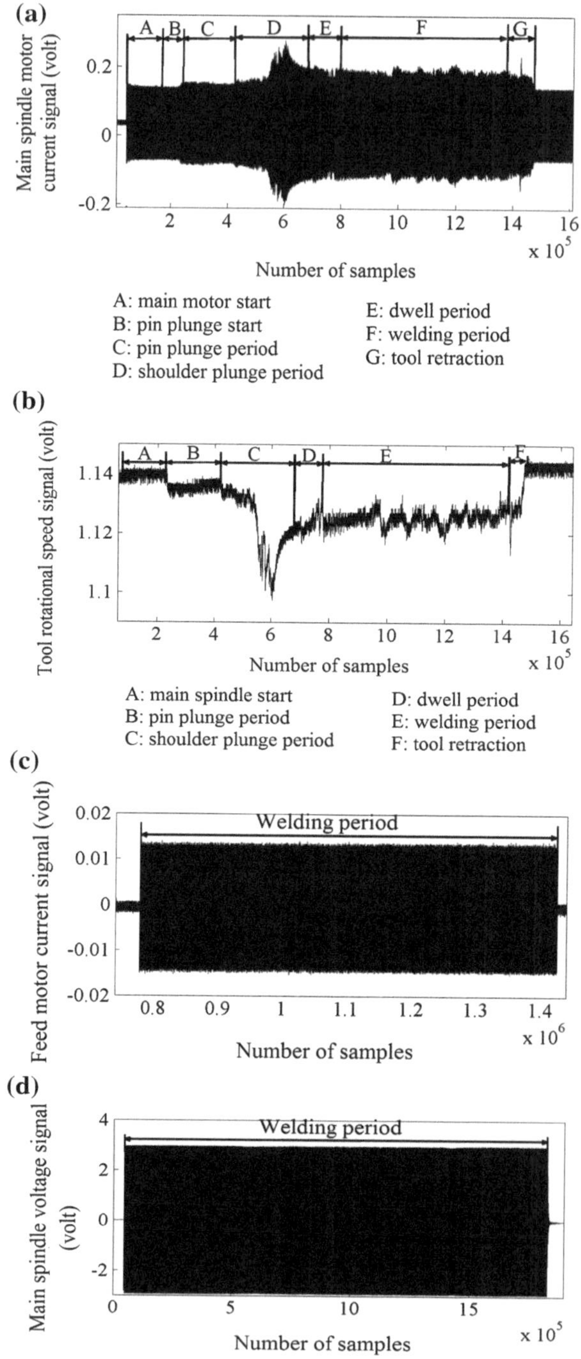

Fig. 4 Signals acquired during friction stir welding **a** main spindle motor current signal, **b** tool rotational speed signal, **c** feed motor current signal, and **d** voltage signal

As shown in Fig. 4a, main spindle motor current signal has the potential to encapsulate the significant stages during FSW process both in trend and magnitude of the signal. As the motor starts, it draws more current. This can be seen in the figure as a sudden rise on the signal amplitude. The amplitude of the signal increases further when there is a contact between the rotating tool and the rigidly clamped material which is due to the resistance offered by the workpiece material to the rotating spindle. To maintain the same rotational speed, main spindle motor draws more current resulting in increasing magnitude of the current signal. This increasing trend in the amplitude of the signal continues, and a sharp rise in the amplitude of the signal can be noticed where amplitude of the signal grows to maximum. This is because of the shoulder penetration to the material which offered more resistance as compared to the tool pin due to the size difference between the pin and the shoulder of the tool. Rotation of the tool introduces rubbing action between the shoulder material and workpiece material which aid toward the generation of frictional heat, causing the material softening, and plasticization. As the plasticization increases, resistance offered by the material decreases and this drop in resistance can be noticed as a drop in signal amplitude in main spindle motor current signal. During dwell period, the signal comes to a stable position and during welding maintains more or less the same level of amplitude as in dwell period. When the tool is retracted back after the welding, the resistance decreases gradually and this gradual change can be seen as a gradual drop in current amplitude to the level of starting current value.

The above analysis holds good for the TRS signal too where all these significant stages were effectively seen both in signal amplitude and trend of the signal. In Fig. 4b where TRS signal is shown, during the welding period, some irregular drops in the signal can be noticed in the trend as well as magnitude of the signal. More interestingly, these trends were also retained by the main spindle motor current signal as shown in Fig. 4a. These trends of the signals give a clear indication that there must be some oddity in the process during welding, which were effectively captured by these two signals. Thus, the analysis of signals would provide more useful information regarding the process and monitoring, by which a better and precise control can be gained over the process. It is surprising to observe that the main spindle motor voltage signal, even though it is acquired from the same main spindle motor, does not reflect any of the significant stages during welding. The probable reason could be that voltage signal is acquired between two phases of the motor, which is supposed to be at a fixed value as the supply voltage. It is not quite astonishing that the significant stages involved in the FSW process were not captured by the feed motor current signal as this motor is switched on only during the welding period, so any prior physical change cannot be expected here. Even though it is anticipated that the abruptions as seen in main spindle motor current signal and TRS signal during the period of welding will be enfolded by feed motor current signal, no such concluding remarks could be made by simply looking at the signal as in other two signals. The aforementioned analysis appraised the capability of different signals acquired during FSW process other than those being considered by different researchers to explore a new avenue for the monitoring of the process.

4 Monitoring of FSW Process

Monitoring in a generic way can be defined as the procedure of observation of some physical process or phenomenon and recording some valuable information through which the performance of the process is evaluated periodically or continuously and the output from it is predicted. Thus, this involves two crucial steps: one is recording of valuable information through observation, and the other one is the estimation or prediction of the output. Monitoring in case of FSW process can be implemented for the process output prediction. The output from the process can be represented in terms of the quality of the welds. Quality of the welds relates to the strength of the welds and hardness distribution of the welds as compared to the base material properties. To have better control over the outcome of the process, the monitoring should be very effective to note every possible change in the process. As mentioned in the earlier section, signals can effectively detect changes occurred in the FSW process over time which gives a clear indication that monitoring for the prediction of weld quality in FSW can be achieved with signals acquired during FSW process. Moreover, acquired signals give a more direct representation of the physical process. Since signals are sampled over time, time information of the signals can play an important role in detecting any unusual event at some point of time occurred during the process, which will undoubtedly enhance the monitoring process.

5 Feature Extraction from Signals

Signals are some sensor data collected with a specific sampling frequency and for a finite period of time and cannot fetch any useful information on their own. The information contained in signals should be presented in a comprehensive way to work for any monitoring system based on signals. Most primitive threshold values from signals can be obtained, and a database can be maintained through rigorous experimentation for the evaluation of the process. Thresholds can be either amplitude of the signals or the frequency of the signals. With this approach, difficulty arises when unknown operating conditions arrive for which database is not available to compare with. To handle such cases, different techniques need to be implemented in order to extract most salient and informative quantities referred as features from the signal.

Features from a signal can be extracted in different context as well as depending on the requirement and roughly can be categorized as follows: (i) features in the time domain of a signal which is about computing average values, the shape of enveloping signal, changes in signal levels such as increments or decrements, crest factor (ratio of peak to the root mean square value of the signal), and polynomial approximation of the signal. These quantities can reveal overall information of the signals and can be computed with lesser efforts. (ii) Frequency domain where spectral features were computed from a signal: Typical features in this category can be total power level of a signal and the power in different spectral bands. The advantage of frequency-domain approach is that it has the ability to render information that is not possible with time-domain approach such as a signal can have multiple frequency components more likely due to the presence of the noise which cannot be observed in time-domain approach. This has the disadvantage that time information of the signal is lost once converted to frequency domain which is essential for detecting the occurrence of some undesired event in the process. For the frequency-domain analysis of signals, most widely used technique is FT, where the transformation of signals from time to frequency domain is achieved using a very well-known algorithm called the fast Fourier transformation (FFT). Short-time FT was the advancement proposed to this method for the elimination of the drawback of losing time information of the signal, but it too suffers from other limitations of fixed window size, etc. (iii) Features in the statistical domain which is applicable to time-, frequency-, and time–frequency-domain approaches: This domain includes features such as root mean square, variance, skewness, kurtosis, coefficients of a time series model used to describe the measured signal (autoregressive, moving average, autoregressive moving average, etc.), and cross-correlation with a measuring cut, etc. (iv) Features in time–frequency domain of the signal are exclusively performed using WT and are more robust and flexible in feature extraction from a signal by retaining its time information with frequency content.

The difficulties faced in FFT during signal analysis were eliminated by the introduction of WT of signals. Wavelet analysis has been proved to be more flexible and robust method for signal processing of non-stationary signals due to the characteristics of the wavelet basis function as irregular, asymmetric, and of finite time length. Wavelet transform of a signal is a function of two parameters, namely time and scale, the latter being a key point of the WT. WT is a technique of extracting coefficients comparing the original signal with the wavelet basis functions. These wavelet basis functions were called as the mother wavelets. The mother wavelet is contracted or dilated by changing the scale parameters. This variation in the scale parameter not only changes the central frequency of the signals but also the window length. In WT, a signal is split into two parts: approximation which contains low-frequency, high-scale components, and details which contain high-frequency, low-scale components. The approximation part is again split into two new approximations and details, and the process is repeated. Schematic representation of WT of a signal is shown in Fig. 5 up to third level of decomposition. In WT, as the approximation part is only split for further approximation and details. This does not guaranty that it will lead to effective features extraction from the signals as details part might also contain some valuable information. Another form of WT is the wavelet packet transformation. In wavelet packet transformation, both the approximation and details part are split into further approximation and details, which give a more rich way of analyzing a signal. This

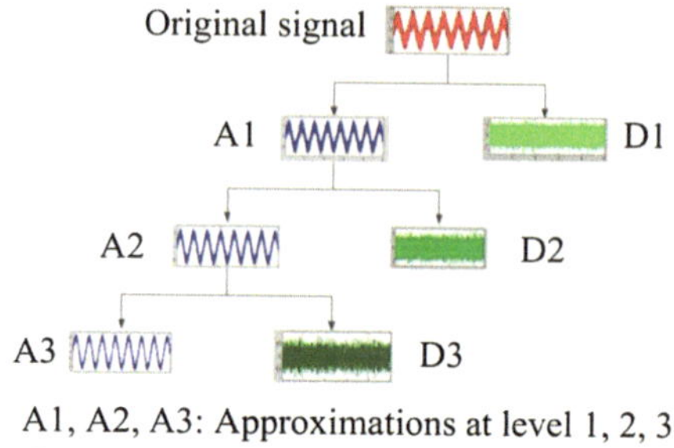

Fig. 5 Third-level wavelet transformation of a sine wave corrupted with white noise

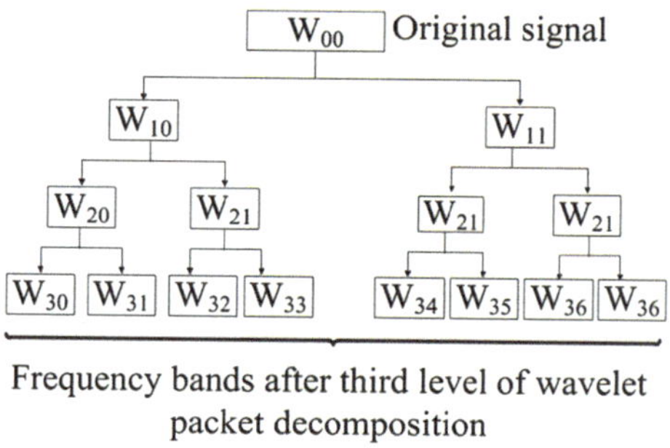

Fig. 6 Schematic of wavelet packet tree at third level of decomposition

generates a tree composed of different nodes at the end of the final level of decomposition where each node represents a particular frequency band of the signals. Schematically, a three-level wavelet packet tree is shown in Fig. 6.

6 Correlation of Process Parameters with Signal Features

Signal features to be used for monitoring of the process should have some correlation with the process parameters, as an indication that any change in the physical behavior of the process is reflected through the extracted features. Figure 7a–d shows the variation of four statistical domain features, namely root mean square, variance, skewness, and kurtosis of the TRS signal, and the TRS and variation with WS are shown in Fig. 8a–d. The variation of signal features with the TRS shows an increasing trend. However, similar conclusive remarks cannot be drawn for the variation of the statistical features with WS. The reason behind this could be that these features were extracted from TRS signal, where the effect of WS is obviously less as compared to TRS. The variation as depicted by the figures intimates a vivid indication of definite correlation between signal features and process parameters. This information conveyed through the above analysis reveal that signal features can be effectively incorporated in monitoring of FSW process.

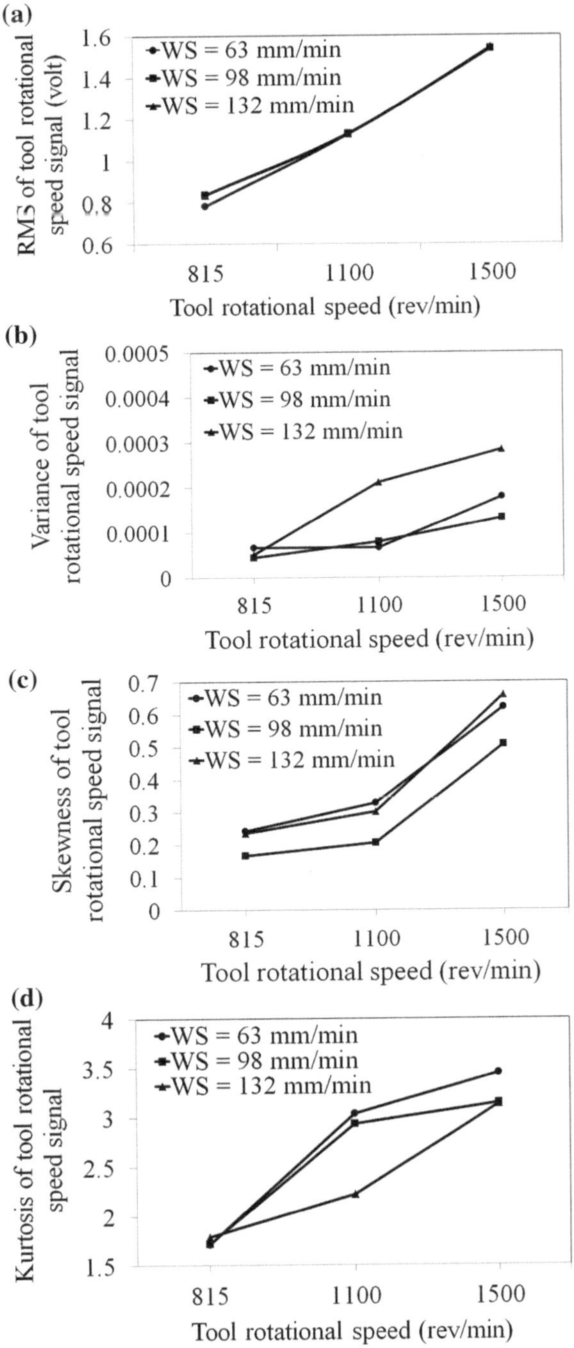

Fig. 7 Variation of **a** RMS, **b** variance, **c** skewness, and **d** kurtosis with tool rotational speed

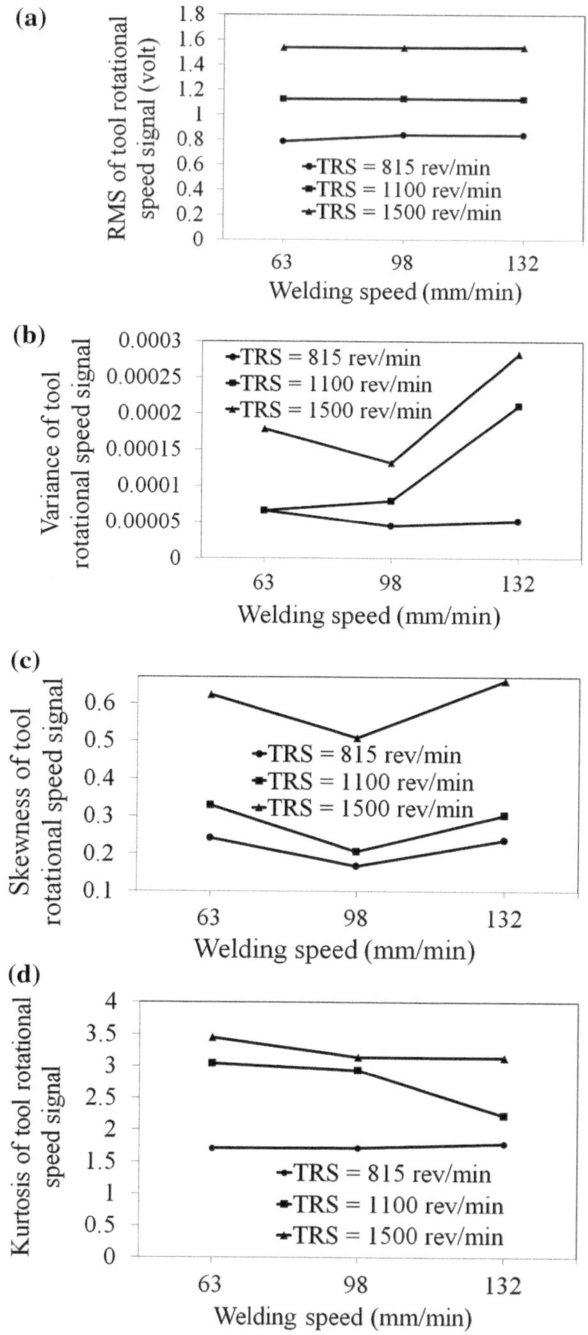

Fig. 8 Variation of **a** RMS, **b** variance, **c** skewness, and **d** kurtosis with welding speed

7 Correlation of Process Output with Signal Features

For any monitoring system, the goal is to observe the performance of the system to achieve desired output through better control over the process. For FSW process, the output from the process is the quality of the welds in terms of tensile strengths and hardness distribution as compared to the base material properties. It is inevitable that process parameters will surely have influence over these outputs from the process. But the question is whether the extracted signal features will have definite correlation with these outputs or not. Until and unless signal features show some correlation with the process output, it would be difficult for any modeling system to predict the process output using the signal features. Figure 9a–d shows the variation of UTS with the four statistical domain features already mentioned in the earlier section. UTS of the welds shows an increasing trend up to a certain value of the signal features of the TRS signal, but after a certain value, it decreases dramatically. This could be explained as with the increase in the rotational speed of the tool, heat input to the material increases and the material undergoes more plasticization. Thus, even at a low WS, material extrusion increases and the welds could not reach to a perfect condition required for joining the workpiece material, which leads to deterioration of the weld quality. Thus, presence of correlation between the signal feature and process output from FSW process is proved and can be implemented in modeling of the process.

8 Modeling of Weld Quality

Different approaches have been tested and effectively implemented by researchers for the efficient modeling of the FSW process for the prediction of different weld quality. Regression analysis (Jayaraman et al. 2009; Mohanty et al. 2012; Rajakumar et al. 2011a), response surface methodology (Lakshminarayanan and Balasubramanian 2009; Elatharasan and Kumar 2012; Palanivel and Mathews 2012; Rajakumar et al. 2011b), and ANN (Ocuyucu et al. 2007; Boldsaikhan et al. 2011; Ghetiya and Patel 2014; Chiteka 2014; Fahd 2014; Shojaeefard et al. 2013; Tansel et al. 2010; Rezgui et al. 2013) are different tools used for the modeling of weld strength. Among these, ANN was mostly used because of its robustness and flexibility in modeling complex processes where precise mathematical formulation is not available as FSW process.

ANN is data-driven modeling technique, which is more robust in mapping any input–output relationship without much simplification and precise mathematical formulation. It can be considered as nonlinear statistical data modeling tool. Various types of neural network are in use such as feed-forward neural network, recurrent neural network, stochastic neural network, and modular neural network. The most widely used neural network is multilayer feed-forward neural network. In Fig. 10, typical multilayer feed-forward neural network architecture is shown. The

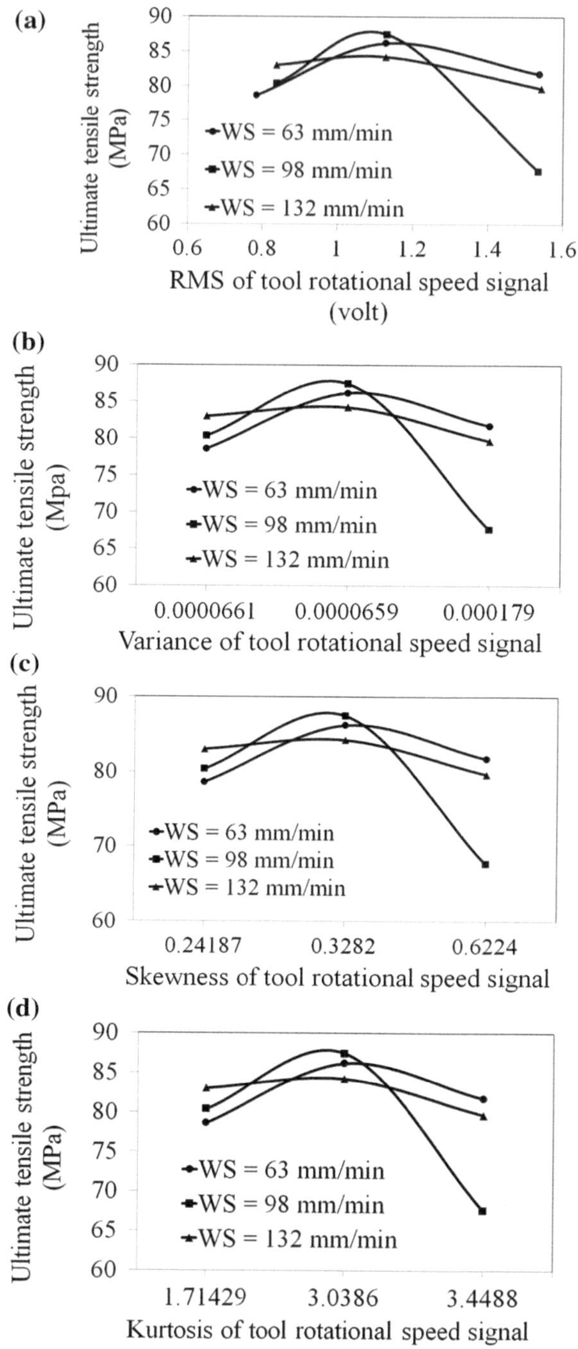

Fig. 9 Variation of ultimate tensile strength with **a** RMS, **b** variance, **c** skewness, and **d** kurtosis

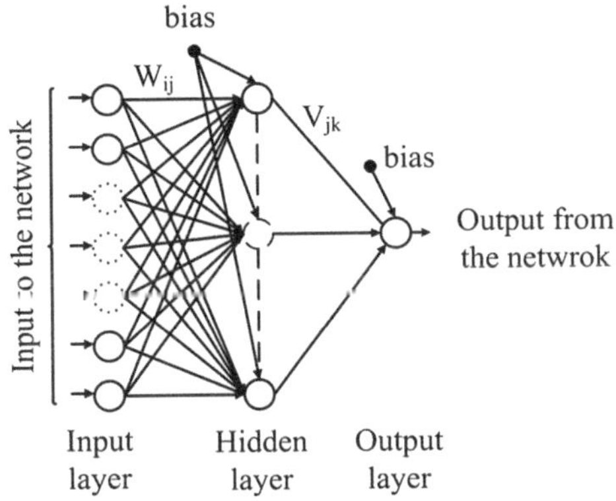

Fig. 10 Schematic representation of multilayer feed-forward neural network with single output and single hidden layer

input and output layer neurons are connected to hidden layer neurons through some weights. The universal approximation theorem of neural networks states that a standard multilayer feed-forward neural network with a single hidden layer containing a finite number of neurons with arbitrary activation function is a universal approximator (Hornik 1991). Each working unit of a neural network is called as the neuron, and it is activated through some functions called as the activation function. This function maps the data from one space to other by some nonlinear mapping. Linear, log–sigmoid, and tan–sigmoid are some activation functions; among them, log–sigmoid is the most popular activation function. Neural network helps in predicting output where input space data are not well correlated and cannot be mapped linearly and where some precise mathematical relationship does not exist between the input and output variables.

A network can be trained using supervised as well as unsupervised learning algorithm. For training of neural network, incremental mode of training and batch mode of training are the two methods for presenting the data to the network during training. For a large data set, batch mode is preferred over incremental mode of training. One of the notable limitations with neural network relies in its working principle. It works on minimizing the error function generated during the training of the network. Since for the minimization, gradient descent method is generally implemented, the chances of solution to get trapped in local minima are high. This can be reduced by selecting multiple starting solutions and adding momentum term. For the training of the network, one most important and essential parameter is learning rate which the user has to select. The more is the value of learning rate, the faster is the convergence, but the network may become unstable, and the lower is the value of learning rate, the slower is the convergence, but the network parameters might get overtrained. So a trade-off has to be maintained while selecting this parameter. Details of ANN can be found in an illustrative manner elsewhere (Haykin 1999). To develop an ANN model, the data set needs to be divided into training, validation, and testing before presenting it to the network. The training data set is used to train the network during which weights' values were updated through any of the methods as gradient descent, conjugate gradient descent,

Lavenberg–Marquardt back-propagation, or quasi-Newton method by minimizing the mean square error. Testing data set is presented to the network once the network is fully trained. Validation data set is used for avoiding the overfitting as neural networks are prone to overfit (Srivastava et al. 2014). The statistical features extracted from the signals acquired during FSW process show prominent correlation with the process output, so inclusion of these features will surely lead to more accurate prediction of the weld quality.

9 Case Study

To represent the usefulness of different signal features in the modeling of UTS of the joints, four statistical features computed from main spindle motor current signal were used and regression models were developed. A total of 64 data sets containing tool rotational speed (*R*), welding speed (*W*), and shoulder diameter (*S*) as process parameters and RMS, variance (*V*), kurtosis (*K*), and skewness (Sk) as the computed features were used to develop the regression models for modeling the UTS. Equation 1 involves all the process parameters and the computed signal features. Here, the goodness of fit obtained is 53.26. Equation 2 represents the UTS model considering only the process parameters which yield goodness of fit as 24.59, and Eq. 3 represents the UTS model considering the signal features only, which yield goodness of fit as 16.49. Though the goodness of fit by including process parameters and signal features together was less, this gives a clear indication that inclusion of the signal features in the regression model leads to the improvement of prediction accuracy of the developed models.

$$\begin{aligned}\mathrm{UTS(MPa)} = {} & -1301.43 - 0.09 \times R - 3.68 \times W + 14.37 \times S + 9138.77 \\ & \times \mathrm{RMS} - 28{,}808.3 \times V + 1248.63 \times K - 34{,}358.8 \times \mathrm{Sk} \\ & - 0.00016 \times R \times W + 0.0027 \times R \times S - 2.260 \times R \times \mathrm{RMS} \\ & + 10.16 \times R \times V + 0.10 \times R \times K - 2.68 \times R \times \mathrm{Sk} + 0.0027 \\ & \times W \times S + 50.85 \times W \times \mathrm{RMS} - 238.59 \times W \times V + 0.47 \\ & \times W \times K + 25.40 \times W \times \mathrm{Sk} + 319.56 \times S \times \mathrm{RMS} - 2465 \\ & \times S \times V - 20.75 \times S \times K - 129.19 \times S \times \mathrm{Sk} + 288{,}955 \\ & \times \mathrm{RMS} \times V - 12{,}228.3 \times \mathrm{RMS} \times K + 1.05594e + 006 \\ & \times \mathrm{RMS} \times \mathrm{Sk} + 47{,}544.2 \times V \times K - 6.04652e + 006 \times V \\ & \times \mathrm{Sk} - 10{,}247.1 \times K \times \mathrm{Sk}\end{aligned} \tag{1}$$

$$\begin{aligned}\mathrm{UTS(MPa)} = {} & 131.8 - 0.02 \times R - 0.07 \times W - 2.45 \times S - 6.84264e - 005 \\ & \times R \times W + 0.00105 \times R \times S + 0.0065 \times W \times S\end{aligned} \tag{2}$$

$$\begin{aligned}\text{UTS(MPa)} = {} & 731.50 - 8996.79 \times \text{RMS} + 32{,}004.3 \times V - 405.09 \times K \\ & - 5784.49 \times Sk - 82{,}987.8 \times \text{RMS} \times V + 5266.47 \\ & \times \text{RMS} \times K + 144{,}327 \times \text{RMS} \times \text{Sk} - 0795.5 \times V \\ & \times K - 1.40944e + 006 \times V \times \text{Sk} + 1235.54 \times K \times \text{Sk}\end{aligned} \tag{3}$$

10 Conclusion

Possibilities of monitoring of FSW process with the acquired signals during welding were presented in this research work. The capability of main spindle motor current signal and TRS signal to capture significant stages involved in FSW process was demonstrated, which gives a clear notion that signals can be considered for effective monitoring of the system to achieve desired output through better control over the process. The correlation of four time-domain features extracted from TRS signal with the UTS of the welds reflects a vivid intimation that inclusion of signal features to modeling tools can yield more satisfactory results as compared to modeling only with the process parameters. At the end, ANN model is presented that can be implemented as strength prediction mechanism which will lead to more accurate prediction that will lead toward the development of tools for efficient prediction of real-time weld quality, which is always desired in any manufacturing process.

References

Boldsaikhan E, Corwin E, Arbegast W (2009) Detecting wormholes in friction stir welds from welding feedback data. In: Midwest instruction and computing symposium 2009, South Dakota, USA

Boldsaikhan E, Corwin EM, Logar AM, Arbegast WJ (2011) The use of neural network and discrete Fourier transform for real-time evaluation of friction stir welding. Appl Soft Comput 11:4839–4846

Buffa G, Fratini L, Micari F (2012) Mechanical and microstructural properties prediction by artificial neural networks in FSW processes of dual phase titanium alloys. J Manuf Process 14:289–296

Buffa G, Campanella D, Fratini L (2013) On tool stirring action in friction stir welding of work hardenable aluminium alloys. Sci Technol Weld Joining 18(2):161–168

Chen C, Kovacevic R, Jandgric D (2003) Wavelet transform analysis of acoustic emission in monitoring friction stir welding of 6061 aluminum. Int J Mach Tools Manuf 43:1383–1390

Chiteka K (2014) Artificial neural networks in tensile strength and input parameter prediction in friction stir welding. In J Mech Eng Rob Res 3(1):145–150

D'Urso G, Ceretti E, Giardini C, Maccarini G (2009) The effect of process parameters and tool geometry on mechanical properties of friction stir welded aluminum butt joints. Int J Mater Form 2(1):303–306

Elatharasan G, Kumar VS (2012) Modelling and optimization of friction stir welding parameters for dissimilar aluminum alloys using RSM. Procedia Eng 38:3477–3481

Fahd SM (2014) Artificial neural network model for friction stir processing. Int J Eng Res 3 (6):396–397

Fleming P, Lammlein D, Wilkes D, Fleming K, Bloodworth T, Cook G, Strauss A, DeLapp D, Leinert T, Bement M, Prater T (2008) In-process gap detection in friction stir welding. Sens Rev 28(1):61–67

Ghetiya ND, Patel KM (2014) Prediction of tensile strength in friction stir welded aluminium alloy using artificial neural network. Procedia Technol 14:274–281

Gibson BT, Lammlein DH, Prater TJ, Longhurst WR, Cox CD, Ballun MC, Dharmaraj KJ, Cook GE, Strauss AM (2014) Friction stir welding: process, automation, and control. J Manuf Process 16:56–73

Hamilton C, Dymek S, Senkov O (2012) Characterisation of friction stir welded 7042-T6 extrusions through differential scanning calorimetry. Sci Technol Weld Joining 17(1):42–48

Haykin S (1999) Neural networks: a comprehensive foundation, 2nd edn. Prentice Hall International Inc, New Jersey

Hornik K (1991) Approximation capabilities of multilayer feedforward networks. Neural Netw 4:251–257

Huang W, Kovacevic R (2011) A neural network and multiple regression method for the characterization of the depth of weld penetration in laser welding based on acoustic signatures. J Intell Manuf 22:131–143

James M, Mahoney M, Waldron D (1999) Residual stress measurements in friction stir welded aluminum alloys. In: Proceedings of 1st international symposium on friction stir welding, 1999. Rockwell Science Center, Thousand Oaks, CA, USA, TWI, June 14–16

Jayaraman M, Sivasubramanian R, Balasubramanian V, Lakshminarayanan AK (2009) Mechanical and microstructural properties prediction by artificial neural networks in FSW processes of dual phase titanium alloys. J Sci Ind Res 68:36–43

Kamp N, Reynolds AP, Robson JD (2009) Modelling of 7050 aluminium alloy friction stir welding. Sci Technol Weld Joining 14(7):589–596

Lakshminarayanan AK, Balasubramanian V (2009) Comparison of RSM with ANN in predicting tensile strength of friction stir welded AA7039 aluminium alloy joints. Trans Nonferrous Met Soc China 19:9–18

Li JQ, Liu HJ (2013) Effects of tool rotation speed on microstructures and mechanical properties of AA2219-T6 welded by the external non-rotational shoulder assisted friction stir welding. Mater Design 43:299–306

Lim S, Kim S, Lee CG, Kim S (2004) Tensile behavior of friction-stir-welded al 6061-T651. Metall Mater Trans A 35A:2829–2835

Liu H, Maeda M, Fujii H, Nogi K (2013) Tensile properties and fracture locations of friction-stir welded joints of 1050-H24 aluminum alloy. J Mater Sci Lett 22:41–43

Mishra RS, Ma ZY (2005) Friction stir welding and processing. Mater Sci Eng R 50:1–78

Miyazawa T, Iwamoto Y, Maruko T, Fujii H (2012) Friction stir welding of 304 stainless steel using Ir based alloy tool. Sci Technol Weld Joining 17(3):207–212

Mohanty HK, Venkateswarlu D, Mahapatra MM, Kumar P, Mandal NR (2012) Modeling the effects of tool probe geometries and process parameters on friction stirred aluminium welds. J Mech Eng Autom 2(4):74–79

Ocuyucu H, Kurt A, Arcaklioglu E (2007) Artificial neural network application to the friction stir welding of aluminum plates. Mater Des 28:78–84

Palanivel R, Mathews PK (2012) Prediction and optimization of process parameter of friction stir welded AA5083-H111 aluminum alloy using response surface methodology. J Cent S Univ 19:1–8

Rajakumar S, Muralidharan C, Balasubramanian V (2011a) Optimisation and sensitivity analysis of friction stir welding process and tool parameters for joining AA1100 aluminium alloy. Int J Microstruct Mater Prop 6(2):132–156

Rajakumar S, Muralidharan C, Balasubramanian V (2011b) Predicting tensile strength, hardness and corrosion rate of friction stir welded AA6061-T6 aluminium alloy joints. Mater Des 32:2878–2890

Ramulu PJ, Narayanan RG, Kailas SV, Reddy J (2013) Internal defect and process parameter analysis during friction stir welding of Al 6061 sheets. Int J Adv Manuf Technol 65:1515–1528

Rezaei H, Mirbeik MH, Bisadi H (2011) Effect of rotational speeds on microstructure and mechanical properties of friction stir-welded 7075-T6 aluminium alloy. In: Proceedings of the institution of mechanical engineers, part C: journal of mechanical engineering science, pp 1761–1773

Rezgui M, Trabelsi A, Bouzaiene H, Ayadi M (2013) Predictive models for the ultimate tensile and yield stresses occurring in joints of untreated friction stir welded 2017AA (ENAW-AlCu4MgSi) plates. Open J Met 3:7–18

Sato YS, Kokawa H, Enomoto M, Jogan S (1999) Microstructural evolution of 6063 aluminum during friction stir welding. Metall Mater Tran A 30A:2429–2437

Shannon CE (1998) Communication in the presence of noise. Proc IEEE 86(2):447–457

Shojaeefard MH, Akbari M, Tahani M, Farhani F (2013) Sensitivity analysis of the artificial neural network outputs in friction stir lap joining of aluminum to brass. Adv Mater Sci Eng 2013:1–7

Srivastava N, Hinton G, Krizhevsky A, Sutskever I, Salakhutdinov R (2014) Dropout: a simple way to prevent neural networks from overfitting. J Mach Learn Res 15:1929–1958

Tansel IN, Demetgul M, Okuyucu H, Yapici A (2010) Optimizations of friction stir welding of aluminum alloy by using genetically optimized neural network. Int J Adv Manuf Technol 48:95–101

Threadgill PL (2007) Terminology in friction stir welding. Sci Technol Weld Joining 12:357–360

Yang Y, Kalya P, Landers RG, Krishnamurthy K (2008) Automatic gap detection in friction stir butt welding operations. Int J Mach Tools Manuf 48:1161–1169

Yen CL, Lu MC, Chen JL (2013) Applying the self-organization feature map (SOM) algorithm to AE-based tool wear monitoring in micro-cutting. Mech Syst Signal Process 34:353–366

Zeng WM, Wu HL, Zhang J (2006) Effects of tool wear on microstructure, mechanical properties and acoustic emission of friction stir welding of 6061 aluminium alloy. ACTA Metall Sin (English Letters) 19:9–19

Coupled Electromagnetic–Structural Simulation of Magnetic Pulse Welding

Angshuman Kapil and Abhay Sharma

Abstract The present study deals with the coupled magnetic–structural analysis of magnetic pulse welding (MPW) process applied on a tubular workpiece. The study investigates the various criteria required for a successful weld between the mating members through a finite element model. The transient electromagnetic field phenomenon is coupled sequentially with mechanical phenomenon. The coupling between the magnetic field and the electrical circuit is formulated in the electromagnetic part of the model, whereas in the structural part, the impact velocities, the effective plastic strain and the shear stress induced in the workpiece are found from the numerical simulations. A viscoplastic material model with rate-dependent material properties is considered in the structural part. The effect of varying process parameters: input voltage and air gap between the two mating members on weld quality are computed through numerical simulations. Based on the results of the numerical simulations, an optimal weldability window is suggested.

Keywords Magnetic pulse welding · Coupled magnetic–structural analysis · Finite element model · Weldability window

1 Introduction

Recent studies and advances in the welding fraternity have been made in order to develop a newer process which enables us to achieve better quality and productivity and is environment friendly. Magnetic pulse welding (MPW) is a step forward into the future of welding. In the recent years, the welding of dissimilar and lightweight

A. Kapil · A. Sharma (✉)
Department of Mechanical and Aerospace Engineering, Indian Institute of Technology Hyderabad, ODF Campus, Yeddumailaram 502205, Telangana, India
e-mail: abhay@iith.ac.in

A. Kapil
e-mail: me13m0001@iith.ac.in

R.G. Narayanan and U.S. Dixit (eds.), *Advances in Material Forming and Joining*, Topics in Mining, Metallurgy and Materials Engineering,
DOI 10.1007/978-81-322-2355-9_13

materials has been a topic of utmost interest for the researchers and industries. MPW is a potential candidate for joining of dissimilar materials (Raoelison et al. 2013). The process is quite effective and reliable in terms of cost and performance, respectively, having analogy to explosive welding (EXW) (Zhang et al. 2011). The entire process is contactless wherein a magnetic field helps generate the required magnetic pressure to drive the flyer tube/plate towards the base tube/plate by change of energy in electrical form to its mechanical form. The flyer workpiece driven with high velocities by the magnetic forces impacts against the base workpiece, and a weld is created as shown in Fig. 1. The feasibility of the welding depends upon the velocity of the flyer (Kore et al. 2010), the direction and magnitude of the shear stress in the weld zone and severe plastic deformation occurring on the surface of the mating members (Mousavi and Al-Hassani 2008). The weld length in MPW process is usually less than 1 cm, time needed for the formation of the bond is several microseconds, and the collision speed is of the order of several

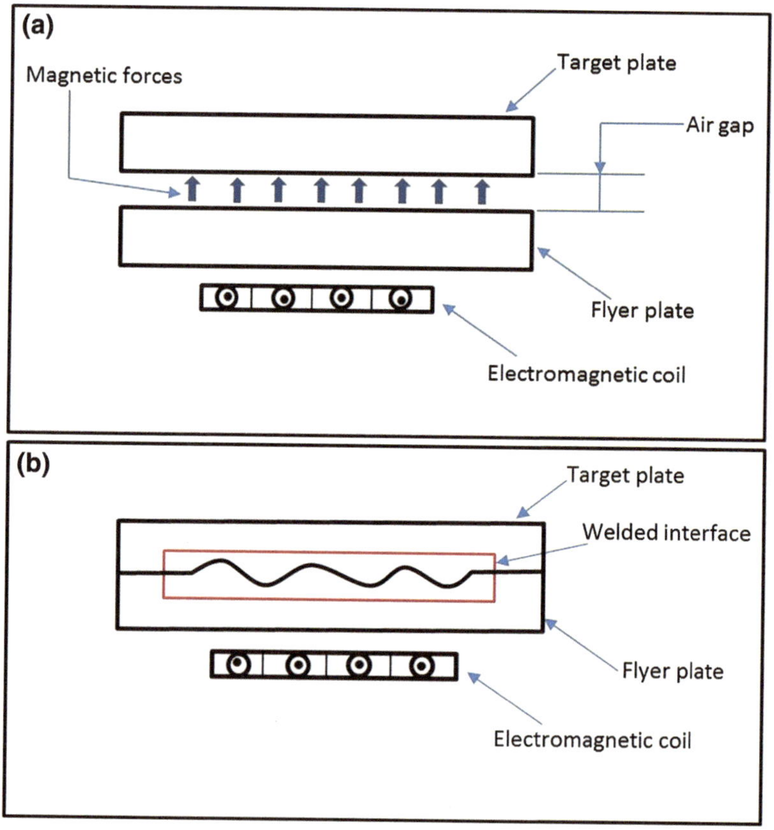

Fig. 1 Schematic illustration of MPW process: **a** initial MPW set-up for flat plates, **b** deformed geometry after the application of magnetic forces

hundreds of metres per second (Stern and Aizenshtein 2002). The temperature at the interface is always less than the melting point of the mating members.

There lies quite a few challenges for the modern day manufacturers; one of them being able to produce joints of high strength economically and without violating the strict environmental rules and regulations (Shribman 2007). One of the major problems faced by the manufacturers is the difficulty in joining dissimilar materials by the use of conventional joining processes. Dissimilarities in material properties such as temperature of melting, specific heat and coefficient of thermal expansion all prove to be a hindrance while attempting to join dissimilar materials. MPW is a substitute to conventional welding processes as the achievable weld joint strength is within the range of the strength of the weakest joining member (Weddeling et al. 2011).

With rapid advances in technology and demand for introduction of lightweight material concepts, the practice of using a single material does not seem feasible in many cases. Hence, a combination of different materials delivers the preferred technological characteristics required for the industries (Tomas 2010). Even though the process has been identified long back, there is a big scope for additional growth and application.

Successful application of MPW for tubular geometries dates back to 1969 (Zhang et al. 2011). Development of the process has made welding of tubular geometry highly successful and less cumbersome. In order to optimise the essential process parameters such as frequency, voltage, discharge energy and equivalent inductance, an electromagnetic field analysis is required. A finite element model (FEM) can allow the user to select optimum values of the parameters and help design the coil with suitable dimensions. The finite element modelling of MPW process requires coupling between electromagnetic and structural models. Very few works have been reported in the literature relating numerical modelling of tubes. Haiping et al. (2014) studied the influence of field shaper on quality of the weld as well as other process parameters such as air gap and input voltage. Zhidan et al. (2013) numerically computed the impact velocity during MPW of Al–Fe tubular geometry. Shim et al. (2010) investigated welding features taking into account the distributions of magnetic forces on the weldment using FEM to identify the ideal process parameters.

The present study deals with the numerical modelling of the MPW process applied to the tubes. The electromagnetic and structural models were coupled using FEM software ‘COMSOL Multiphysics’. The feasibility of welding was investigated by studying three factors, viz. impact velocity, effective plastic strain and shear stress acting in the welded zone/impact zone based on published literature (Kore et al. 2010; Mousavi and Al-Hassani 2008). The impact velocity, effective plastic strain and shear stress along the weld zone were all determined from the numerical simulations. Based on the available literature, analytical calculations were done to determine the minimum impact velocity required for a successful weld to occur. The published literature also suggests minimum values of effective plastic strain and shear stress along the weld zone to determine the weldability of the joint. The feasibility of MPW of tubes was then ascertained by comparing the analytical values of impact velocities, effective plastic strain and shear stress with the numerically computed values. The foregoing three criteria were calculated at various levels of process parameters, viz. input voltage and air gap between the mating members. Based on the varying process parameters and subsequent values of impact velocity, effective plastic strain and shear stress along the weld zone crossing a threshold value required for successful welding, a suitable weldability window for producing high weld length was reached upon.

2 Magnetic Pulse Welding: Process Components and Operating Principle

The MPW set-up consists mainly of four main units (Shribman 2007) with each unit having one or more than one component. Table 1 lists all the units and components of a general MPW set-up.

MPW is based on the principles given by Ampere which state that the force prevailing between two long parallel conductors of infinite length, separated from each other by distance 'd' and carrying currents I_1 and I_2 may be written as follows (Kore et al. 2007):

$$F = \frac{\mu_0}{2\pi d} I_1 I_2 \quad (1)$$

This force between the coil and the workpiece is repulsive in nature when the currents I_1 and I_2 move in directions opposite to each other and vice versa.

MPW is a very fast process with usual time taken for a cycle to complete being in microseconds. The process begins by charging the capacitors to the desired energy level. The workpieces are then positioned in the electromagnetic coil, and the trigger system operates the gap switch. Due to the closing of the switch, current

Table 1 Units and components of the MPW set-up (Shribman 2007)

Units	Components
Pulse generator	• Capacitor bank (inductance–capacitance circuit)
Control cabinet	• Power supply (DC power supply)
	• Spark gap (high voltage switch)
Workstation	• Work coils (actuators)
	• Workpieces
	- Flyer plate/tube
	- Target plate/tube
	• Field shaper (optional)
	• Electrical cables
Operational unit	• PLC

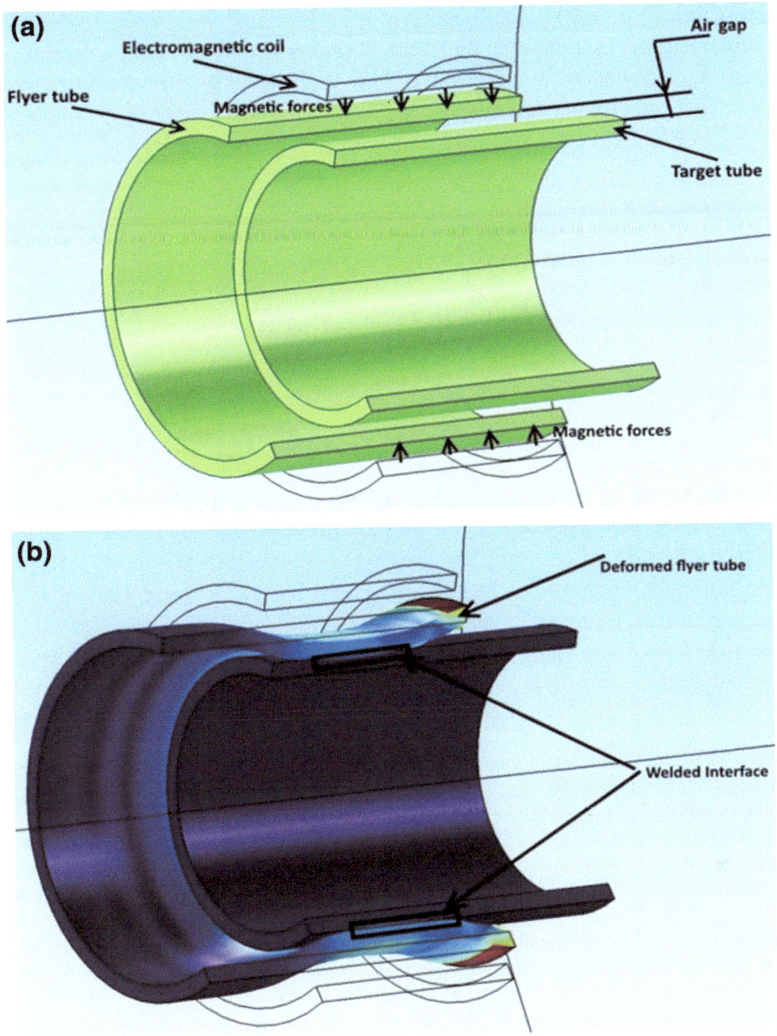

Fig. 2 MPW set-up for two tubes: **a** before the application of pulsed current, **b** after application of pulsed current

starts flowing rapidly through the coil, causing the magnetic flux to grow quickly from the coil winding and outward. This generates a magnetic field around the electromagnetic coil, and due to the secondary currents, i.e. eddy currents on the workpiece surface, a second magnetic field of opposite nature is generated as shown in Fig. 2.

The generated magnetic fields of opposing nature then interact with one other leading to the creation of a magnetic force field in between the coil and the workpiece. These magnetic forces, i.e. Lorentz forces, apply a magnetic pressure and cause the flyer workpiece to collide with the base workpiece under very high velocities ranging in between 250 and 500 m/s (Aizawa et al. 2007; Epechurin 1974, 1978; Chudakov et al. 1980). The high velocity impact leads to plastic deformation of the mating members, and in certain circumstances, a solid-state weld is generated between the same.

3 Modelling for Analysis

3.1 Materials, Geometry and Process Parameters

In order to examine the process of MPW, structural steel ASTM A36 tubes were simulated as a 2D axisymmetric problem in the FEM. Table 2 shows the chemical composition of the material used in the model.

The investigated welding tests were typically composed of a hollow flyer tube and a hollow cylindrical base tube. The outer and inner diameters of the flyer tube were adjusted to calibrate the required air gap and are denoted as D2 and D1, respectively. The air gap between the tubes has been denoted by G. The arrangement of the flyer and the base tubes along with specific dimensions is shown in Fig. 3.

The outer and inner diameters of the base tube were kept fixed at 19 and 15 mm, respectively, for each of the test cases. The simulations were carried out at varying

Table 2 Chemical composition of structural steel ASTM A36 (ASTM Standard A36/A36M 2012)

Element	Composition (%)
Carbon, C	0.25–0.290
Copper, Cu	0.20
Iron, Fe	98.0
Manganese, Mn	1.03
Phosphorous, P	0.040
Silicon, Si	0.280
Sulphur, S	0.050

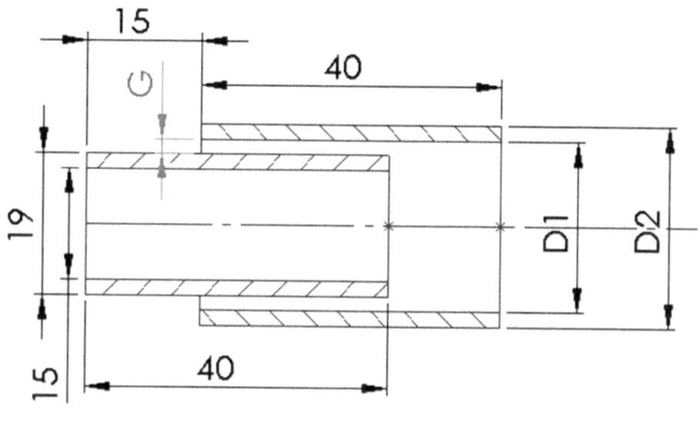

Fig. 3 Configuration of flyer and base tubes for different air gaps

Table 3 Material properties and dimensions

Properties of flyer tube and target tube	Density (kg/m^3)		7850
	Ultimate tensile strength (MPa)		400–550
	Tensile yield strength (MPa)		250
	Modulus of elasticity (GPa)		200
	Bulk modulus (GPa)		140
	Shear modulus (GPa)		79.3
	Poisson's ratio		0.26
	Speed of sound (m/s)		4512
	Threshold value of impact velocity (m/s) (see Eq. 6)		297.7
	Threshold value of plastic strain Mousavi and Al-Hassani (2008)		0.35
	Threshold value of shear stress (GPa) Mousavi et al. (2005)		0.5
Properties of electromagnetic coil—copper	Relative permeability		1
	Resistivity (Ω m)		3.4×10^{-8}
	Inductance (H)		10^{-7}
Dimensions—flyer tube	Air gap (mm)	Outer diameter (mm)	Inner diameter (mm)
	0.5	24	20
	1	25	21
	1.5	26	22
	2	27	23
	2.5	28	24

process parameters. The process parameters which were varied for performing the simulations were air gap and input voltage. The input voltage was varied in steps of 0.5 kV from 6 to 9.5 kV and the air gap being varied from 0.5 to 2.5 mm in steps of 0.5 mm. The simulations were performed with a multi-turn copper coil. Properties of the copper coil are listed in Table 3. Table 3 also lists the material properties and dimensions of the flyer tube.

3.2 FEA Model of the MPW Process

The flowchart for a sequentially coupled Electromagnetic–Structural analysis is illustrated in Fig. 4. The physical environments, viz. electromagnetic and structural (electromagnetic coil, and flyer and base tubes, respectively) were established at first. The electromagnetic environment was then solved, which calculated the transient magnetic forces, i.e. Lorentz forces. These forces were fed as input load in the structural module to calculate the flyer tube's deformation at subsequent time steps. Based on the updated geometry of the tube, the time-dependent magnetic

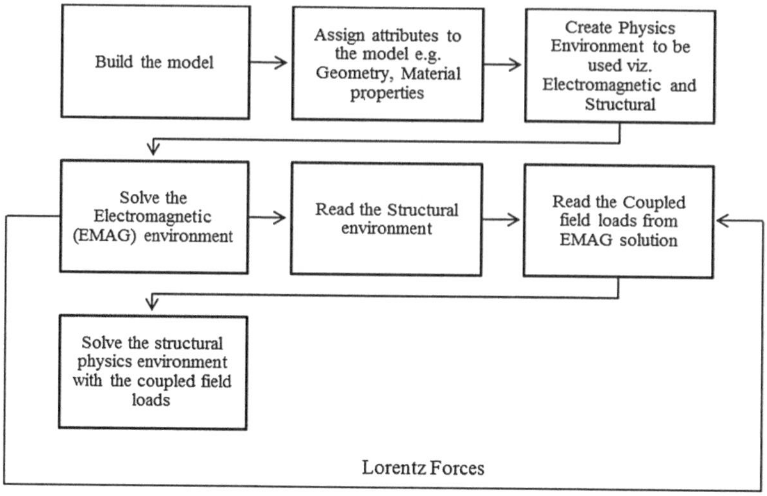

Fig. 4 Simulation flow chart for a sequentially coupled electromagnetic—structural analysis

forces were found out at subsequent time steps. The electromagnetic module consisted of the flyer and target tubes, the electromagnetic coil and a surrounding air region, whereas the structural module was related to the flyer, the target tubes and the coil. The structural module took into consideration the inertial effects because of time-dependent stress.

3.2.1 Governing Equations

In the tube region, Eq. (2) stated below is reached upon by substituting the constitutive equations into the equations given by Maxwell (Haiping and Chunfeng 2009):

$$\nabla \times \left(\frac{1}{\mu} \nabla \times \vec{A} \right) = -\Upsilon \frac{\partial \vec{A}}{\partial t} \tag{2}$$

where μ represents the permeability of the medium (H/m), Υ represents the conductivity of the medium (S/m), $-\Upsilon(\partial \vec{A}/\partial t)$ represents the current density (A/m^2) and $\vec{A}$ represents the magnetic vector potential.

The magnetic force $\vec{f}$ in unit volume of medium, i.e. the magnetic force density, is given by Maxwell's equation as follows (Haiping and Chunfeng 2009):

$$\vec{f} = \vec{J} \times \vec{B} = \frac{1}{\mu} (\nabla \times \vec{B}) \times \vec{B} \tag{3}$$

where $\vec{J}$ represents the coil current density (A/m^2) and $\vec{B}$ represents magnetic flux density (T).

The forces applied on the tube due to the generated magnetic fields can thus be calculated by substituting $\nabla \times \vec{A} = \vec{B}$ and Eq. (2) into Eq. (3) and the input body load in the structural module.

The load in the electromagnetic module is the current which passes through the electromagnetic coil, and this current is approximately expressed as follows (Haiping and Chunfeng 2009):

$$I = U\sqrt{\left\{\frac{C}{L}\exp(-\beta t)\sin(\omega t)\right\}} \tag{4}$$

where U represents the input voltage, C represents the capacitance, L represents the inductance, β represents the damping exponent and ω is the angular frequency.

The constitutive behaviour of the tube material is described by the default constitutive relation built in COMSOL, i.e. the Cowper–Symonds constitutive model (Haiping and Chunfeng 2009).

$$\sigma = \sigma_y\left[1 + \left(\frac{\dot{\varepsilon}}{P}\right)^m\right] \tag{5}$$

where σ_y represents the quasi-static flow stress, $\dot{\varepsilon}$ represents the plastic strain rate (s^{-1}), and P and m are specific material parameters.

3.3 Weld Validation Criteria

Three different criteria were investigated to ascertain the weldability of the joint, namely:

(a) Impact velocity
(b) Effective plastic strain
(c) Shear stress

(a) Impact velocity

The determination of impact velocity is a very essential step during MPW. Impact velocity determines the success of the weld. Available literature suggests analytical relations that give threshold values of impact velocity. The weld occurs as an when the threshold limit is crossed. Researchers have carried out extensive studies regarding the effect of impact velocity on the occurrence of weld. Separate relations have been suggested for similar and dissimilar material combinations.

Due to the absence of any specific criterion for MPW and the process being closely analogous to EXW, the criterion suggested by Kore et al. (2010) was used to obtain the minimum impact velocity required for a successful weld to occur in case of similar material combination. The relation is given as follows (Kore et al. 2010):

$$U = \left(\frac{\sigma_{\mathrm{TU}}}{S}\right)^{1/2} \tag{6}$$

where U is the threshold velocity (m/s), σ_{TU} is the ultimate tensile stress (MPa) and S is the velocity of sound in the material considered (m/s).

Similarly for the case of dissimilar material combination, a different analytical relation was used to calculate the threshold value of impact velocity. The relation is given as follows (Botros and Groves 1980):

$$P = \frac{1}{2} Z_{\mathrm{eq}} U \cos \Phi \tag{7}$$

where U is the required impact velocity for the occurrence of a successful weld, Z_{eq} is the equivalent acoustic impedance of the mating members, P is the critical impact pressure for jet formation and Φ is the critical angle for jet formation.

$$Z_{\mathrm{eq}} = \frac{2}{{}^{1}/_{Z_1} + {}^{1}/_{Z_2}} \tag{8}$$

where

$$Z_1 = \rho_1 s_1 \text{ is the flyer tube acoustic impedance} \tag{9}$$

$$Z_2 = \rho_2 s_2 \text{ is the base tube acoustic impedance} \tag{10}$$

where ρ_1 and ρ_2 are the densities of the material considered and s_1 and s_2 are the speed of sound in the flyer and base tubes, respectively.

Apart from crossing the threshold impact velocity, there is a minimum value of impact pressure that also needs to be crossed to attain a successful weld. This is termed as the critical impact pressure and is determined by the relation (Botros and Groves 1980):

$$P = 5 \times \text{Hugoniot elastic limit (HEL)} \tag{11}$$

where HEL is given by the relation (Botros and Groves 1980):

$$\mathrm{HEL} = \frac{1}{2}\left(\frac{K}{G} + \frac{4}{3}\right) Y_0 \tag{12}$$

where K is the bulk modulus, G is the shear modulus and Y_0 is the tensile yield stress.

In MPW, higher impact velocity leads to an increase in the impact pressure which causes severe plastic deformation at the weld interface of the mating members. The fundamental need of a successful weld is that the workpiece surfaces are free from the presence of any contaminants and oxide layers. Velocity of impact is the main parameter that causes variations in the bonding. Too high a velocity leads to formation of intermetallics at the weld interface which subsequently leads to brittle damage of the mating members. On the other hand at a very low velocity, the jet formed is unable to eliminate the contaminants and oxide layer from the workpiece surface leading to an unsuccessful weld.

The impact velocity is directly related to input voltage which in turn is related to discharge energy. The relation between discharge energy and input voltage is given in Eq. (13) (Raoelison et al. 2012). Increase in discharge energy leads to an increase in the energy of the workpiece (flyer) which subsequently increases the impact velocity.

$$E = CU^2/2 \tag{13}$$

where E represents the energy stored in the capacitors, C is the capacitance of the MPW system and U is the input voltage.

For every material arrangement, there is a certain level of energy essential for joining the metals and beyond a certain energy level weld does not occur (Kore et al. 2008).

The impact velocity required to conduct a successful weld between two structural steel ASTM A36 tubes was calculated to be 297.7 m/s from Eq. (6).

(b) Effective Plastic Strain

Available literature (Mousavi and Al-Hassani 2008) on MPW as well as EXW suggests that effective strain can be used as a possible criterion for bonding. Literature suggests a threshold value of effective plastic strain for bonding to take place. A value of 0.35 was suggested as the limiting value for bonding of steel to steel (Mousavi and Al-Hassani 2008).

(c) Shear Stress

Available literature (Mousavi et al. 2005; Mousavi 2008) suggests that shear stresses in the base and flyer tubes at the impact zone should be of opposite sign for successful welding. If the shear stresses in the weld interface were of the identical sign in the two mating members, welding was doubtful. A threshold value of 0.5 GPa was suggested to decide the weldability for stainless steel joints (Mousavi et al. 2005).

4 Results and Discussions

The numerical simulations to assess the weldability of the structural steel ASTM A36 tubular assembly was carried out at various process parameters as mentioned in the preceding section. The flyer tube was affected by both radial and axial forces at a time, and thus, it suffered complicated stresses. The surface Von Mises stress developed in the flyer tube increased continuously with time (Fig. 5a–d) and

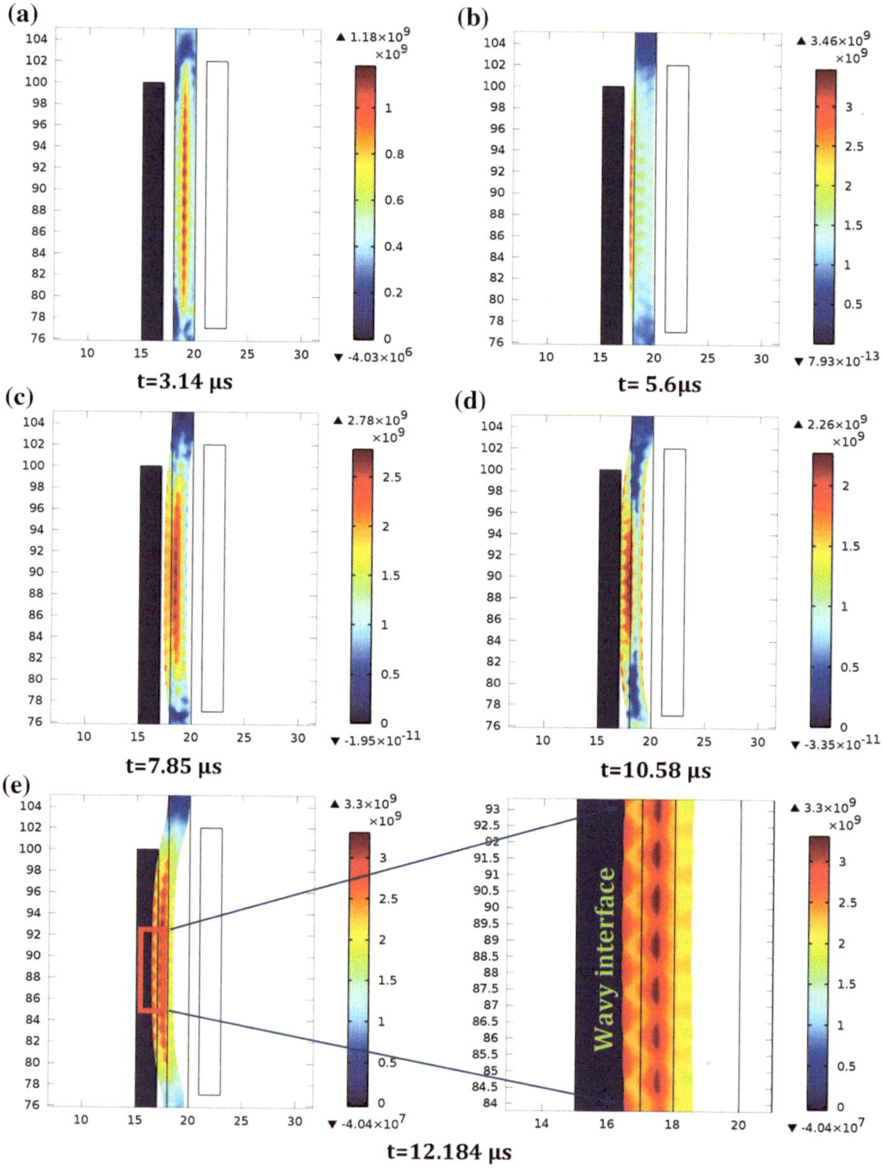

Fig. 5 Surface Von Mises stress at different times (input voltage—8.5 kV, air gap—1 mm)

reached the maximum value when the flyer tube collided with the target tube Fig. 5e. A wavy interface at the impact zone can be seen at the time of impact. The impact velocities, effective plastic strain and the shear stress at the time of impact were recorded for the 40 cases. The resulting weldability window and process mechanism are discussed in the succeeding section.

Figure 6a–c shows the comparison of three weldability criteria along the arc length at varying input voltages and a constant air gap of 1 mm. The arc length is the distance along the edge of the flyer plate at the impact zone. It can be observed that the impact velocity criterion could not be satisfied at 8 kV (Fig. 6a), whereas the effective plastic strain crossed the threshold at all the three input voltages (Fig. 6b). On the contrary, the shear stress criterion crossed the threshold at 8 kV but failed to do the same at 9 kV (Fig. 6c). The input voltage of 8.5 kV satisfied all the three criteria. Thus, the existence of process parameters which can satisfy all the three criteria in the MPW process was non-trivial.

The foregoing description was further extended to the remaining test cases and consolidated in Fig. 7. The pairs of the input voltage and the air gap are marked for the cases when an individual weldability criterion crossed the respective threshold value. It is seen that the plastic strain was the most versatile criteria that crossed the threshold limit followed by the impact velocity and the shear stress. The impact velocity crossed the threshold value at moderate and higher values of the investigated input voltage, whereas the same happened at moderate and lower values for the shear stress. The moderate input voltage except the minimal air gap of 0.5 mm was successful in crossing the threshold of the three criteria. Based on the above discussion, a weldability window was reached upon, as shown in Fig. 7. This window encompassed only those values of process parameters, where the impact velocity, effective plastic strain and shear stress together crossed their respective threshold values. The window identified the particular process parameters suitable for conducting a successful weld.

The process parameters that characterised the three weldability criteria were interrelated. The interrelation is explained through Fig. 8a–c that depicts the influence of process parameters on the impact velocity, the effective plastic strain and the shear stress, respectively.

From Fig. 8a, it is seen that the impact velocity increased with an increase in the input voltage. However, a minimum input voltage of 7.5 kV was essential to cross the threshold impact velocity. This is in agreement with the previously published observations (Zhidan et al. 2013) that one of the simplest ways to increase the quality of MPW is to increase the input voltage. At lower input voltages such as 6 and 6.5 kV, maximum impact velocity was obtained with the lowest air gap of 0.5 mm. A further increase in the input voltage at 0.5 mm air gap did not result in a significant amount of change in the impact velocity and eventually the flyer plate damaged at voltages ranging from 8 to 9.5 kV. The damage might be a result of low strain rate and high stress level that occurred at the time of discharge leading to crack initiation and propagation in the flyer (Raoelison et al. 2013). At a particular value of the air gap, the velocity and kinetic energy of the tubes reached a maximum value. Below this particular value, the tubes were unable to attain the

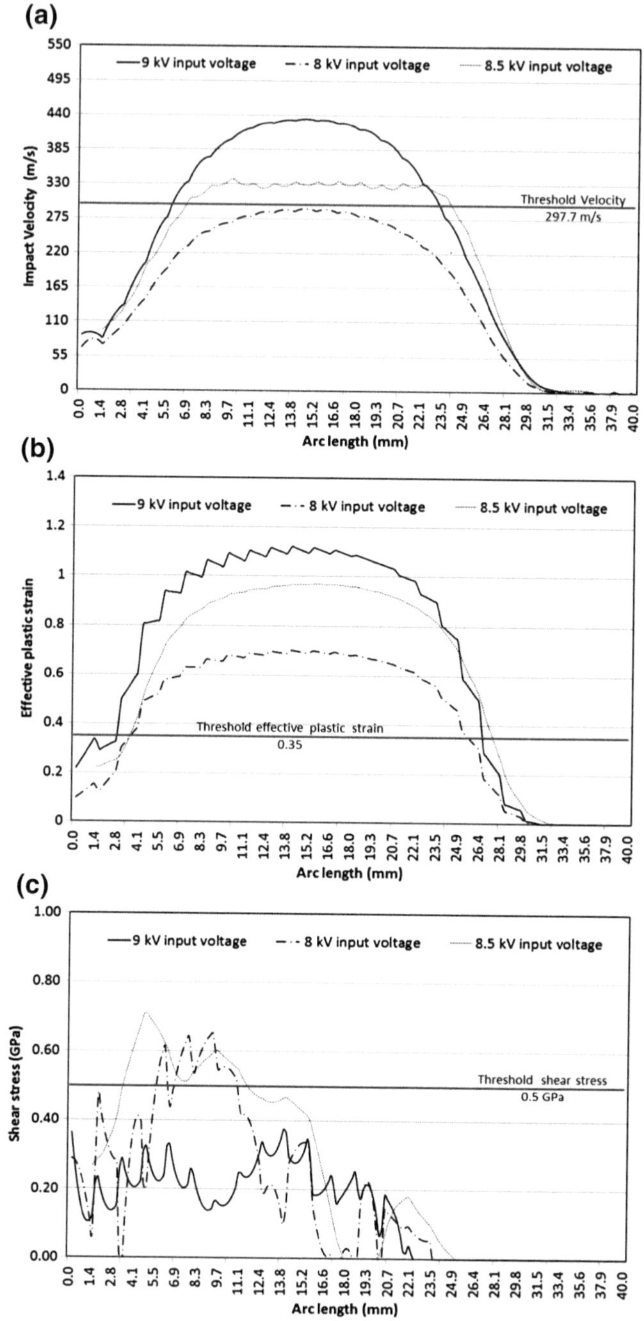

Fig. 6 Comparison of weldability criteria **a** impact velocity, **b** effective plastic strain, **c** shear stress (air gap = 1 mm)

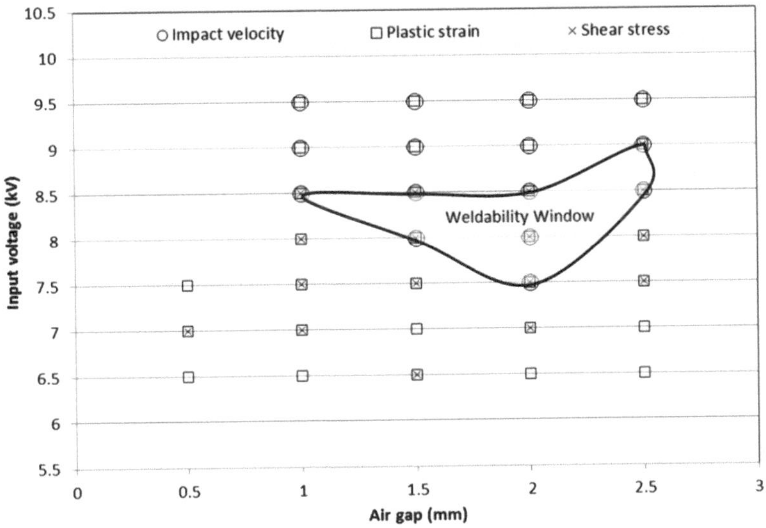

Fig. 7 Weldability window for MPW of structural steel ASTM A36

maximum possible velocity. At lower values of the air gap, e.g. 0.5 mm in the present case, the tubes collided well before the flyer attained the maximum velocity, whereas in case of larger air gaps, the velocity reached a value lower than the maximum at the time of the impact. The previously published experimental results (Kore et al. 2007) are in agreement with the observed numerical results in the present study.

As the input voltage was increased, the plastic strain induced in the members as well as the shear stress, increased up to an extent as shown in Fig. 8b, c, respectively. The increase of the input voltage led to an increase in the pressure acting upon the flyer tube. The pressure in the impact zone was mainly due to two phenomena: pressure induced due to the magnetic field and a pressure due to the impact intensity (Raoelison et al. 2012). High impact velocities produced high plastic deformation which subsequently resulted in higher levels of effective plastic strain at the impact zone. The numerically computed results suggest the existence of a plastic strain band, as well as a severely deformed impact zone with high values of plastic strain. The values of plastic strain crossed the threshold value for input voltages ranging from 6.5 to 9.5 kV suggesting that below 6.5 kV, the bonding would not take place (Fig. 8b). As the air gap was increased, the plastic strain values showed an increasing trend up to a certain value of input voltage and subsequently the value decreased. This confirms the presence of an optimum air gap (around 1.5 mm in the present case) between the members to achieve a good weld. At the lower air gap, it would not be possible to create a weld due to pressure deficiency, whereas at higher gaps, the impact would not take place.

The foregoing observation was also supported by the shear stress distribution shown in Fig. 8c. The shear stress values had opposite signs for the flyer and target

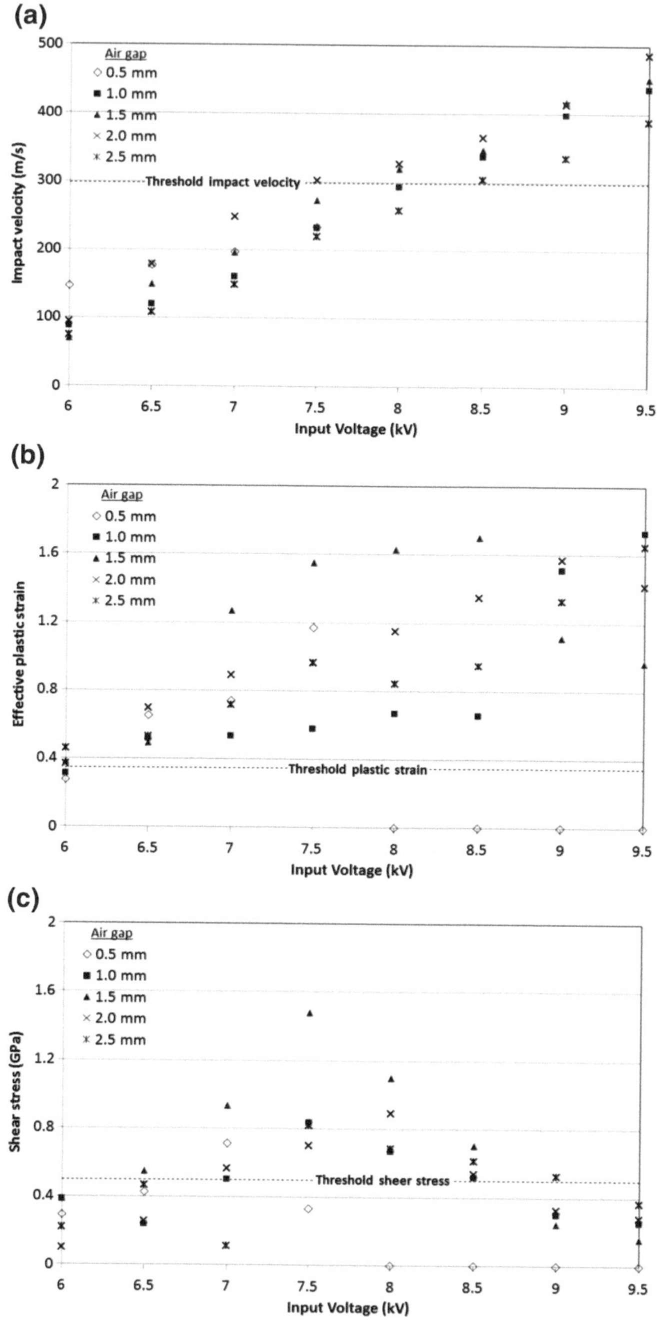

Fig. 8 Weldability criteria at different air gaps and input voltages **a** impact velocity, **b** effective plastic strain, **c** shear stress

tubes, respectively, for the cases where it crossed the threshold value of 0.5 GPa. At an input voltage of 6 kV, the shear stress value was below the threshold value for all the air gaps. From Fig. 8c, it can be observed that the shear stress crossed the threshold value for almost all air gaps at voltages ranging from 7 to 8.5 kV. Beyond this value of input voltage, the shear stress started to decrease and was unable to cross the threshold value. This would limit the allowable range of input voltage in a manner the impact velocity and the effective plastic strain would limit the allowable air gap as mentioned earlier.

The results of the present investigation emphasise on the need for an all-inclusive approach towards weldability criteria for MPW. Considering all, the three criteria would provide a more reliable range of process parameters to work with. Furthermore, use of FEM simulation would save the cost and the time spent in development of a product.

5 Conclusions

1. The three weldability criteria, namely impact velocity, effective plastic strain and the direction and magnitude of the shear stress studied in this investigation have a significant role in MPW of tubular joint.
2. Existence of process parameters which can simultaneously satisfy the three foregoing criteria is non-trivial. A comprehensive approach considering each of the foregoing weldability criteria for MPW should be adopted.
3. The process parameters in MPW are interrelated. A moderate input voltage at an optimum air gap could achieve a sound joint.
4. The demonstrated methodology of developing a weldability window through FEM simulation would save cost and time spent in production of product using MPW.

References

Aizawa T, Kashani M, Okagawa K (2007) Application of magnetic pulse welding for aluminum alloys and SPCC steel sheet joints. Weld J 86:119–124

Mousavi SAAA, Al-Hassani STS (2008) Finite element simulation of explosively-driven plate impact with application to explosive welding. Mater Des 29:1–19

Akbari Mousavi SAA, Burley SJ, Al-Hassani STS (2005) Simulation of explosive welding using the williamsburg equation of state to model low detonation velocity explosives. Int J Impact Eng 31:719–734

Akbari Mousavi SAA (2008) Numerical studies of explosive welding of three-layer cylinder composites-part 2. Mater Sci Forum 580–582:327–330

ASTM Standard A36/A36 M (2012) Specification for carbon structural Steel. ASTM International, West Conshohocken, PA doi:10.1520/A0036_A0036M-12

Botros KK, Groves TK (1980) Fundamental impact-welding parameters-an experimental investigation using a 76-mm powder cannon. J Appl Phys 51:3706–3714

Chudakov VA, Khrenov KK, Sergeeva YA, Gordan GN (1980) The effect of temperature to which the material is heated on the process of formation of intermetallic compound in magnetic pulse welding. Weld Prod 27(9):23–25

Epechurin VP (1978) Determination of the forces of the compressive effect in nonmagnetic multilayer conductors. Power Eng 16(3):115–120

Epechurin VP (1974) Properties of bimetal joints by magnetic pulse welding. Weld Prod 21(5):4–12

Haiping YU, Chunfeng LI (2009) Effects of current frequency on electromagnetic tube compression. J Mater Process Tech 209:1053–1059

Haiping YU, Zhisong F, Chunfeng LI (2014) Magnetic pulse cladding of aluminum alloy on mild steel tube. J Mater Process Tech 214:141–150

Kore SD, Date PP, Kulkarni SV (2008) Electromagnetic impact welding of aluminum to stainless steel sheets. J Mater Process Tech 208:486–493

Kore SD, Date PP, Kulkarni SV (2007) Effect of process parameters on electromagnetic impact welding of aluminum sheets. Int J Impact Eng 34:1327–1341

Kore SD, Dhanesh P, Kulkarni SV, Date PP (2010) Numerical modeling of electromagnetic welding. Int J Appl Electrom 32:1–19

Raoelison RN, Buiron N, Rachik M, Haye D, Franz G (2012) Efficient welding conditions in magnetic pulse welding process. J Manuf 14:372–377

Raoelison RN, Buiron N, Rachik M, Haye D, Franz G, Habak M (2013) Study of the elaboration of a practical weldability window in magnetic pulse welding. J Mater Process Tech 213:1348–1354

Shim JY, Kang BY, Kim IS, Kang MJ, Park DH, Kim IJ (2010) A study on distributions of electromagnetic force of the dissimilar metal joining in MPW using a FEM. Adv Mat Res 83–86:214–221

Shribman V (2007) Magnetic pulse welding of automotive HVAC parts. Pulsar—Magnetic pulse solutions, pp 1–31

Stern A, Aizenshtein M (2002) Bonding zone formation in magnetic pulse welds. Sci Technol Weld Joi 7(5):339–342

Tomas BMC (2010) Magnetic pulse welding MPW. Ph. D. thesis, Department of Mechanical Engineering Universidade Nova de Lisboa, Portugal

Weddeling C, Woodward ST, Marre M, Nellesen J, Psyk V, Tekkaya AE, Tillmann W (2011) Influence of groove characteristics on strength of form-fit joints. J Mater Process Tech 211:925–935

Zhang Y, Babu SS, Prothe C, Blakely M, Kwasegroch J, LaHa M, Daehn GS (2011) Application of high velocity impact welding at varied different length scales. J Mater Process Tech 211:944–952

Zhidan XU, Junjia C, Haiping YU, Chunfeng LI (2013) Research on the impact velocity of magnetic impulse welding of pipe fitting. Mater Des 49:736–745

Numerical Analysis of Heat Transfer of Arc Welded Plate

Aniruddha Ghosh, Pawan Kumar and Arvind Kumar

Abstract In submerged arc welding process, the understanding of temperature distribution is essential in order to control the dimension of heat affected zone and to get the required weld bead shape and size. Moreover, the temperature profile is required to estimate the stress distribution in thermo mechanical analysis of the process. In this work, a numerical solution for moving heat source with Gaussian distribution of heat flux density over the volume of oval shape is derived using finite difference method. Heat transfer in welded plates during welding from fusion zone to heat affected zone (HAZ) is assumed to be conductive heat transfer. Convective and radiative heat losses are also considered for remaining zone of welded plate. With the help of the numerical solution, transient temperature distribution is estimated. HAZ widths are also measured experimentally. Decent agreements between predicted and experimental values are achieved.

Keywords Submerged arc welding · Gaussian heat distribution · Oval heat source · Finite difference method

List of Symbols

a, b, c, m	Heat source parameters, mm
A, B, C	Bead geometry parameters, mm
C_p	Specific heat of work piece, J/kgK
h	Convective heat transfer coefficient W/m^{2}K
I	Current, amp
k	Thermal conductivity of work piece, W/mK
l	Mesh length, mm

A. Ghosh
Department of Mechanical Engineering, Government College of Engineering and Textile Technology, Berhampore, West Bengal, India

P. Kumar · A. Kumar (✉)
Department of Mechanical Engineering, Indian Institute of Technology, Kanpur, Uttar Pradesh, India
e-mail: arvindkr@iitk.ac.in

R.G. Narayanan and U.S. Dixit (eds.), *Advances in Material Forming and Joining*, Topics in Mining, Metallurgy and Materials Engineering, DOI 10.1007/978-81-322-2355-9_14

q	Heat density, W/m^3
Q_0	Total heat input, W
t	Time, sec
T	Temperature, °C
T_∞	Ambient temperature, °C
U	Travel Speed (cm/min)
V	Voltage, Volt
x, y, z	Coordinates, mm

Greek letters

α	Thermal diffusivity, m^2/sec
Δ	Increment
ε	Emissivity of the work piece
H	Arc efficiency
P	Density of workpiece, kg/m^3
σ	Stefan–Boltzmann constant (=5.67 × 10^{-6} W/m^{2}K^4)
T	Mesh Fourier number

1 Introduction

Critical investigation of the transient temperature distribution is important for maintaining quality of the submerged arc welding of plates. Temperature history of welded components has a significant influence on temperature distribution in the welded plate. Such transient heat transfer problem having simple geometries with simplified boundary conditions can be solved analytically. But many problems encountered in practice involve complicated geometries with complex boundary conditions or variable thermo-physical properties and cannot be solved analytically. In such cases, sufficiently accurate approximate solutions can be obtained using numerical method.

Goldak et al. (1984) proposed double ellipsoidal heat source, and the welding problem was analyzed by ASGARD, a nonlinear transient finite element (FEM) heat flow program. However, they have not considered the convective heat loss. A three-dimensional, steady-state thermal model of the GMA welding process with a moving coordinate framework was formulated by Pardo and Weckman (1989) and was solved by the finite element method. Kumar and Bhaduri (1994) developed a 3D finite element model to predict the transient temperature distribution in the work piece for GMA welding process. Their results were validated with experimental data. Ohring and Lunt (1999) investigated the gas metal arc weld pool considering radiation, evaporation, and viscous stress in the deformable free surface boundary conditions. Sabapathy et al. (2000) showed that numerical methods have ability to assess the welding conditions for a safe welding of high-pressure gas pipelines.

A new mathematical description of a heat source was obtained with the help of experimental data. Lindgren (2001) discussed complexities involved in simulation of welding process. Considering constant thermal properties, Yeh et al. (2003) investigated numerically and experimentally the temperature distribution of aluminum plates welded by gas tungsten arc.

Klobcar et al. (2004) developed mathematical modeling with the help of finite element method solving by ABAQUS computer code of GTA weld-surfacing process. It was found that double ellipsoidal heat source assumption prevailed over ellipsoidal heat source assumption in comparison with numerical results and experimental data for welded shape and thermal welding cycles. Bianco et al. (2004a) carried out two numerical methods for two- and three-dimensional models evaluating transient conductive fields with moving heat source. Later, Bianco et al. (2004b) investigated transient three-dimensional temperature distributions numerically by COMSOL. Radiation and convection modes of heat transfer from work piece surfaces as well as variable thermo-physical properties were taken into account. Cho (2006) investigated 3D numerical simulation of arc welding by using volume of fluid technique. Comparisons of weld bead hump and transient radius and temperature distribution on welded plates were made with experimental data. Mahapatra et al. (2006) used three-dimensional finite element analyses to predict the effect of submersed arc welding process parameters on temperature distribution and angular distortions in single-pass butt joints with top and bottom reinforcements. Elsen et al. (2007) described the analytical and numerical solutions of the heat conduction equation for a localized moving heat source for use in laser material processing viz. layered manufacturing and laser alloying. Yeh et al. (2007) employed Douglas–Gunn scheme to solve 3D transient heat conduction problem in gas tungsten arc welded plates. In addition, experiment on aluminum plates was conducted to verify the theoretical results. Scutelnicu et al. (2008) described several results of theoretical and experimental investigations on heat transfer in copper-low carbon steel welded joints of thin plates. Convection and radiation heat lost, and temperature-dependent thermo-physical properties are considered in their model. Measurements of the temperatures distribution have been made during welding through infrared thermography. Using the finite element code ANSYS, Kazemi and Goldak (2009) developed a three-dimensional finite element model to simulate the bead geometry of welded plates. The transient temperature distribution as well as the dimensions of the weld bead geometry during the welding process was calculated. Biswas and Mandal (2010) considered a numerical model and solved by the finite element package ANSYS for single-pass single-side submerged arc welding of square butt joints.

Esmaily and Shokuhfar (2010) claimed that finite element method with Eulerian and Lagrangian formulation, namely Arbitrary Lagrangian Eulerian (ALE), is the most suitable method for various simulation methods of processes having severe plastic deformation. They analyzed the heat transfer process in friction stir welding using ANSYS software and compared with the obtained measured data. In this study, welding made of 7075-T6 Aluminum alloy and high carbon steel. Temperature measurement for this process in the main sections was carried out by an accurate and special method during welding. They achieved very good agreement between measured and calculated data.

Pang et al. (2010) presented a parallel numerical study of the heat transfer and fluid flow of the weld pool; however, most of the cases current numerical studies are only limited to serial simulations running on a single CPU. In present numerical simulation effect of Marangoni force, buoyancy force, friction force of the mushy zone region, and the effect of keyhole were considered. A combined keyhole volume and surface heat source model was also considered. The coupled transient heat transfer and Navier–Stokes equations were solved with a high-order accuracy parallel projection method. It was found that 200 % speedup can be achieved on a shared memory quad-core CPU using the presented parallel simulation system. They achieved very good agreement between measured and calculated data.

Bag and De (2010) successfully demonstrated the heat source term in the frame of conduction heat transfer analysis for a number of laser spot and linear welds. They validated the quantitative estimation of the fluid velocity through order of magnitude analysis. From their study, an improved physical understanding of laser welding process that served as a basis for the design of welding process was found out. Isiklar and Girgin (2011) developed a numerical method for modeling underwater welding; they calculated transient temperature distribution of 3D plate considering point heat source and simulating the arc and convective, radiative, and boiling surface boundary conditions. To calculate transient temperature distribution, they used finite volume method as numerical scheme. Good agreement was achieved between predicted and measured data. Mohassela et al. (2013) described a thermal model which was developed to investigate the effects of welding parameters on the distribution of temperature in wide gap aluminothermic rail welds. Finite difference method (FDM) was employed to find out this. During numerical solution, phase change and convective heat transfers in weld pool were considered. A reasonable agreement was found between numerical and experimental results. The effects of welding variables, such as preheating, initial liquid temperature and weld gap on temperature distribution were studied. It was found from the modeling results that the magnitude of weld gap had the greatest influence on thermal behavior of the joint. Bag et al. (2012) described their work in two approaches. One is an intrinsic mapping of the heat source geometry with the numerically computed melt pool dimensions within the framework of modeling calculations in an iterative manner. It was validated for the prediction of final weld dimensions in autogenously gas tungsten arc welds with butt joint geometry. Another approach is an analytical estimation of the volumetric source term as function of only the welding conditions and the initial weld joint geometry and it was examined for the prediction of weld pool dimensions and thermal cycles in tandem submerged arc welds with typical groove joint geometry successfully. Bendaoud et al. (2014) described the numerical simulation of an industrial case of hybrid laser–MIG welding of high thickness duplex steel UR2507Cu with Y-shaped chamfer geometry using COMSOL multi physics.

Yaduwanshi et al. (2014) aimed to develop a heat transfer model of friction stir welding process and to study on effects of influential variables on transient thermal analysis. In their study, effect of transverse tool speed was neglected. It was revealed from experimentation that the temperature distributions are not symmetric with respect to welding line and maximum temperature occurs behind the tool pin.

In the present work, heat transport in submerged arc welding of steel plates is investigated numerically, through finite difference method, and experimentally. It is assumed that heat is distributed on welded plate in Gaussian distribution and shape of heat source is oval type. The oval shape arises due to heat source and convection in fusion zone. The aim of present work is to study how this heat from this oval shape weld pool (fusion zone) is being transferred to remaining of the welded work piece that forms heat affected zone. The mode of heat transfer is conduction for heat transfer from fusion zone to heat affected zone. Parameters controlling the oval heat source are obtained through the measurement of weld pool geometry. Variable thermo-physical properties and convective and radiative heat losses are taken into account in the proposed model. Previously no work was attempted to find out transient temperature distribution through finite difference method and considering convective and radiative heat loss, oval heat source,, and variable thermo-physical properties. Practically during welding process, thermo-physical properties are not constant and convective and radiative heat transfers have a vital role. Finally, a simplified experiment is conducted to validate the numerical results.

2 Experiment

A semi-automatic welding machine with constant voltage and rectifier-type power source with a 1200-A capacity was used to join two C–Mn steel plates as butt joint. The C–Mn steel work piece (300 × 150 × 20 mm) is cut and V groove of angle 60° as per the standards are prepared. The chemical composition of work piece material is described in Table 1. One mm root opening is selected to join the plates in the flat position keeping electrode positive and perpendicular to the plate. The job was firmly fixed to a base plate by means of tack welding and then the submerged arc welding was finally carried out. The welding parameters were recorded during actual welding to determine their fluctuations, if any. The slag was removed and the job was allowed to cool down. Welding is carried out for the square butt joint configuration. Temperatures were measured at different points of the welded plates

Table 1 Chemical composition of C–Mn steel work piece (in weight %)

C	Sn	Mn	P	S
0.18	0.36	1.58	0.023	0.027
Cr	Ni	Mo	Cu	Al
0.06	0.03	0.01	0.04	0.05

Table 2 Submerged arc welding conditions and HAZ

S. no.	Heat input (kJ/cm)	HAZ width (mm)
1	30.88	2.37
2	43.24	3.45
3	39.71	2.63
4	55.59	3.93
5	17.50	2.19
6	24.50	2.27
7	22.50	2.25
8	31.50	2.39

except welding line by infrared thermometers (OMEGA SCOPE OS 524E, temperature range 2482 °C, accuracy is ±1 % or 2 °C whichever is greater, resolution 1 °C, response time 10 ms). Finally, the welded plate was cut at the center of the bead to obtain 10-mm-wide test specimens. The standard metallographic process (i.e., metal polishing with a series of empty sheets and disk polishing using diamond paste from 5 μm down to 0.5 μm) was carried out. The established color etching procedure for steel was employed to identify different regions of the weldment. An optical research microscope (NEOPHOT-32) was used for this purpose. With the help of the microscope, HAZ widths were measured (Table 2).

3 Theoretical Modeling

3.1 Heat Source

For this study, it is assumed that heat is distributed on welded plate through Gaussian manner (as shown in Fig. 1) and shape of heat source is oval type (as shown in Fig. 2).The equation of this 3D oval shape becomes

$$ax^2 + \left(by^2 + cz^2\right)e^{mx} = 1 \tag{1}$$

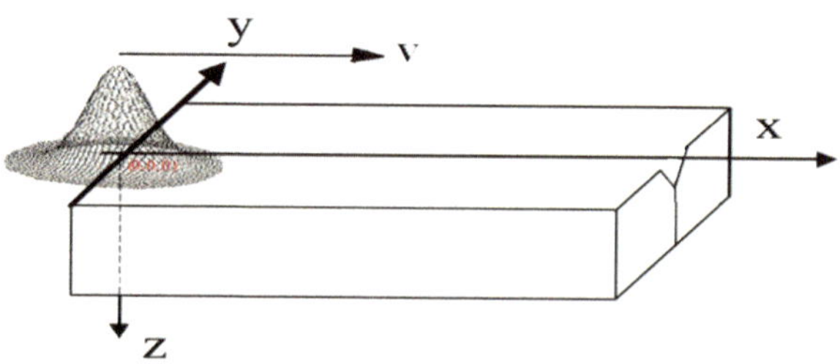

Fig. 1 Sketch of the work piece for submerged arc welding process and Gaussian heat distribution

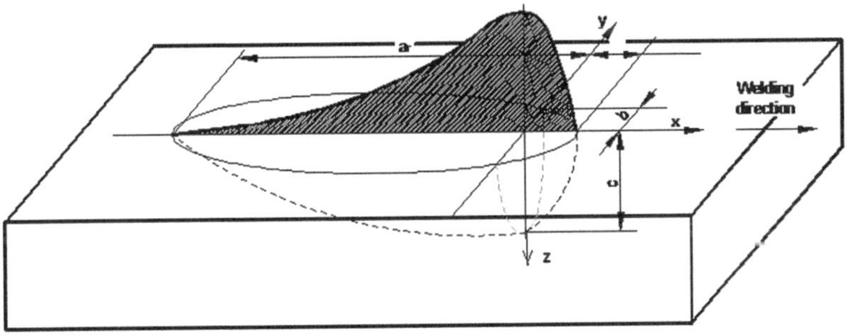

Fig. 2 Oval heat source shape

where $m = 0.2$, measured from experimental data. The heat density $q(x, y, z)$ at a point (x, y, z) within oval shape is given by the following equation:

$$q(x, y, z) = q_o e^{-\left[ax^2 + \left(by^2 + cz^2\right)e^{mx}\right]} \tag{2}$$

In Eq. (2), q_0 is Gaussian heat distribution parameter and a, b, c, and m are oval heat source parameters. If Q_0 is the total heat input, then

$$2Q_0 = \int_{-\infty}^{\infty} \int_{-\infty}^{\infty} \int_{-\infty}^{\infty} q(x, y, z) \mathrm{d}x \, \mathrm{d}y \, \mathrm{d}z \tag{3}$$

Substituting Eq. (2) into Eq. (3) yields

$$q_0 = \frac{2\sqrt{abc}}{\pi^{3/2}} \cdot e^{\frac{-m^2}{4a}} \cdot Q_0 \tag{4}$$

Then, the oval shape heat distribution equation can be rewritten as follows:

$$q(x, y, z) = Q_0 \frac{2\sqrt{abc}}{\pi^{3/2}} e^{-\left[\frac{m^2}{4a} + ax^2 + \left(by^2 + cz^2\right)e^{mx}\right]} \tag{5}$$

In this study, Q_0 is the heat input of the heat source which describes the heat flux from the arc. It equals to ηVI, where η is the efficiency of the arc, V is the arc voltage, and I is the welding current. Arc efficiency is taken as unity for submerged arc welding process.

Let A, B, and C are the Oval Shape Bead Geometry parameters. It should be pointed out that A is half of the major axis of oval shape, B stands for half of the bead width, and C represents penetration. In addition, values of A, B, and C can be measured from weld bead geometry. It is defined that the heat source energy at the boundary between the molten pool and the base material equals 0.05 $q(0)$. Thus, in the x-direction:

$$q\,(A,\,0,\,0) = q\,(0)e^{-aA^2} = 0.05\,q(0) \tag{6}$$

Hence,

$$a = \frac{\ln 20}{A^2} \cong \frac{3}{A^2} \tag{7}$$

Similarly, it is derived that

$$b \cong \frac{3}{B^2} \text{ and } c \cong \frac{3}{C^2} \tag{8}$$

Considering a rectangular region in which heat conduction is significant in the x- , y- , and z-directions. Heat is distributed on the welded plates. Let heat density expression be

$$\dot{q}(x, y, z, t) = \mathrm{e}^{-[a(x-vt)^2+(by^2+cz^2)\mathrm{e}^{mx}]} \tag{9}$$

where x, y, and z are the position of electrode at time t and v denotes velocity of the moving heat source. Note that ax^2 in Eq. (2) is replaced by $a(x - vt)^2$ in Eq. (9) since the electrode is moving.

3.2 Thermal Properties

In this study, thermal conductivity, specific heat, thermal expansion coefficient, and Young's modulus of the work piece are all temperature dependent. Due to the wide temperature interval involved in the welding of C–Mn steel, from 30 to 2500 °C, it is required to take this dependence into account. To facilitate computation, polynomial relations were fitted from available experimental data (Biswas and Mandal 2010). The relations were derived in the temperature intervals indicated below:

$$k(T) = 2 \times 10^{-5}T^2 - 0.0135T + 52.2(\text{W/m K}) \tag{10a}$$

when 20 °C $\leqq$ T $\leqq$ 800 °C

$$k(T) = 0.0037T + 23.8\;(\text{W/m K}) \tag{10b}$$

when 800 °C < T $\leqq$ 5000 °C

For specific heat capacity,

$$C_p(T) = 4 \times 10^{-6}T^3 - 0.0034T^2 + 1.0143T + 442.5 \tag{11a}$$

when 20 °C ≦ T ≦ 720 °C

$$C_p(T) = 22{,}543T^{-0.501} \tag{11b}$$

when 720 °C < T ≦ 5000 °C.

For thermal expansion coefficient,

$$\beta(T) = 0.0068T + 10.094 \tag{12a}$$

when 20 °C ≦ T ≦ 550 °C

$$\beta(T) = -2 \times 10^{-7}T^2 + 0.0016T + 13.04 \tag{12b}$$

when 550 °C < T ≦ 5000 °C.

The density of welded plates, ρ, is 7806 kg/m^3, convective heat transfer coefficient, h, is 15 W/m^2 K, Stefan–Boltzmann constant, σ, is equals to 5.67×10^{-8} W/m^2 K^4, and the emissivity of the work piece, ε, is assumed to be 0.95. In addition, arc efficiency, η, is taken as unity for this submerged arc welding process.

3.3 Boundary Conditions

A specified initial temperature covering the entire plate surface is

$$T - T_\infty \text{ for } t = 0 \tag{13}$$

where T_∞ is the ambient temperature. During the welding process, heat is dissipated into environment through convection and radiation heat losses on the top surface of the welded plate. Convection and radiation heat losses are also assumed on the bottom surface. The heat loss due to convection and radiation over these surfaces is given by

$$-k(T)\frac{\partial T}{\partial n} = h(T - T_\infty) + \varepsilon\sigma\left(T^4 - T_\infty^4\right) \quad \text{for } t > 0 \tag{14}$$

only convection heat transfer is considered for the lateral surface of the work piece, i.e.,

$$-k(T)\frac{\partial T}{\partial n} = h(T - T_\infty) \quad \text{for } t > 0 \tag{15}$$

3.4 Finite Difference Formulation

Generally reasonable accurate results can be obtained by replacing differential quantities by sufficiently small difference.

From Fig. 3, it can be written as

$$\frac{\mathrm{d}f(x)}{\mathrm{d}x} \cong \frac{f(x + \Delta x) - f(x)}{\Delta x} \tag{16}$$

From Fig. 4, it can be written that

$$\left.\frac{\mathrm{d}T}{\mathrm{d}x}\right|_{m-\frac{1}{2}} = \frac{T_m - T_{m-1}}{\Delta x} \tag{17}$$

$$\left.\frac{\mathrm{d}T}{\mathrm{d}x}\right|_{m+\frac{1}{2}} = \frac{T_{m+1} - T_m}{\Delta x} \tag{18}$$

$$\left.\frac{\mathrm{d}^2T}{\mathrm{d}x^2}\right|_m = \frac{\left.\frac{\mathrm{d}T}{\mathrm{d}x}\right|_{m+\frac{1}{2}} - \left.\frac{\mathrm{d}T}{\mathrm{d}x}\right|_{m-\frac{1}{2}}}{\Delta x} = \frac{T_{m-1} - 2T_m + T_{m+1}}{\Delta x^2} \tag{19}$$

So steady-state two-dimensional heat conduction (without heat generation) equation will be (comp. Fig. 5)

$$\frac{T_{m+1,n} - 2T_{m,n} + T_{m-1,n}}{\Delta x^2} + \frac{T_{m,n+1} - 2T_{m,n} + T_{m,n-1}}{\Delta x^2} = 0 \tag{20}$$

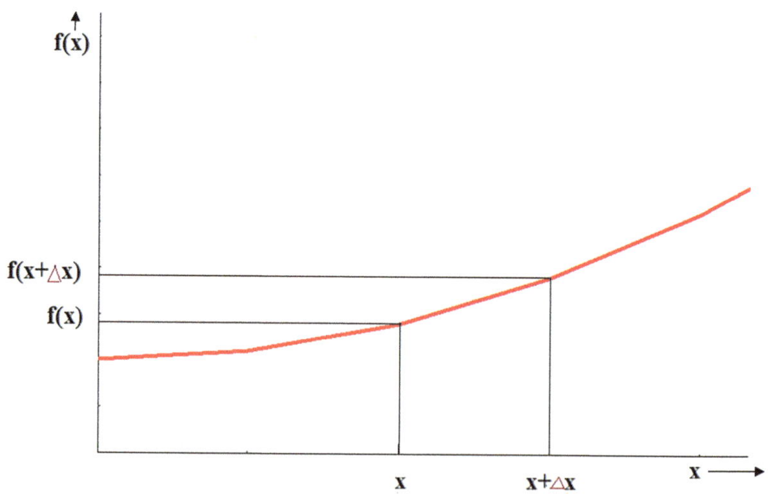

Fig. 3 The derivative of a function at a point represents the slope of the function at that point

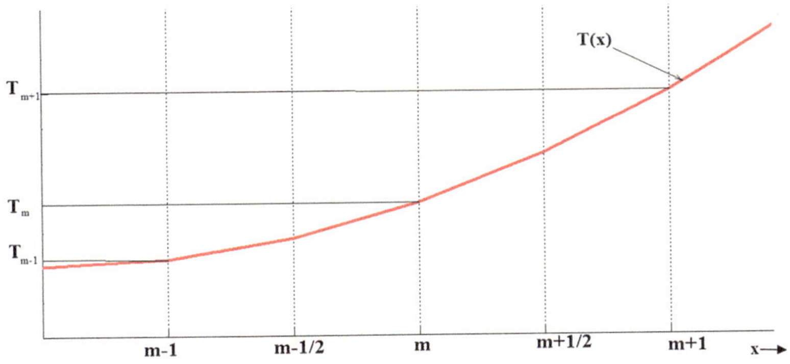

Fig. 4 Schematic of the nodes and the nodal temperatures used in the development of the finite difference formulation of heat transfer in a plane wall

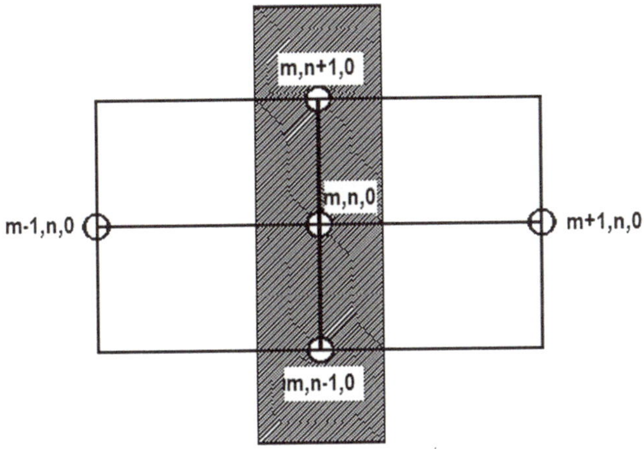

Fig. 5 Representation of finite difference mesh

The explicit finite difference equations are determined on the basis of the energy balance for the transient case expressed as follows:

$$\sum_{\text{allsides}} \dot{Q}^i + e(x, y, z, \dot{t})V_{\text{element}} = \rho V_{\text{element}} C_p \frac{T^{i+1}{}_{m,n,o} - T^i{}_{m,n,o}}{\Delta t} \tag{21}$$

In the present problem mode of heat transfer are conduction, radiation, and convection. So, Eq. (21) can be written in the form:

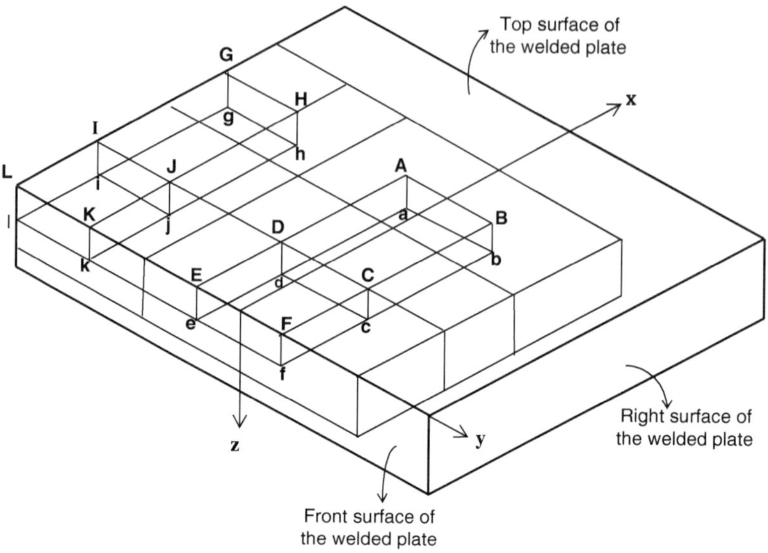

Fig. 6 Description of meshes of the welded plate

When heat is generated from the block "ABCDabcd" (comp. Fig. 6), equation will be written as follows:

$$\begin{aligned}
&h \times \Delta x \times \Delta y \times (T_\infty - T^i_{m,n,o}) + \sigma \times \varepsilon \times \Delta x \times \Delta y \times (T^4_\infty - (T_{m,n,o})^4) \\
&\quad + k(T) \times \Delta x \times \frac{\Delta z}{2} \times \left(\frac{T^i_{m-1,n,o} - T^i_{m,n,o}}{\Delta y}\right) + k(T) \times \Delta y \times \frac{\Delta z}{2} \times \left(\frac{T^i_{m,n+1,o} - T^i_{m,n,o}}{\Delta x}\right) \\
&\quad + k(T) \times \Delta x \times \frac{\Delta z}{2} \times \left(\frac{T^i_{m+1,n,o} - T^i_{m,n,o}}{\Delta y}\right) + k(T) \times \Delta x \times \frac{\Delta z}{2} \times \left(\frac{T^i_{m,n-1,o} - T^i_{m,n,o}}{\Delta x}\right) \\
&\quad + k(T) \times \Delta x \times \Delta y \times \left(\frac{T^i_{m,n,o+1} - T^i_{m,n,o}}{\Delta z}\right) + \dot{q}(x, y, z, t) \times \Delta x \times \Delta y \times \frac{\Delta z}{2} \\
&= \rho \times \Delta x \times \Delta y \times \frac{\Delta z}{2} \times C_p(T) \times \left(\frac{T^{i+1}_{m,n,o} - T^i_{m,n,o}}{\Delta t}\right)
\end{aligned} \tag{22a}$$

Let $\Delta x = \Delta y = \Delta z = 1$, τ = Mesh Fourier Number = $\frac{\alpha \Delta t}{\Delta x^2}$, α = Thermal Diffusivity = $\frac{k(T)}{\rho C_p(T)}$

So Eq. (17) becomes

$$T^{i+1}_{m,n,o} = \left(1 - \frac{2hl\tau}{k(T)} - 6\tau\right) \times T^{i}_{m,n,o} + \frac{2hl\tau}{k(T)} \times T_{\infty} + 2 \times \sigma \times \varepsilon \times \frac{1\tau}{k(T)} \times \left(T^{4}_{\infty} - (T_{m,n,o})^{4}\right)$$
$$+ \tau\left(T^{i}_{m-1,n,o} + T^{i}_{m,n+1,o} + T^{i}_{m,n+1,o} + 2T^{i}_{m,n,o+1}\right) + \frac{\dot{q}^{i}(x,y,z)1^{2}\tau}{k(T)} \tag{22b}$$

Stability criteria for Eq. (22b) is

$$\left(1 - \frac{4hl\tau}{k(T)} - 6\tau\right) \geq 0$$
$$\text{or } \Delta t \leq \frac{l^{2}}{\left(\frac{4hl\tau}{k(T)} + 6\right)\alpha} \tag{22c}$$

as, τ = Mesh Fourier Number = $\frac{\alpha \Delta t}{\Delta x^{2}}$; $\Delta x = \Delta y = \Delta z = 1$.

When heat is generated from the block “GHJIghji” (comp. Fig. 6), equation will be written as follows:

$$h \times \Delta x \times \Delta y \times \left(T_{\infty} - T^{i}_{m,n,o}\right) + \sigma \times \varepsilon \times \Delta x \times \Delta y \times \left(T^{4}_{\infty} - (T_{m,n,o})^{4}\right)$$
$$+ k(T) \times \Delta x \times \frac{\Delta z}{2} \times \left(\frac{T^{i}_{m-1,n,o} - T^{i}_{m,n,o}}{\Delta y}\right) + k(T) \times \Delta y \times \frac{\Delta z}{2} \times \left(\frac{T^{i}_{m,n+1,o} - T^{i}_{m,n,o}}{\Delta x}\right)$$
$$+ k(T) \times \Delta x \times \frac{\Delta z}{2} \times \left(\frac{T^{i}_{m+1,n,o} - T^{i}_{m,n,o}}{\Delta y}\right) + k(T) \times \Delta x \times \frac{\Delta z}{2} \times \left(\frac{T^{i}_{m,n-1,o} - T^{i}_{m,n,o}}{\Delta x}\right)$$
$$+ k(T) \times \Delta x \times \Delta y \times \left(\frac{T^{i}_{m,n,o+1} - T^{i}_{m,n,o}}{\Delta z}\right) + \dot{q}(x,y,z,t) \times \Delta x \times \Delta y \times \frac{\Delta z}{2}$$
$$= \rho \times \Delta x \times \Delta y \times \frac{\Delta z}{2} \times C_{p}(T) \times \left(\frac{T^{i+1}_{m,n,o} - T^{i}_{m,n,o}}{\Delta t}\right) \tag{23a}$$

or

$$T^{i+1}_{m,n,o} = \left(1 - \frac{4hl\tau}{k(T)} - 6\tau\right) \times T^{i}_{m,n,o} + \frac{2hl\tau}{k(T)} \times T_{\infty} + 2 \times \sigma \times \varepsilon \times \frac{l\tau}{k(T)} \times \left(T^{4}_{\infty} - (T_{m,n,o})^{4}\right)$$
$$+ \tau\left(T^{i}_{m+1,n,o} + T^{i}_{m,n+1,o} + T^{i}_{m,n-1,o} + 2T^{i}_{m,n,o+1}\right) + \frac{\dot{q}^{i}(x,y,z)l^{2}\tau}{k(T)} \tag{23b}$$

Stability criteria for Eq. (23b) is

$$\begin{aligned} &\left(1 - \frac{4hl\tau}{k(T)} - 6\tau\right) \geq 0 \\ &\text{or } \Delta t \leq \frac{l^2}{\left(\frac{4hl\tau}{k(T)} + 6\right)\alpha} \end{aligned} \tag{23c}$$

When heat is generated from the block "CDEFcdef" (comp. Fig. 6), equation will be written as follows:

$$\begin{aligned} &h \times \frac{\Delta x}{2} \times \Delta y \times \left(T_\infty - T^i_{m,n,o}\right) + \sigma \times \varepsilon \times \frac{\Delta x}{2} \times \Delta y \times \left(T^4_\infty - (T_{m,n,o})^4\right) \\ &\quad + h \times \Delta y \times \frac{\Delta z}{2} \times \left(T_\infty - T^i_{m,n,o}\right) k(T) \times \frac{\Delta x}{2} \times \frac{\Delta z}{2} \times \left(\frac{T^i_{m-1,n,o} - T^i_{m,n,o}}{\Delta x}\right) \\ &\quad + k(T) \times \Delta y \times \frac{\Delta z}{2} \times \left(\frac{T^i_{m,n+1,o} - T^i_{m,n,o}}{\Delta z}\right) + k(T) \times \frac{\Delta x}{2} \times \frac{\Delta z}{2} \times \left(\frac{T^i_{m+1,n,o} - T^i_{m,n,o}}{\Delta y}\right) \\ &\quad + k(T) \times \Delta y \times \frac{\Delta z}{2} \times \left(\frac{T^i_{m,n,o+1} - T^i_{m,n,o}}{\Delta z}\right) + \dot{q}^i(x,y,z,t) \times \Delta y \times \frac{\Delta x}{2} \times \frac{\Delta z}{2} \\ &= \rho \times \Delta y \times \frac{\Delta x}{2} \times \frac{\Delta z}{2} \times C_p(T) \times \left(\frac{T^{i+1}_{m,n,o} - T^i_{m,n,o}}{\Delta t}\right) \end{aligned} \tag{24a}$$

or,

$$\begin{aligned} T^{i+1}_{m,n,o} = &\left(1 - \frac{4hl\tau}{k(T)} - 6\tau\right) \times T^i_{m,n,o} + \frac{2hl\tau}{k(T)} \times T_\infty + 2 \times \sigma \times \varepsilon \times \frac{l\tau}{k(T)} \times \left(T^4_\infty - (T_{m,n,o})^4\right) \\ &+ \tau\left(T^i_{m-1,n,o} + T^i_{m+1,n,o} + T^i_{m,n+1,o} + 2T^i_{m,n+1,o}\right) + \frac{\dot{q}^i(x,y,z) l^2 \tau}{k(T)} \end{aligned} \tag{24b}$$

Stability criteria for Eq. (24b) is

$$\begin{aligned} &\left(1 - \frac{4hl\tau}{k(T)} - 6\tau\right) \geq 0 \\ &\text{or } \Delta t \leq \frac{l^2}{\left(\frac{4hl\tau}{k(T)} + 6\right)\alpha} \end{aligned} \tag{24c}$$

When heat is generated from the block "IJKLijkl" (comp. Fig. 6), equation will be written as follows:

$$
\begin{aligned}
& h \times \frac{\Delta x}{2} \times \frac{\Delta y}{2} \times \left(T_\infty - T^i_{m,n,o}\right) + \sigma \times \varepsilon \times \frac{\Delta x}{2} \times \frac{\Delta y}{2} \times \left(T^4_\infty - (T_{m,n,o})^4\right) \\
& \quad + h \times \frac{\Delta x}{2} \times \frac{\Delta z}{2} \times \left(T_\infty - T^i_{m,n,o}\right) k(T) \times \frac{\Delta y}{2} \times \frac{\Delta z}{2} \times \left(\frac{T^i_{m,n+1,o} - T^i_{m,n,o}}{\Delta x}\right) \\
& \quad + k(T) \times \frac{\Delta x}{2} \times \frac{\Delta z}{2} \times \left(\frac{T^i_{m+1,n,o} - T^i_{m,n,o}}{\Delta y}\right) + k(T) \times \frac{\Delta x}{2} \times \frac{\Delta y}{2} \times \left(\frac{T^i_{m,n,o+1} - T^i_{m,n,o}}{\Delta z}\right) \\
& \quad + h \times \frac{\Delta z}{2} \times \frac{\Delta y}{2} \times \left(T^i_{m,n-1,o} - T^i_{m,n,o}\right) + \dot{q}^i(x,y,z,t) \times \Delta y \times \Delta x \times \frac{\Delta z}{2} \\
& = \rho \times \frac{\Delta y}{2} \times \frac{\Delta x}{2} \times \frac{\Delta z}{2} \times C_p(T) \times \left(\frac{T^{i+1}_{m,n,o} - T^i_{m,n,o}}{\Delta t}\right)
\end{aligned}
\tag{25a}
$$

or

$$
\begin{aligned}
T^{i+1}_{m,n,o} = & \left(1 - \frac{6hl\tau}{k(T)} - 6\tau\right) \times T^i_{m,n,o} + \frac{2hl\tau}{k(T)} \times T_\infty + 2 \times \sigma \times \varepsilon \times \frac{l\tau}{k(T)} \times \left(T^4_\infty - (T_{m,n,o})^4\right) \\
& + \tau\left(T^i_{m+1,n,o} + T^i_{m,n+1,o} + 2T^i_{m,n+1,o}\right) + \frac{\dot{q}^i(x,y,z) l^2 \tau}{k(T)}
\end{aligned}
\tag{25b}
$$

So, stability criteria for Eq. (25b) is

$$
\begin{aligned}
& \left(1 - \frac{6hl\tau}{k(T)} - 6\tau\right) \geq 0 \\
& \text{or } \Delta t \leq \frac{l^2}{\left(\frac{6hl\tau}{k(T)} + 6\right)\alpha}
\end{aligned}
\tag{25c}
$$

3.5 Solution Procedure

In this work, the governing Eq. (21) and the boundary conditions Eqs. (13)–(15) are transformed into a set of explicit finite difference Eqs. (22a, 22b, 22c)–(25a, 25b, 25c) and were solved numerically. The input data for this calculation included the ambient temperature, the initial temperature of the specimen, the size of the specimen and of the elements of the mesh, the initial position of the arc, the heat transfer coefficient, and the parameters which characterized the arc. In order to ensure stability of the calculations, the mesh size and time step has to be limited. Considering Eqs. (22a, 22b, 22c)–(25a, 25b, 25c), the dominant stability criterion for all nodes is expressed as follows:

$$\Delta t \leq \frac{l^2 \rho C_p(T)}{6[k(T) + hl]} \quad (26)$$

The stability condition is further given by

$$\Delta t \leq \frac{l^2 \rho C_{p,\min}}{6(k_{\max} + hl)} \quad (27)$$

where $C_{p,\min}$ and $k_{\max}$ are, respectively, the minimum value of heat capacity and the maximum value of thermal conductivity over the entire temperature range of interest. The solutions of Eq. (16) subject to boundary conditions Eqs. (13)–(15) give the temperature profile at each time step explicitly in terms of known information from the previous time step. To assure the proposed explicit method to be conditionally stable, the stability criteria Eq. (27) must be satisfied.

3.6 Grid Independence

A careful check for the grid independence of numerical solutions has been made to ensure the accuracy and validity of the numerical schemes.

For this purpose, three grid systems, $\Delta t = 0.01$ s and $\Delta x = \Delta y = \Delta z = 2$ mm, $\Delta x = \Delta y = \Delta z = 1$ mm, and $\Delta t = 0.002$ s and $\Delta x = \Delta y = \Delta z = 0.5$ mm are tested. It is found the maximum relative error in the temperature at the center of the work piece between the latter two solutions is within 1 %. Spatial increments, Δx, Δy, and Δz of 1 mm, and a time step $\Delta t = 0.01$ s are thus adopted in the calculation.

3.7 Study on Thermal Stress Distribution and Deformation of Welded Plates at the Instant When Position of Moving Heat Source (15, 0, 0)

Study on stress distribution and deformation of welded plates was also been made with the easy finite element method software and by calculating temperatures on different node points with the help of equation No. 1–27 and EASY F E M V1.0 software. Graphical representations of thermal stress distribution (unit of stress N/m^2) are shown in Fig. 7.

From Fig. 8, it has been revealed that when moving heat source is in midpoint on the welded plates then deformation direction (shown by arrow) away from the center of moving heat source that means maximum elongation or thermal stress, thermal strain in the mid-position.

In an elastic body that is subject to a system of loads in three dimensions, a complex three-dimensional system of stresses is developed. That is, at any point

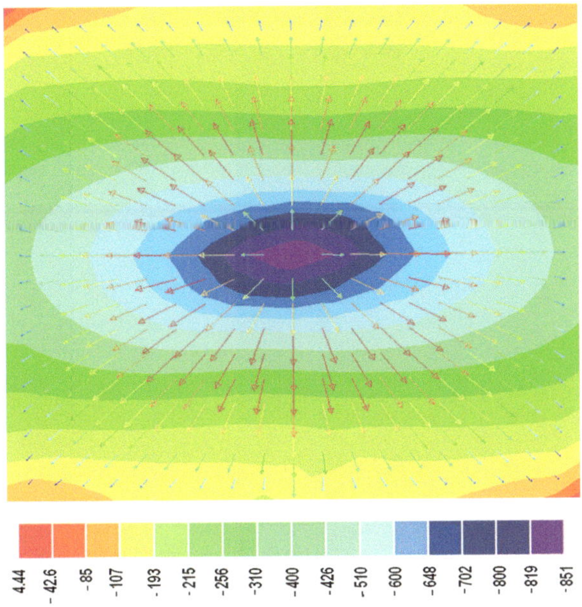

Fig. 7 Stress YY on the welded plates and vector plot of nodal displacements when electrode position is on (15, 0, 0)

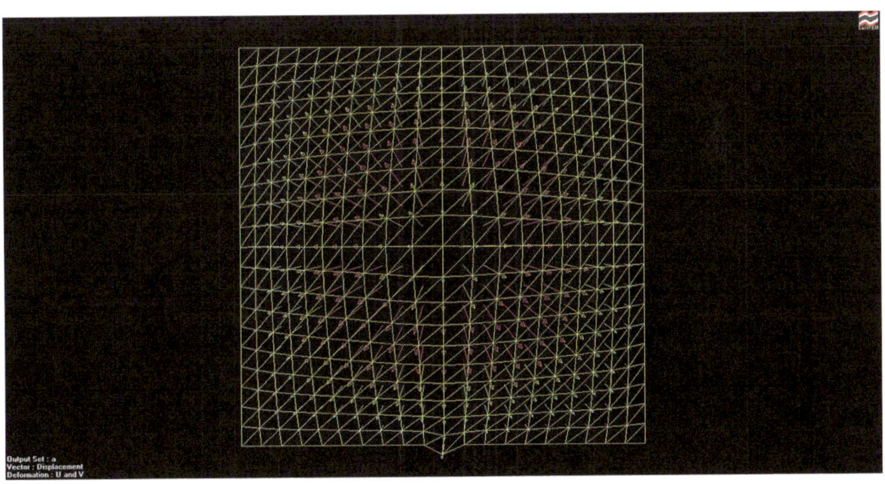

Fig. 8 Deformed meshes and vector plot of nodal displacements when electrode position is on (15, 0, 0)

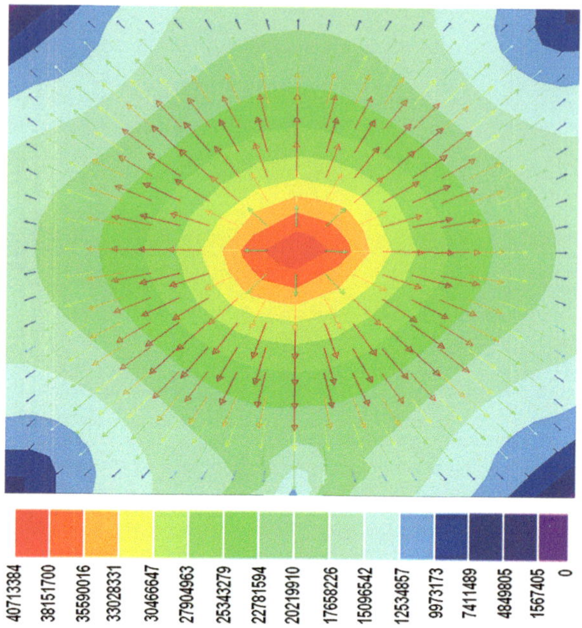

Fig. 9 Von-Mises stress distribution on welded plates and vector plot of nodal displacements

within the body, there are stresses acting in different directions, and the direction and magnitude of stresses changes from point to point. The von-Mises criterion is a formula for calculating whether the stress combination at a given point will cause failure. The equivalent stress is often called the von-Mises stress as a shorthand description. It is not really a stress, but a number that is used as an index. If the von-Mises stress exceeds the yield stress, then the material is considered to be at the failure condition.

With the help of EASY F E M V1.0 software, von-Mises stress distribution on welded plates is also found out (as shown in Fig. 9).

4 Results and Discussion

Heat is transferred from fusion zone to heat affected zone by conduction during the welding process. But during heat flow from heat affected zone to remaining zone, convective heat loss has also a vital role. It was found from experiment (refer Fig. 10) that near weld line (i.e., at point $x = 110$ mm, $y = 15$ mm), temperature was 1437 °C (when electrode was nearest to this point) but after 11 s from reaching of peak temperature (i.e., 1437 °C) temperature of this point was 1045 °C, after 29 s the temperature was 958 °C, and after 38 s temperature was 919 °C; so rate of cooling was initially high but it was decreasing with decrease of temperature of welded plate. The convection heat loss is proportional to temperature difference

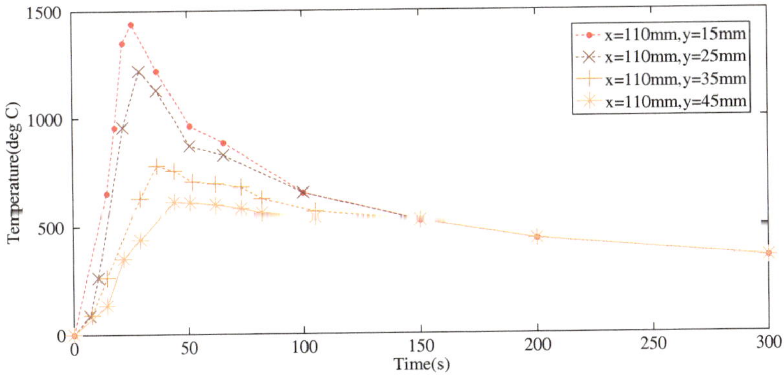

Fig. 10 Measured temperature variation with respect to time on welded plate

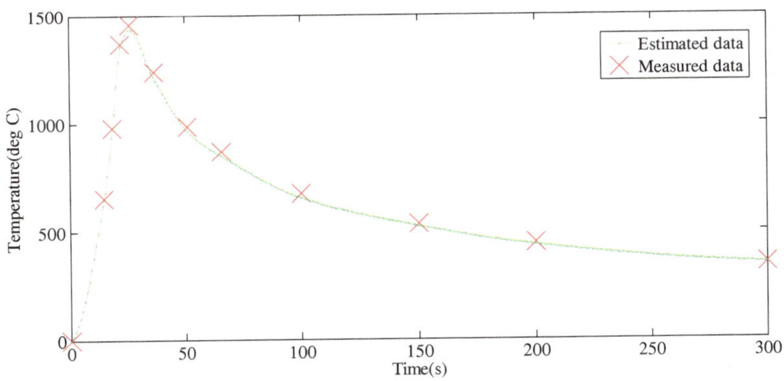

Fig. 11 Comparison between measured and estimated data at x = 110 mm, y = 15 mm

between welded plate and atmospheric temperature. Initially, temperature difference between welded plate and atmospheric temperature was larger but after some time temperature difference was smaller; so, convective heat loss was also low. It was found from time–temperature curve of Fig. 5 that the rate of change of temperature is high near the welding line as compared to that at a location away from the welding line for the time range 0–50 s. This is due to convective heat loss and as the electrode was reached at point x = 110 mm, y = 0 mm, z = 0 mm, just after 22 s from starting point (0, 0, 0) (as shown in Fig. 1). Decent agreement between predicted data and measured data has been achieved (refer Fig. 11).In present work, measured range of heat affected zone width is 2.19–3.93 mm (as described in Table 2).

5 Conclusions

To estimate the transient temperature distribution on welded plate in submerged arc welding process, the heat source was assumed as oval in shape (oval shape raised due to heat source and convection in fusion zone) and thermo-physical properties of materials were considered as function of temperature. Convective and radiative heat losses were also considered. The experimental measurement shows that the peak temperature was higher and the cooling rate was larger for a work piece closer to welding path with respect to other part of welded plate as convection heat loss is proportional to temperature difference between welded plate and atmospheric temperature. The numerical solution for oval shape heat source was used to calculate transient temperatures at selected points on C–Mn steel plates which were welded. The estimated temperature distribution obtained through finite difference method and those obtained from experimental measurements compared fairly well with variation up to 3 % for the peak temperatures. It was observed that the HAZ widths increase with increase in heat input or arc energy. Because of high heat input the heat removal rate is insufficient through convection cooling, and as a result temperature near the welded zone remains higher than the recrystallization temperature of welded plate for a sufficient time.

The validated numerical predictions indicate that the present solution could offer a good prediction for transient temperatures near the weld pool, as well as away from welding path. The newly developed heat source model has predictive potential for various welding simulations.

References

Bag S, De A (2010) Computational modelling of conduction mode laser welding process, laser welding. In: Stone XN (ed.) InTech. doi:10.5772/9861, ISBN 978-953-307-129-9

Bag S, Kiran DM, Syed A, De A (2012) Efficient estimation of volumetric heat source in fusion welding process simulation. Weld World 56(11–12):88–97

Bianco N, Manca O, Naso V (2004a) Numerical analysis of transient temperature fields in solids by a moving heat source. In: 3rd international conference on heat transfer, fluid mechanics and thermodynamics, Paper no. BN2, 21–24 June 2004 (HEFAT2004), Capetown, South Africa

Bianco N, Manca O, Nardini S (2004b), Two dimensional transient analysis of temperature distribution in a solid irradiated by a Gaussian lesser source. In: Proceeding of ESDA2004-58286 7th biennial conference on engineering systems design and analysis, 19–22 July 2004, Manchester, UK

Biswas P, Mandal NR (2010) Thermo mechanical finite element analysis and experimental investigation of single-pass single-sided submerged arc welding of C-Mn steel plates. Proc IMechE Part B J Eng Manuf 224:627–639

Cho MH (2006) Numerical simulation of arc welding process. Ph.D. thesis, School of the Ohio State University, USA

Elsen VM, Baelmans M, Mercelis P, Kruth JP (2007) Solutions for modeling moving heat sources in a semi-infinite medium and applications to laser material processing. Int J Heat Mass Transf 50:4872–4882

Esmaliy M, Shokuhfar A (2010) Numerical simulation of heat transfer in friction stir welding of 7075-T6 aluminium alloy and high carbon steel using Arbitrary Lagrangian Eulerian technique. Materialwissens craft und Werksofftechnik (Special issue: Advanced Computational Engineering and Experimenting) 41(5):350–355

Goldak JA, Chakraborty A, Bibby M (1984) A new finite element model for welding heat sources. Metall Trans B 15B:299–305

Isiklar YV, Girgin I (2011) Numerical modelling of underwater welding. J Naval Sci Eng 7(2):11–29

Bendaoud I, Matteï S, Cicala E, Tomashchuk I, Andrzejewski H, Sallamand P, Mathieu A, Bouchaud F (2014) The numerical simulation of heat transfer during a hybrid laser–MIG welding using equivalent heat source approach. Opt Laser Technol 56:334–342

Kazemi K, Goldak JA (2009) Numerical simulation of laser full penetration welding. Comput Mater Sci 44:841–849

Klobcar D, Tusek J, Taljat B (2004) Finite element modeling of GTA weld surfacing applied to hot-work tooling, Comput Mater Sci 31:368–378

Kumar S, Bhaduri SC (1994) Three-dimensional finite element modeling of gas metal-arc welding. Metall Trans B 25B:935–941

Lindgren LF (2001) Finite element modeling and simulation of welding Part-1: increased complexity. J Therm Stress 24:141–192

Mahapatra MM, Datta GL, Pradhan B, Mandal NR (2006) Three-dimensional finite element analysis to predict the effects of SAW process parameters on temperature distribution and angular distortions in single-pass butt joints with top and bottom reinforcements. Int J Press Vessels Pip 83:721–729

Mohassela A, Kokabi AH, Davamia P, Ranjbarnodeh E, Movahedia M (2013) Heat transfer modelling in wide gap rail thermite welding. J Adv Mater Process 1:29–38 (Article 5, Spring 2013)

Ohring S, Lunt HJ (1999) Numerical simulation of a time dependent 3-D GMA weld pool due to a moving arc. Weld J 79:416-s–424-s

Pang SY, Chen LL, Yin YJ, Duan AQ, Zhou JX, Hu LJ (2010) A parallel numerical study of transient heat transfer and fluid flow of weld pool during laser Keyhole welding. Adv Mater Res 97–101:3001–3004

Pardo E, Weckman DC (1989) Prediction of weld pool and reinforcement dimensions of GMA welds using a finite-element model. Metall Trans B 20B:937–947

Sabapathy PN, Wahab MA, Painter MJ (2000) The prediction of burn-through during in-service welding of gas pipelines. Int J Press Vessels Pip 77:669–677

Scutelnicu E, Iordachescu M, Blasco M, Iordachescu D (2008) Arc welding of dissimilar metals: FEA and experiments. In: Proceedings of the 8th international conference on trends in welding research, 1–6 June 2008. Callaway Gardens Resort, Pine Mountain, Georgia, USA, pp 241–246

Yaduwanshi DK, Bag S, Pal S (2014) Heat transfer analyses in friction stir welding of aluminium alloy. Proc Inst Mech Eng B J Eng Manuf. doi:10.1177/0954405414539297

Yeh RH, Liaw SP, Yu HB (2003) Thermal analysis of welding on aluminum plates. J Mar Sci Technol 11:213–220

Yeh RH, Liaw SP, Tu YP (2007) Transient three-dimensional analysis of gas tungsten arc welding plates. Numer Heat Transf Part A 51:573–592

Development and Analysis of Butt and Lap Welds in Micro-friction Stir Welding (μFSW)

Shuja Ahmed, Abhishek Shubhrant, Akash Deep and Probir Saha

Abstract Manufacturing of the ever smaller components, be it mechanical, electronic, etc., is being chased now a days. Likewise in friction stir welding (FSW), when the thickness of the joining plates is reduced to less than 1 mm, i.e. micro-friction stir welding (μFSW), the application areas relate to joining thin-walled structures in electrical, electronic and micro-mechanical assemblies and in the packaging industry. The advantages of FSW like it being a solid state process, it not requiring any shielding gases or fluxes and its possibility to weld dissimilar and different thickness alloys emphasize the use of it to weld micro-thickness plates. In the present work, commercial grade AA6XXX series aluminium alloys of thickness 0.44 mm have been welded together in both butt and lap fashion. Tests have been conducted to measure the joint's tensile strength (by conducting both transverse and longitudinal tension tests) and the micro-hardness. The use of lap welds in application areas of μFSW has been established.

Keywords Micro-friction stir welding (μFSW) · Butt and lap fashion · Tensile strength · Micro-hardness

S. Ahmed · A. Shubhrant · A. Deep · P. Saha (✉)
Department of Mechanical Engineering, Indian Institute of Technology, Patna, India
e-mail: psaha@iitp.ac.in

S. Ahmed
e-mail: shuja.pme13@iitp.ac.in

A. Shubhrant
e-mail: abhishek.me10@iitp.ac.in

A. Deep
e-mail: akash.me10@iitp.ac.in

R.G. Narayanan and U.S. Dixit (eds.), *Advances in Material Forming and Joining*,
Topics in Mining, Metallurgy and Materials Engineering,
DOI 10.1007/978-81-322-2355-9_15

1 Introduction

In the year 1991, The Welding Institute (TWI), UK, invented a new technology for joining materials using the frictional heat. Therein, the heat is generated using a rotating tool that aids in joining which is unlike the process of friction welding where the rotating shafts are themselves joined. Termed as 'friction stir welding' (FSW), this technology very soon gained importance for its ability to join all series of aluminium alloys which are used to a large extent in the aerospace, shipping, railroad and automobile industries.

For aluminium alloys, fusion welding process is not suitable due to its high thermal expansion coefficient. The high heat transfer rate makes the laser welding difficult, and due to high electrical transfer, the use of resistant welding is also a restriction. The weldability for aluminium alloys can be improved by using FSW. The process being a solid state eliminates problems related to fusion welding. Recrystallized microstructure having better mechanical properties can be obtained. The environment friendly nature of the process, which being the need of the time, goes on adding to its advantages.

In FSW, a non-consumable rotating tool with a probe is plunged into the joining area. The tool after reaching the required plunge depth undergoes dwell (rotation at the plunge depth for few seconds) which softens the workpiece material ahead in translation, making welding easier. The heating is achieved due to friction between the tool and the workpiece, and by the plastic deformation of the workpiece. Next is the rotation plus translation of the tool along the weld line which leads to the stirring action all along the joint line. During stirring, the material moves from the advancing side to the retreating side of the tool and undergoes plastic deformation at high temperature which produces fine and recrystallized grains. Figure 1 shows the schematic representation of the process and the parameters it is associated with.

Joint configurations for FSW include square butt, edge butt, T-butt, lap, multiple lap, T-lap, etc. In the application areas of μFSW, butt as well as lap joints can be used. Research works done in joining of micro-thickness plates have mainly focused on the butt joint configurations. However, lap joints can also be used as a replacement of fasteners and specifically at places where modifications in the current parts are not necessary. Workpieces of different thickness can also be easily friction stir welded when in lap joint configuration.

Tool rotation speed and tool traverse speed are two important process parameters that govern the quality of the joint. Using multi-objective optimization and analysis of variance (ANOVA), it can be shown that the tool rotational speed is a highly significant factor (Vijayan et al. 2010). The alloy used by the authors was AA 5083, and it was welded in 'butt' joint configuration. Further it is observed that there is an optimum value of FSW process parameters (spindle speed, welding speed, plunge force/depth) for which the weld quality is the best (Lakshminarayanan and Balasubramanian 2008; Peel et al. 2003; Hirata et al. 2007). At other combinations of these parameters, the resulting weld is comparatively poor. The analysis in the

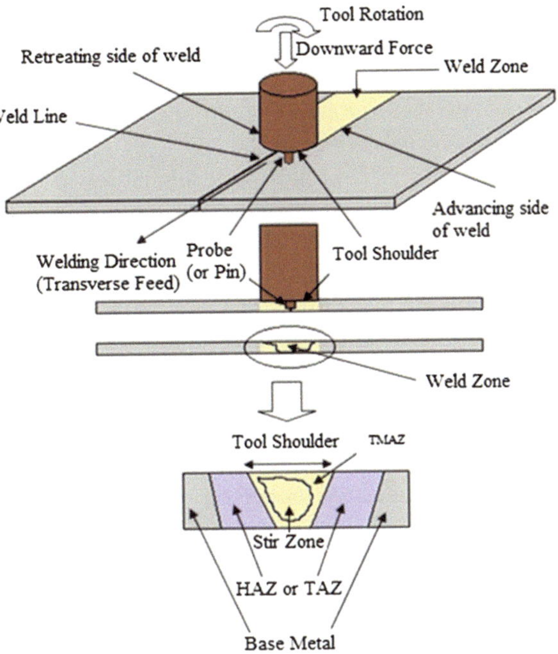

Fig. 1 Schematic representation of the FSW process (*TMAZ* thermo-mechanically affected zone; *HAZ* heat-affected zone; *TAZ* thermally altered zone)

present work has been carried out using such combinations of process parameters that gave good quality of the weld.

Limited work has been done in joining of micro-thickness plates by using μFSW. Scialpi et al. (2007) did a mechanical analysis of ultra-thin FSW and joined sheets with dissimilar and similar materials. μFSW was used for producing only 'butt' welds on 0.8-mm-thick sheets of 2024-T3, 6082-T6 materials and of dissimilar 2024-6082 alloys. The results showed that the joints had excellent mechanical properties, and the tensile failure occurring in the welded zone was because of irregularities in the thickness rather than by any defect. A successful feasibility study of μFSW was taken up by Nishihara and Nagasaka (2004) on AZ31 magnesium alloy. Sattari et al. (2012) welded sheets of AA 5083, 0.8 mm thick, in 'butt' joint configuration and studied its mechanical properties as well as the temperature distribution during welding. It was concluded that with higher speed, the resulting weld was having microscopic defects and welds obtained between 430 and 510 °C were defect free. On welding 0.8-mm-thick alloys of 2024-T3 and 6082-T6, Scialpi et al. (2008) observed the formation of two distinct stirred zones, one for each alloy.

The study of lap joints in μFSW is still an open area of research. For applications as in joining of lid to make an airtight (hermetically) sealed package, which is one common requirement of high-reliability electronics, 'butt' as well as 'lap' weld geometries can be used to join materials having similar or dissimilar cross sections. On investigating the influence of process parameters on the metallurgical and mechanical properties of friction stir welded lap joints of AA 2198-T4 aluminium alloy, it was shown that to produce effective FSW lap joints, the nugget area should be kept as wide as possible considering at the same time, the nugget integrity (Buffa et al. 2009). Further, the influence of tool geometry on the structural and mechanical properties of the lap joint of 5456 aluminium alloy was studied by Salari et al. (2014), and the result indicated that the stepped conical thread pin improved the joint's integrity and mechanical properties by improving the material's flow during FSW. The mounting of specimens for lap joint configuration was also taken care of by the authors during welding. To investigate the effect of pin profile on the friction stir processed zone formation in AA 2219 aluminium alloy, 'straight cylindrical, tapered cylindrical, threaded cylindrical, triangular and square' were the five profiles that were used to fabricate the joints at three different rotational speeds (Elangovan and Balasubramanian 2007). It was found that the square tool pin profile produces mechanically sound and metallurgically defect free welds as compared to other tool pin profiles. Fujii et al. (2006) studied the effect of tool shape on mechanical properties and microstructures of welded aluminium plates. Based on the results, an optimal tool shape was suggested for high melting temperature metals such as steels. Parida et al. (2014) surveyed that complicated tool profile wears out after a few number of runs specially while working with higher tensile strength materials. In the present work, the study was carried out using simple tool geometry—a straight cylindrical pin profile with flat shoulder.

2 Experimental Procedure

Commercial grade aluminium alloys of thickness 0.44 mm were joined using a rotating tool made of H13. The composition of the aluminium alloy and the tool is as shown in Tables 1 and 2, respectively. The rotating tool was plunged into and traversed perpendicular (zero tilt-angle) to the base material along the joint line of the faying surfaces. To begin with, a large number of joints were prepared at certain feeds (traverse speed) and rotations (spindle speed). Comparatively, better joints (according to visual inspection) were recorded at some particular parameters. Further tests have been conducted on samples prepared only at these parameters and which are as shown in Table 3.

Table 1 Composition of Al alloy

Element	Al	Si	Mg	Cr	Fe	Cu
%Weight	97.62	0.80	0.42	0.06	0.72	0.38

Table 2 Composition of H13 tool

Element	Cr	Mo	Si	V	C	Mn
% Weight	4.98	1.32	1.05	0.98	0.40	0.38

Table 3 Process parameters used in μFSW joints

Process parameters / Joint configuration	Tool rotation speed (rpm)	Tool traverse speed (mm/min)
Butt weld	1650	25
Lap weld	1700	25

2.1 Development of Tool

FSW at macro level requires a sufficiently hard welding tool which can soften the material without itself being deformed. Similarly, while making an approach to weld materials at micro-level, hardness play a vital role. The tool material must also retain its strength and hardness at the temperature of welding. The temperature reached during welding is up to 90 % of the melting point temperature of the material being joined. In the past, researchers have used tool materials such as H13 (52 HRC) and 56NiCrMoV7-KU (24.2 HRC) for welding aluminium alloys (Scialpi et al. 2007, 2008; Sattari et al. 2012). The tool material used for the present study was H13. Development of the tool geometry having dimensions in hundred of microns was achieved through micro-turning on DT-110 (high precision multi-purpose machine tool). The micro-turning operation was performed at 100 rpm spindle speed and a manual feed in sub-millimetres. Some deburring operations (sanding, facing and straight turning) were also performed after welding three to four samples. This was necessary as there were some burr formations observed on the tool, which would hamper the joint quality if proper care was not taken. Figure 2 shows the tool geometries that were used for welding the aluminium alloys in 'butt' and 'lap' configurations. The tool dimensions were chosen as per the literature survey.

2.2 Development of Fixture

μFSW requires very high precision. The fixture plays a very crucial role in this technique. The sheets should be aligned very accurately to avoid any deflection at the top surface. It has to be held very tightly so that by nullifying its movement during welding we get joints without defects. Also due to high heat conduction rate of aluminium, the heat loss can make it very difficult to soften the sheets. So, the fixture must be made up of heat-resistant materials. The fixture consisted of two major components—one was backing plate and the other was top clamps. To begin with, acrylic sheets of 10 mm thickness were used both as backing plate and as top clamping. But, it was observed that the welding samples got deflected at 1200 rpm of tool rotation speed. Also, due to high heat dissipation during welding operation, low-quality welds were obtained. Some improper welded samples and their reasons are as shown in Fig. 3.

Fig. 2 Geometry of the H13 tool used for μFSW

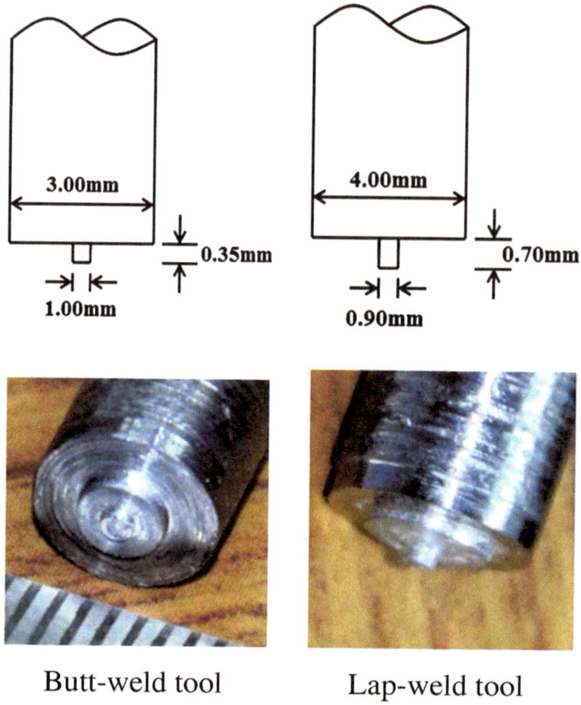

One potential reason among these was poor clamping of aluminium sheets. Therefore, a new fixture design was proposed in order to have better clamping during welding operation. Here, the clamping force was concentrated over the plates using two bolts creating a couple of forces in opposite direction at different lengths, hence providing the net downward moment to the clamps where they touched the sheets. Figure 3 shows the corresponding μFSW experimental setup. The backing plate used was of heat-resistant material Bakelite and this further minimized the effect of heat loss, which was more dominant in case of earlier used acrylic material.

2.3 *Welding Operation*

The aluminium sheets were cut as per the required dimensions on wire-EDM so as to have precise cutting, which would aid in better clamping of the samples at and around the faying surface. The welding operation was performed using the micro-machine DT-110. This machine provides high precision motion transmission with negligible backlash, low deformation and negligible vibration in order to achieve high accuracy in the welded component. The rotating tool was mounted on the spindle using a suitable collet. The sheets to be joined were clamped onto the

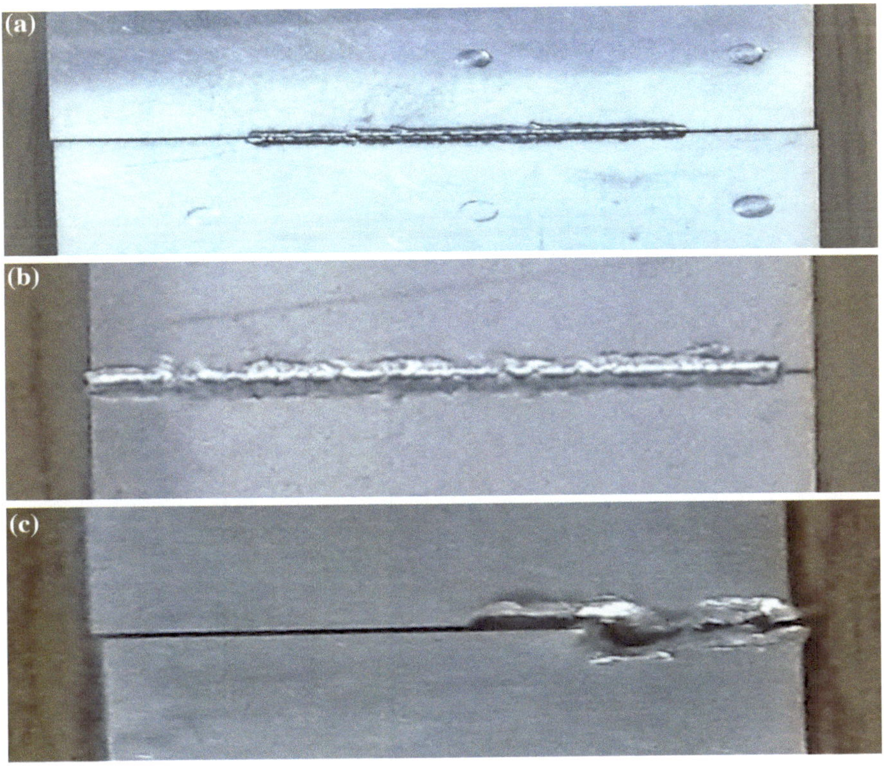

Fig. 3 Some failures observed during the welding. **a** Failure due to improper contact of shoulder and sheets. **b** Failure due to excessive thinning. **c** Failure due to deflection of sheets from the fixture

backing plate in such a way that any movement of the sheets could be avoided during welding. With the help of 'contact probe', the tool pin was made to touch the sheets at the weld line, and the co-ordinates were set to zero at that point (work reference system). Next, the G- and M-codes, where the rotational speed, plunge depth, transverse speed and weld length, which were set beforehand, were executed serially as mentioned below for performing the welding operation:

1. Choose co-ordinate system (absolute)
2. Select work reference system as defined
3. Start tool rotation at certain RPM
4. Rapid motion of tool to a certain safe height (z)
5. Rapid motion to point (x = 0, y = 0)
6. Linear interpolation of tool to z = 0.1 at certain feed
7. Linear interpolation of tool up to plunge depth at certain feed
8. Dwell for a few seconds
9. Linear interpolation up to weld length at certain traverse feed
10. Exit of tool from workpiece at certain feed
11. Stop tool rotation

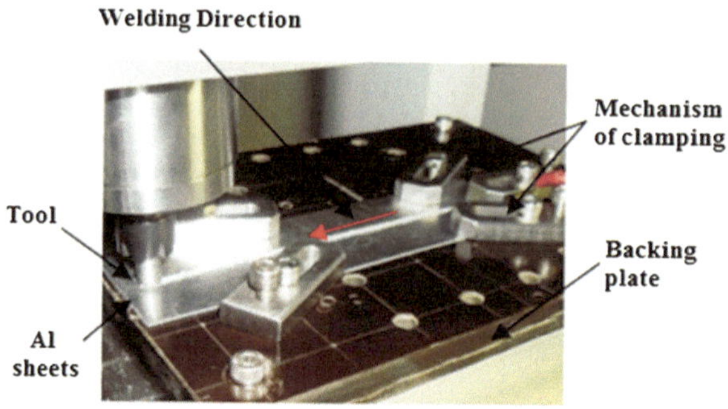

Fig. 4 μFSW experimental set up

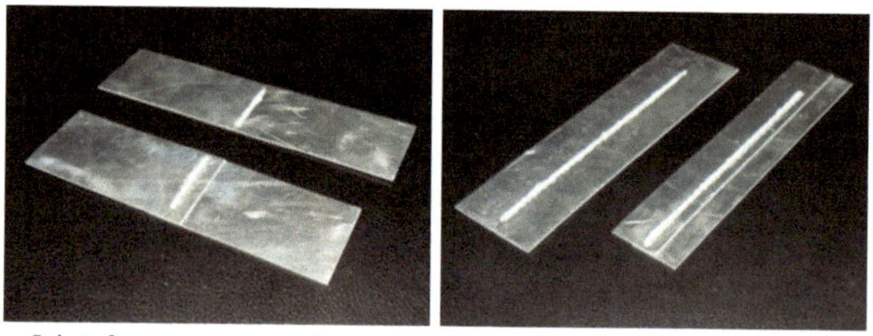

Joints for transverse tension test   Joints for longitudinal tension test

Fig. 5 Weld specimens prepared using μFSW

While performing lap welding, it was taken care that the advancing side of the probe was located near the top sheet edge (Salari et al. 2014). An overlap width of about 10 mm was used. The μFSW experimental set up is shown in Figs. 4 and 5 which shows some of the weld specimens prepared.

3 Results and Discussion

The results of longitudinal tension test, transverse tension test and micro-hardness test on both of 'butt' and 'lap' joint configurations are described in subsequent subsections.

3.1 Tensile Test

Tensile tests were conducted to assess the mechanical properties of the joint. The transverse tension test, where the applied load is orthogonal to the weld direction investigates the joint's efficiency by studying its mechanical behaviour. The

Table 4 Mechanical properties of Al alloy

Property	Value
Yield tensile strength (MPa)	90
Ultimate tensile strength (MPa)	110
% Elongation	8
Vickers hardness (HV)	46

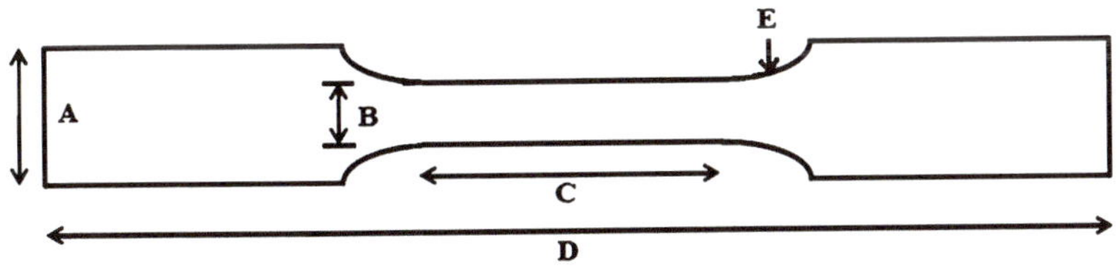

(i) A=10; B=6; C=32; D=100; E=R6.

(ii) A=15; B=4; C=24; D=120; E=R13.

Fig. 6 (i) Dimensions (in mm) of the transverse tension test specimen as per ASTM standard (ii) Dimensions (in mm) of the longitudinal tension test specimen as per ASTM standard

longitudinal tension test, where the applied load is along the direction of the weld, is used to study the mechanical behaviour of the weld zone subjected to stirring. The results obtained from these tests were compared to with that of the base material. Table 4 gives some mechanical properties of the aluminium alloy we have used.

To ensure the accuracy of the measurement, the tests were conducted on three welded samples. First, the welded samples of specific weld length were produced using this technology and then these were cut on wire-EDM as per the specifications of the standard test method for tension testing, ASTM-E8 as shown in Fig. 6.

The tensile tests were performed on UTM machine. Figure 7 shows the engineering stress–strain curve of the base metal, butt weld and lap weld in transverse tension test, and Fig. 8 shows the engineering stress–strain curve of the same in longitudinal tension test.

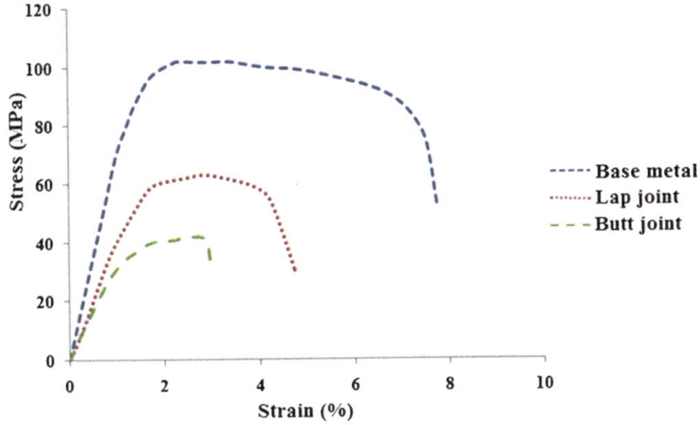

Fig. 7 Stress–strain curve for transverse tension test

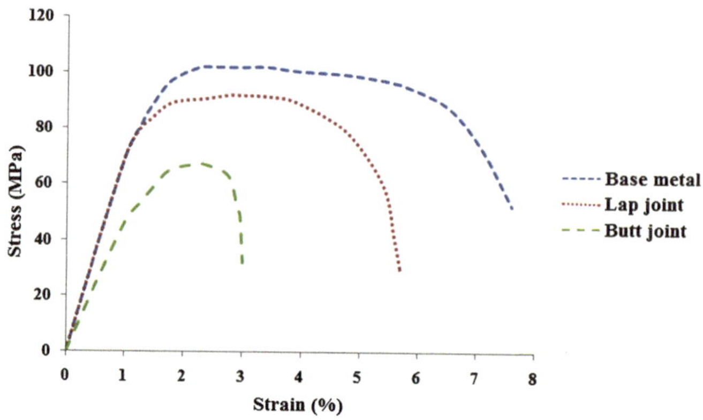

Fig. 8 Stress–strain curve for longitudinal tension test

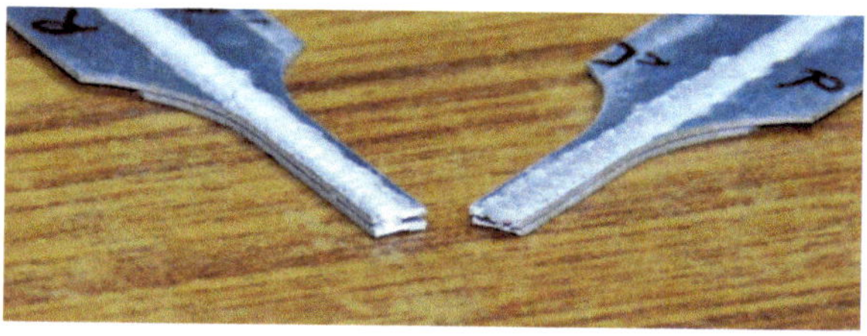

Fig. 9 Failure cross section after longitudinal tensile test on lap weld

Through the plots, it can be seen that the lap welds are having better tensile properties than the butt welds in both transverse and longitudinal tension tests. Failure of the lap joint in transverse tension was observed to be in the thinner upper weld instead of at the nugget between the sheets revealing thinning of material as the reason for failure. Longitudinal tests for lap joints showed convincing results. Higher yield point was observed in longitudinal tension tests as compared to transverse tests. Figure 9 shows the corresponding longitudinal tensile test failure on one of the lap welded samples.

The joint efficiency defined as the tensile strength of the weld divided by the tensile strength of the base material is a measure of the welded joint's strength. The joint efficiency for transverse welds was observed as 65 % for lap joints and 44 % for butt joints, and for longitudinal welds, the observed joint efficiency was 91 and 66 % for lap and butt joints, respectively.

3.2 *Micro-Hardness Test*

Hardness tests are performed to test the material's ability to resist plastic deformation, usually by penetration and at places also due to scratching or cutting. The

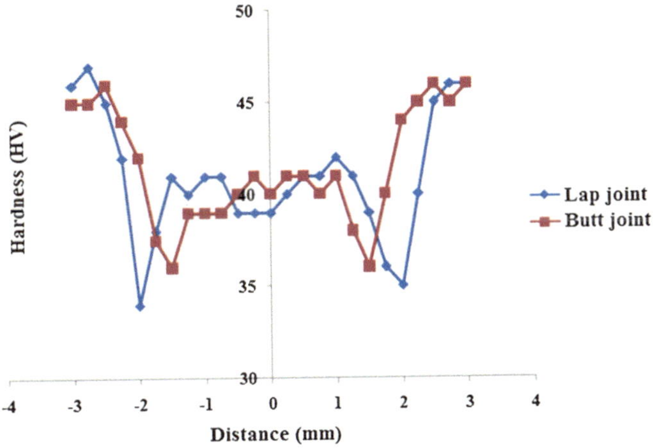

Fig. 10 Cross-sectional micro-hardness of 'butt' and 'lap' μFSW

present micro-hardness test was performed using a Vickers penetrator under a load of 100 gf for 10 s which was as found in some literatures (Scialpi et al. 2007, 2008; Sattari et al. 2012). The samples were polished with emery papers of different grit sizes. Readings were taken on cross sections orthogonal to the welding direction at distances of 0.25 mm apart. The obtained curve for hardness profile is as shown in Fig. 10.

It can be observed that the stir zone (SZ) has more or less a constant value of hardness. A slight drop was found when moving from SZ to TMAZ with a minimum somewhat at the junction between TMAZ and HAZ. Hardness value again increased in the HAZ and became constant for the base metal. Ductile failure near the joint end was observed relating it with lower hardness values there. No significant difference between lap and butt joint hardness trend could be observed other than a little lower minimum value for the lap joint which could be attributed to its having better ductility at the time of tensile test failure.

4 Conclusions

In this work, using μFSW, 0.44-mm-thick aluminium alloys have been successfully welded in both 'butt' and 'lap' joint configurations. In application areas where there is possibility of butt as well as lap configuration of joints, joining the materials in lap fashion would produce better joint strength in the mentioned aluminium alloy. Increase in thickness, near and along the joint line, when producing lap weld is a demerit in its application areas as this occupies comparatively more space. It was observed that the more the material is available for stirring, the stronger is the joint. For the given alloy, having thickness approximately less than 0.5 mm, it is better to opt for lap welding where the thickness increase at the joint would increase the material available for stirring, and hence improve the performance measures of μFSW.

Some of the future challenges that remain are in fabrication of complex tool geometry at such small scale to study its effect on the heat generation and distribution during welding. Further development in fixture is possible which could lead to more precise clamping all along the weld length. This would again improve the joint quality.

References

Buffa G, Campanile G, Fratini L, Prisco A (2009) A friction stir welding of lap joints: influence of process parameters on the metallurgical and mechanical properties. J Mater Sci Eng A 519:19–26

Elangovan K, Balasubramanian V (2007) Influences of pin profile and rotational speed of the tool on the formation of friction stir processing zone in AA2219 aluminium alloy. J Mater Sci Eng A 459:7–18

Fujii H, Cui L, Maeda M, Nogi K (2006) Effect of tool shape on mechanical properties and microstructure of friction stir welded aluminium alloys. J Mater Sci Eng A 419:25–31

Hirata T, Oguri T, Hagino H et al (2007) Influence of friction stir welding parameters on grain size and formability in 5083 aluminum alloy. J Mater Sci Eng A 456:344–349

Lakshminarayanan K, Balasubramanian V (2008) Process parameters optimization for friction stir welding of RDE-40 aluminum alloy using Taguchi technique. Trans Nonferrous Met Soc China 18:548–554

Nishihara T, Nagasaka Y (2004) Development of micro-FSW. In: Proceedings of the 5th international friction stir welding symposium Metz, France, 14–16 Sept 2004

Parida B, Mohapatra MM, Biswas P (2014) Effect of tool geometry on mechanical and microstructural properties of friction stir welding of Al-alloy, Int J Curr Eng Technol, Special Issue 2, pp 88–92

Peel M, Steuwer A, Preuss M, Withers PJ (2003) Microstructure, mechanical properties and residual stresses as a function of welding speed in aluminium AA5083 friction stir welds. Acta Mater 51:4791–4801

Salari E, Jahazi M, Khodabandeh A, Ghasemi-Nanesa H (2014) Influence of tool geometry and rotational speed on mechanical properties and defect formation in friction stir lap welded 5456 aluminium alloy sheets. J Mater Des 58:381–389

Sattari S, Bisadi H, Sajed M (2012) Mechanical properties and temperature distributions of thin friction stir welded sheets of AA5083. Int J Mech Appl 2(1):1–6

Scialpi A, De Giorgi M, De Filippis LAC, Nobile R, Panella FW (2007) Mechanical analysis of ultra-thin friction stir welding joined sheets with dissimilar and similar materials. J Mater Des 29:928–936

Scialpi A, De Fillipis LAC, Cuomo P, Di Summa P (2008) Micro friction stir welding of 2024-6082 aluminium alloys. Weld Int 22(1):16–22

Vijayan S, Raju R, Rao SRK (2010) Multiobjective optimization of friction stir welding process parameters on aluminium alloy AA 5083 using Taguchi-based grey relation analysis. Mater Manuf Process 25(11):1206–1212

Numerical Modeling of Impact and Solidification of a Molten Alloy Droplet on a Substrate

Rajesh Kumar Shukla, Sateesh Kumar Yadav, Mihir Hemant Shete and Arvind Kumar

Abstract Most of the studies reported for droplet impact and spreading on a substrate in a thermal spray coating process assume that droplet material solidifies as a pure substance, i.e., phase change occurs at a fixed temperature. The alloy-type behavior of the droplet impact where it solidifies within liquidus and solidus temperature is not well reported. The role of formation of mushy zone and species composition variation during the coating layer formation while using a multi-constituent alloy material is not established. This work investigates the spreading and solidification characteristics of an alloy droplet impacting on a substrate. Two-dimensional axisymmetric model has been used to simulate the transient flow and alloy solidification dynamics during the droplet impingement process. Volume of fluid (VOF) surface tracking method coupled with the alloy solidification model within a one-domain continuum formulation is developed to describe the transport phenomena during the droplet impact, spreading, and solidification of an alloy droplet on a flat substrate. Using the model, the characteristics of alloy solidification in coating formation are highlighted.

Keywords Droplet impact · Solidification · CFD · Mushy zone · Coating

Nomenclature

c	Specific heat capacity (m^2 s^{-2} K^{-1})
C	Solute concentration (kg kg^{-1})
D	Solute mass diffusivity (m^2 s^{-1})
D_o	Initial droplet diameter (mm)
f_l	Weight fraction of liquid (kg kg^{-1})
f_s	Weight fraction of solid (kg kg^{-1})
F	Volume of fluid function
F_{vol}	Continuum surface tension force source term (N m^{-3})
$\vec{g}_r$	Acceleration due to gravity vector (m s^{-2})

R.K. Shukla · S.K. Yadav · M.H. Shete · A. Kumar (✉)
Department of Mechanical Engineering, Indian Institute of Technology Kanpur, Kanpur 208016, India
e-mail: arvindkr@iitk.ac.in

R.G. Narayanan and U.S. Dixit (eds.), *Advances in Material Forming and Joining*, Topics in Mining, Metallurgy and Materials Engineering,
DOI 10.1007/978-81-322-2355-9_16

k	Thermal conductivity (W m^{-1} K^{-1})
k_p	Equilibrium partition coefficient
K	Permeability of mushy zone (m^2)
L	Latent heat of fusion (J kg^{-1})
m_L	Liquidus slope (°C wt%$^{-1}$)
p	Pressure (N m^{-2})
t	Time (s)
T	Temperature (K)
T_m	Melting point of pure substance (K)
$\vec{u}$	Continuum velocity vector (m s^{-1})
U_0	Initial droplet velocity (m s^{-1})

Greek symbols

ρ	Density (kg m^{-3})
μ	Dynamic viscosity (kg m^{-1} s^{-1})

Subscript

d	Droplet
subs	Substrate
air	Air
alloy	Alloy
eff	Effective
l	Liquid
s	Solid
0	Initial

1 Introduction

The impact of a drop of molten liquid on a solid surface is currently receiving much attention by industries. The study of impact of molten droplet on substrate can be utilized in many industrial applications such as spray cooling of hot surfaces, ink jet printing, spray painting, fire suppression using sprinkler systems, spray forming, deposition of solder bumps on printed circuit boards, thermal spray coating, soil erosion by rain drops, ice accumulation on electric wires, and aircraft. There are many physical phenomena during the molten droplet impact which are still to be addressed. Such phenomena are given as follows: tracking of free liquid surfaces, motion of a liquid–solid–air contact line, wetting of solid surfaces, and fluid instabilities that cause formation of fingers around the droplet periphery, eventually detachment of satellite droplets. The complexity is further increased if droplets solidify while spreading. This study primarily focuses on the thermal spray deposition process. The coating formed by thermal spray is widely used to protect components exposed to corrosion, wear, and heat.

In thermal spraying coating processes, powder materials are injected into a high-temperature flame where they are melted and propelled toward the surface of a substrate. The individual molten particles impact, spread, and solidify to form a coating deposit on the substrate's surface. Prior to the coating development, the process involved is impact of the droplet and its flattening and solidification on the substrate (Fig. 1). The quality of the coating is directly related to the particle impingement process in which individual particles are flattened and solidified on the substrate forming the individual splats (Kamnis and Gu 2005; Kumar et al. 2012a). It is essential to have a clear understanding of the physics of droplet impingement on the surface of substrates for better control of the generation of splats and the structure of coating. Experimental investigations provide a link between factors related to droplet impact and final coating morphology and quality. Numerical simulations provide further insights into the mechanism of droplet impingement and coating layer formation. It also provides an idea about tuning the process parameters to suit different coating applications. Computational fluid dynamics (CFD)-based volume of fluid (VOF) method is utilized in many cases to track the deformation in the surface of molten droplet during its impact on a substrate.

Many studies have been carried out for modeling droplet impact on surfaces. Madejski (1976) developed a simple model that predicted the maximum extent of spread of a molten droplet impacting on a solid surface, which agreed within an order of magnitude with the size of splats formed by dropping tin and lead droplets on a fiat plate. Liu et al. (1993) numerically studied the deformation, interaction, and freezing during impact of molten droplets on a substrate in plasma spray processes and reported that the final diameter and total freezing time of the splat depend on the

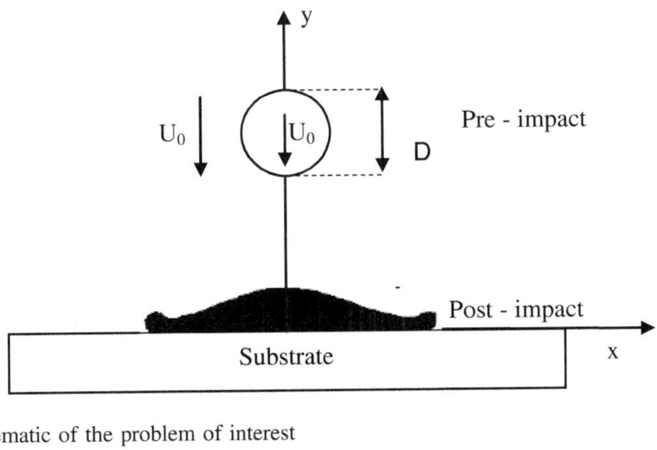

Fig. 1 Schematic of the problem of interest

initial droplet velocity, droplet temperature, and substrate temperature. Bennett and Poulikakos (1994), Kang et al. (1994) studied droplet deposition assuming that solidification starts only after complete spreading of droplet in the form of a disk. Theoretical and experimental studies done by Bennett and Poulikakos (1994) showed that the thermal conductivity of the substrate significantly influences the cooling rate of the splat. Inada and Yang (1994) photographed the underside of a lead droplet as it landed on a quartz plate and observed solidification at the droplet–substrate interface. Waldvogel et al. (1996) deposited a eutectic tin–lead solder droplet with diameter of about 50 mm on a composite substrate. The detailed study of final solidified shapes and flow field at the onset of phase change indicated a strong coupling of the droplet fluid dynamics to the freezing behavior.

Rangel and Bian (1997) developed a heat transfer model to study the mechanism of substrate remelting and the interactions of relevant parameters during the process of single metal droplet deposition and solidification. Fukumoto and Huang (1999) studied the transition behavior of flattening nickel particles sprayed onto a flat substrate. They showed that splashing is less when there is good wetting between the particle and substrate. Wang et al. (1999) conducted numerical studies of the cooling and solidification, calculated the temperature history of the splat and the substrate, and generated the data for maximum melting depth of the substrate for various process parameters. Bussman et al. (2000) used a three-dimensional CFD code to model droplet impact and splashing. They initiated the growth of fingers in their simulations by introducing a sinusoidal perturbation into the velocity field immediately after impact. They speculated that in reality protrusions on the surface disturb liquid flow, with the amplitude of the perturbation proportional to the magnitude of surface roughness. Aziz and Chandra (2000) studied the effect of varying impact velocity and substrate temperature during impact of molten tin droplets on a stainless steel plate. They photographed the deformation of impinging droplets and measured surface temperature variation during impact. Attinger and Poulikakos (2001) studied the molten micro-droplet impact and solidification on a cold flat substrate that can melt and they reported the melted volume as a function of time for various combinations of thermal and fluid dynamics parameters. Shakeri and Chandra (2002) studied the splashing of molten tin droplets on a rough steel surface. Ahmed and Rangel (2002) numerically studied the impingement and solidification of an aluminum droplet on uneven substrates, using a two-dimensional axisymmetric model. Their results show that droplet impact onto an uneven substrate is almost always accompanied by splashing. However, the degree of splashing decreases with the increase in surface roughness height. Li et al. (2004) developed a numerical model to study the integrated effect of droplet solidification, and melting and resolidification of the substrate during thermal spray coating. Hong and Qiu (2005) studied the mm-sized molten droplet impacting on a substrate of the same material. It was found that the thermal contact resistance affects not only the droplet spreading but also the substrate remelting volume and remelting front configurations.

Kamnis and Gu (2005) developed a numerical model to investigate the impingement of tin droplets on a flat stainless steel plate. In this, they reported that the solidification of droplets is significantly affected by the thermal contact resistance/substrate surface roughness. Kamnis et al. (2008) studied the sequential tin droplet impingements on a stainless steel substrate using a 3D numerical simulation. Fukumoto et al. (2011) studied the effects of substrate temperature and ambient pressure on flattening and solidification behavior of free falling droplet. Tabbara and Gu (2012) studied the impingement phenomena for molten tin droplets with low to high velocities on a steel substrate and reported that the solidification process may consist of up to three stages: planar solidification, uneven solidification, and wave mixing. Kumar et al. (2012a) compared the impingement behavior of a hollow ZrO_2 droplet and an analogous continuous droplet onto a substrate.

Most of the studies reported assume that droplet material solidifies as a pure substance, i.e., phase change occurs at a fixed temperature, during its impact and spreading on a substrate. The impingement of an alloy droplet onto a substrate and its spreading behavior can differ fundamentally from a pure metal droplet. In this connection, the investigation of the impact of an alloy droplet onto a solid substrate is of scientific and applied importance. The current work aimed at modeling the impact, spreading, and segregation of species of a Pb–Sn alloy droplet on a stainless steel substrate. A sound understanding of the alloy droplet impingement behavior onto a substrate will improve the knowledge of coating formation by multi-constituents materials in thermal spray coating.

2 Model and Governing Equations

The impact of a molten Sn–10 % Pb alloy droplet with a diameter of 1.1 mm at an initial uniform temperature of 519 K is made on a stainless steel substrate with a velocity of 4 m/s. Figure 2 shows a schematic of the computational domain. Since the physical geometry undertaken is axisymmetric, we have considered a 2D axisymmetric computational domain which has been shown in Fig. 2. The solidification of the binary alloy is considered in this study which occurs by the formation of a mushy zone. The phase diagram of the binary alloy considered is shown in Fig. 3. The thermophysical property data used in the simulations for the Sn–10 % Pb alloy are shown in Table 1 (Bellet et al. 2009).

The numerical methodology is based on a previously validated model for a dense droplet impact on a substrate (Kamnis and Gu 2005). The dense droplet impingement model reported in reference Kamnis and Gu (2005) was validated with the experimental measurement of tin droplets (Shakeri and Chandra 2002). In the present model, unsteady fluid flow dynamics is considered during the impact of droplet and spreading onto the substrate. Heat transfer along with solidification is considered using the volume of fluid surface tracking method (VOF) coupled with a solidification model within a one-domain continuum approach based on the classical mixture theory (Kamnis and Gu 2005; Kumar et al. 2012a, b). However, an alloy solidification model is implemented in the current study as compared to a pure substance solidification model in Kumar et al. (2012a, b). The mushy zone, the solidification kinetics, and the species transport, as applicable for a binary alloy, are

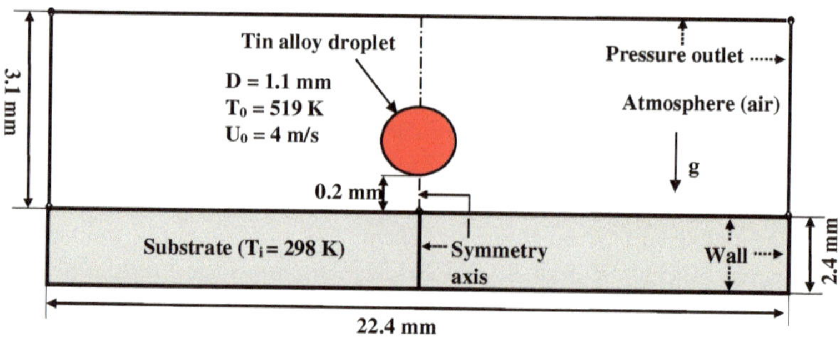

Fig. 2 Schematic diagram of the computational domain

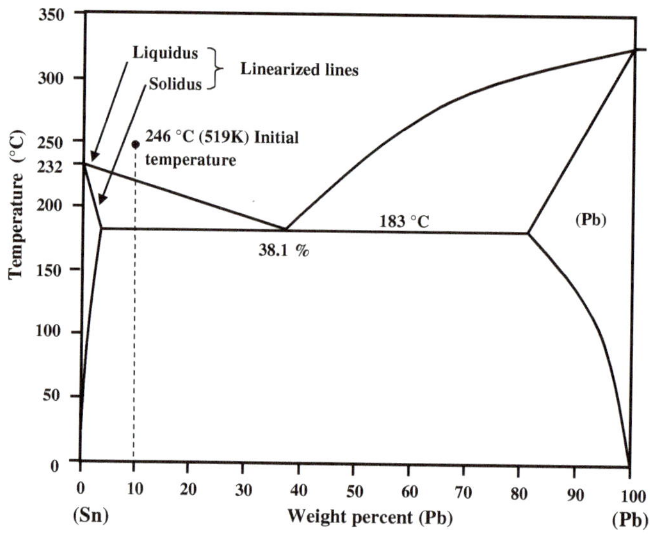

Fig. 3 Phase diagram of Sn–Pb binary alloy

considered in the current model. Shrinkage due to solidification is neglected ($\rho_s = \rho_l$). The density for alloy phase is defined as follows:

$$\rho_{\text{mixture}} = (1 - g_l)\rho_s + g_l\rho_l \tag{1}$$

Due to shrinkage being neglected, the volume fraction (g_k) is equivalent to mass fraction (f_k). The solid phase velocity is zero, $\vec{u}_s = 0$. Accordingly, the mixture velocity is given as follows:

$$\vec{u} = f_l\vec{u}_l + f_s\vec{u}_s = f_l\vec{u}_l \tag{2}$$

Table 1 Thermophysical properties

Properties	Value
Impinging droplet material	Sn–10 % Pb alloy
Substrate material	Stainless steel (SS)
Gas phase (the droplet surrounding medium)	Air (air)
Droplet initial temperature	519 K
Substrate initial temperature	298 K
Melting point of pure substance (Tin)	505 K
Initial concentration (Pb)	10.0 (wt%)
Thermal conductivity	55 W $(\text{m K})^{-1}$
Thermal conductivity (SS)	14.9 W $(\text{m K})^{-1}$
Thermal conductivity (air)	0.0242 W $(\text{m K})^{-1}$
Density (liquid alloy)	7000 kg $(\text{m})^{-1}$
Density (SS)	7900 kg $(\text{m})^{-1}$
Density (air)	1.225 kg $(\text{m})^{-1}$
Droplet surface tension	0.566 N $(\text{m})^{-1}$
Viscosity (liquid droplet)	0.0022 kg $(\text{m s})^{-1}$
Viscosity (air)	1.7894×10^{-5} kg $(\text{m s})^{-1}$
Specific heat capacity (solid and liquid phase)	260 J $(\text{kg K})^{-1}$
Specific heat capacity (SS)	477 J $(\text{kg K})^{-1}$
Specific heat capacity (air)	1006.43 J $(\text{kg K})^{-1}$
Latent heat of fusion	58500 J $(\text{kg})^{-1}$
Eutectic composition	38.1 (wt%)
Equilibrium partition coefficient	0.0656
Liquidus slope	−1.286 (°C $\text{wt\%}^{-1}$)

A VOF method is used to track of the alloy–air interface. The volume of fraction (F) can range from zero to unity. $F = 0$ indicates the cell containing only air, and $F = 1$ indicates the cell containing only alloy. $0 < F < 1$ corresponds to cell containing the alloy–air interface. The volume of fraction function is evaluated using the velocity field and the transport equation.

VOF equation:

$$\frac{\partial F}{\partial t} + \nabla \cdot \left(\vec{u} F\right) = 0 \tag{3}$$

The properties appearing in the transport equation are evaluated by the presence of component phases in each control volume. For example, the density in each control volume is given as follows:

$$\rho = F\rho_{\text{alloy}} + (1 - F)\rho_{\text{air}} \tag{4}$$

The flow is assumed to be Newtonian and incompressible; the governing conservation equations are given below:

Continuity:

$$\nabla \cdot \vec{u} = 0 \tag{5}$$

Momentum conservation:

$$\frac{\partial}{\partial t}(\rho \vec{u}) + \nabla \cdot (\rho \vec{u}\vec{u}) = -\nabla p + \nabla \cdot \left(\mu \frac{\rho}{\rho_l} \nabla \vec{u}\right) + \rho \vec{g_r} + \overrightarrow{F_{vol}} - S\vec{u} \tag{6}$$

$$S\vec{u} = \begin{cases} \frac{\mu_l \vec{u}}{K} & F = 1 \\ 0 & F < 1 \end{cases} \tag{7}$$

$$K = K_0 \frac{f_l^3}{(1 - f_l)^2} \tag{8}$$

For computational cells which are undergoing phase change (solidification), the solid–liquid interaction in the momentum conservation (Eq. 6) is considered using Darcy's model of viscous flow through a porous medium using the source term Su (Eq. 7) (Kamnis and Gu 2005). Permeability is defined by the Carman–Kozeny formula (Eq. 8). Momentum conservation equation (Eq. 6) also accounts for surface tension effects at the free surface, which is considered by a continuum surface force model using the term F_{vol} (Kumar et al. 2012a).

Energy conservation:

$$\frac{\partial}{\partial t}(\rho c_{eff} T) + \nabla \cdot (\rho \vec{u} c_{eff} T) = \nabla \cdot (k_{eff} \nabla T) + S_h \tag{9}$$

$$k_{eff} = F k_{metal} + (1 - F) k_{air} \tag{10}$$

$$c_{eff} = F c_{metal} + (1 - F) c_{air} \tag{11}$$

$$S_h = \begin{cases} -L\left[\frac{\partial}{\partial t}(\rho f_l) + \nabla \cdot (\rho \vec{u} f_l)\right] & F = 1 \\ 0 & F < 1 \end{cases} \tag{12}$$

The momentum and the energy conservation equations (Eqs. 6 and 9) are coupled. The source term S_h (Eq. 12), for handling the solidification phase change, appearing in the energy conservation (Eq. 9) is active only for the computational cells filled with molten droplet ($F = 1$).

Species transport equation:

$$\frac{\partial}{\partial t}(\rho C_l) + \nabla \cdot (\rho \vec{u} C_l) = \nabla \cdot (D^+ \nabla C_l) + S_c \tag{13}$$

where

$$D^{+} = \rho\left(f_{\mathrm{s}}D_{\mathrm{s}}k_{\mathrm{p}} + f_{\mathrm{l}}D_{\mathrm{l}}\right) \tag{14}$$

$$S_{\mathrm{c}} = \frac{\partial}{\partial t}(\rho f_{\mathrm{s}}C_{\mathrm{l}}) - k_{\mathrm{p}}C_{\mathrm{l}}\frac{\partial}{\partial t}(\rho f_{\mathrm{s}}) \tag{15}$$

Species segregation is governed by lever rule model (Eq. 16). The solidification path (according to lever rule) is given by the following equation:

$$f_{\mathrm{l}} = \frac{1}{1-k_{\mathrm{p}}}\left[1 - \frac{m_{\mathrm{L}}C_{0}}{T-T_{\mathrm{m}}}\right] \tag{16}$$

$$T = T_{\mathrm{m}} + m_{\mathrm{L}}C_{\mathrm{l}} \tag{17}$$

The mode of heat transfer in the substrate is considered by conduction only. A constant value of thermal contact resistance of 1.8×10^{-6} m^{2} K W^{-1} is considered corresponding to a substrate roughness of 0.06 μm (Kumar et al. 2012a). The heat transfer in the substrate is governed by the following equation:

$$\frac{\partial}{\partial t}(\rho_{\mathrm{sub}}C_{\mathrm{sub}}T) = \nabla \cdot (k_{\mathrm{sub}}\nabla T) \tag{18}$$

The computational domain, shown in Fig. 2, consists of 220,000 structured computational cells with a refined grid at the substrate surface.

3 Results and Discussion

A 1.1-mm-diameter Sn–10 % Pb droplet is impinging with a velocity of 4 m/s on a stainless steel substrate. The thermophysical properties are given in Table 1. In the following, the characteristics related to impact, flattening, and solidification of the alloy droplet are studied and discussed in detail. The initial flattening and impact is mainly dominated by the inertia of the impacting alloy droplet. The flattening and the spreading of the droplet can be seen in Fig. 4 which show the VOF function of the alloy phase. At the periphery, the spreading droplet solidifies due to heat transfer with the substrate as shown in Fig. 5. The droplets material at the edge of the splat is found to separate from the spreading droplet. This causes the splat to become discontinuous. Such discontinuous splat can influence the coating quality adversely.

Due to solidification occurring between liquidus and solidus, there is a formation of mushy zone as seen in Fig. 5. This also causes an increase in the solidification time as compared to pure metal droplet. The mushy zone formed is disordered and wavy in shape. The disordered mushy zone offers significant resistance to the droplet material to spread during subsequent flattening process. For better clarity of the mushy zone, zoomed images are shown in Fig. 6.

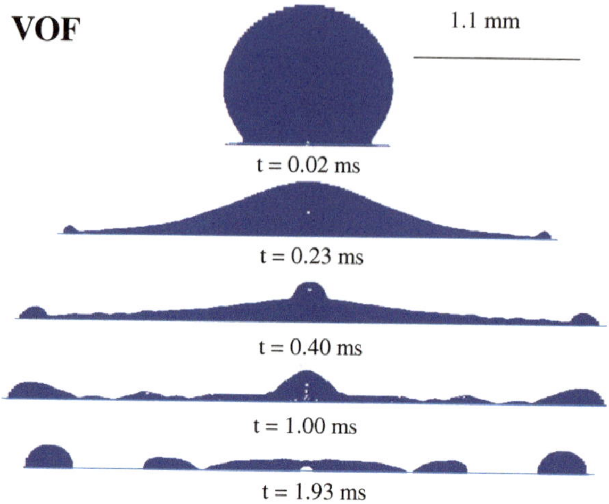

Fig. 4 Droplet impact and spreading pattern

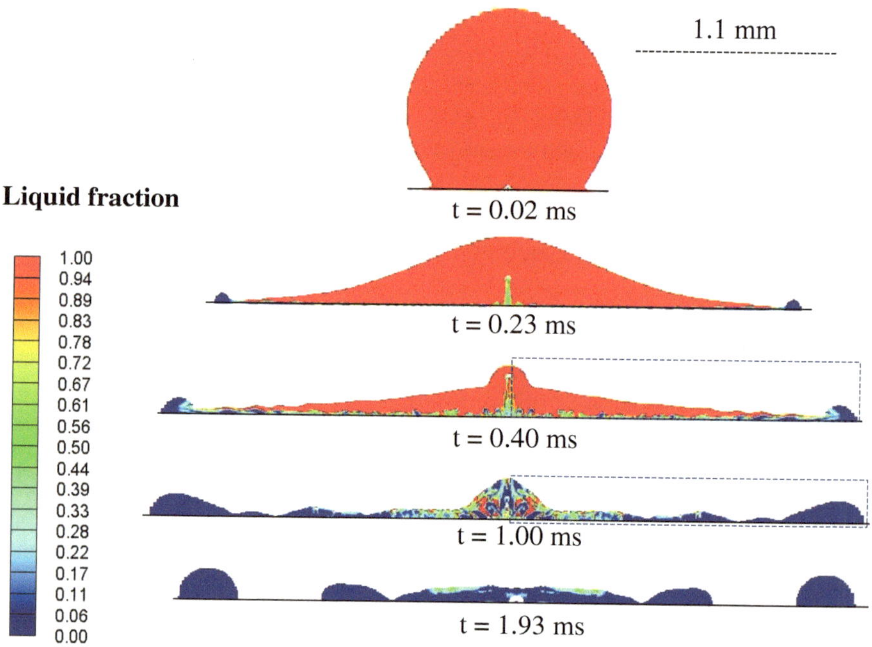

Fig. 5 Solidification pattern and formation of mushy zone at various times

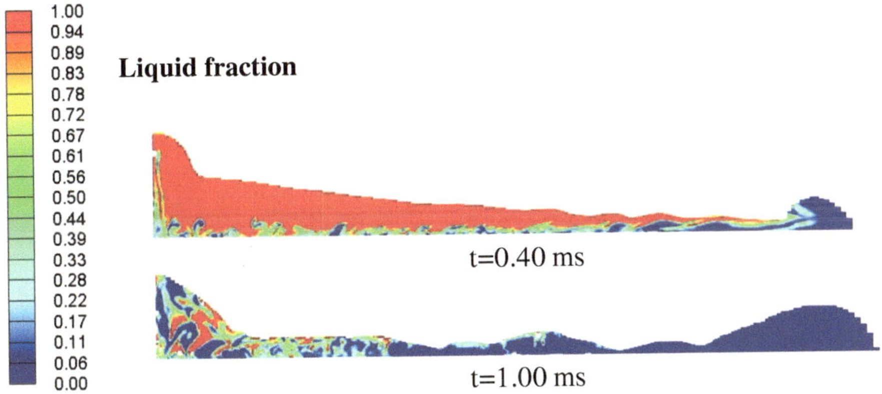

Fig. 6 Zoomed images of mushy zone in rectangular portions marked in Fig. 5

After the initial impact, most of the potential energy of the alloy droplet gets converted into kinetic energy. At the region of initial impact and then toward the periphery during the subsequent flattening process, the droplet velocity becomes significantly high, i.e., around 8 m/s as depicted in Fig. 7. Such high velocities combined with disordered mushy zone and rapid freezing at the advancing edge lead to break up of droplet material along the advancing front creating thereby a discontinuous splat. This phenomenon is known as freezing-induced breakup in the solidifying splat which occurs at the periphery (Chandra and Fauchais 2009).

It can be observed in Fig. 7 that spreading velocity is very high just after the impact. This suggests that the motion just after the impact is primarily governed by inertia forces. Further the velocity decreases due to viscous drag and surface tension. Figure 8 shows the temperature contours during the droplet spreading at different instants. During the spreading of the molten droplet, heat exchanges among surrounding air, droplet, and the substrate. Temperature of the molten droplet decreases and the substrate temperature begins to increase due to heat exchange between the droplet and substrate. This can be seen from the temperature contour (Fig. 8). Figure 8 shows that the droplet temperature is higher in the center region and lower in the vicinity of the spreading front in the entire spreading process. This is because high-temperature fluid is continuously supplied to the center region, whereas at the outer region there is higher heat transfer due to low temperature of the substrate.

Figure 9 shows snapshots of species distribution (mass fraction of Pb). Significant variation in Pb concentration is observed due to macrosegregation caused by the fluid flow in the flattening droplet. The wavy species variation pattern is in tune with the wavy mushy zone as observed in Fig. 5. A clearer look of the species distribution can be seen in Fig. 10 where zoomed images are shown. The prediction of species distribution in our model is possible by solving the species transport

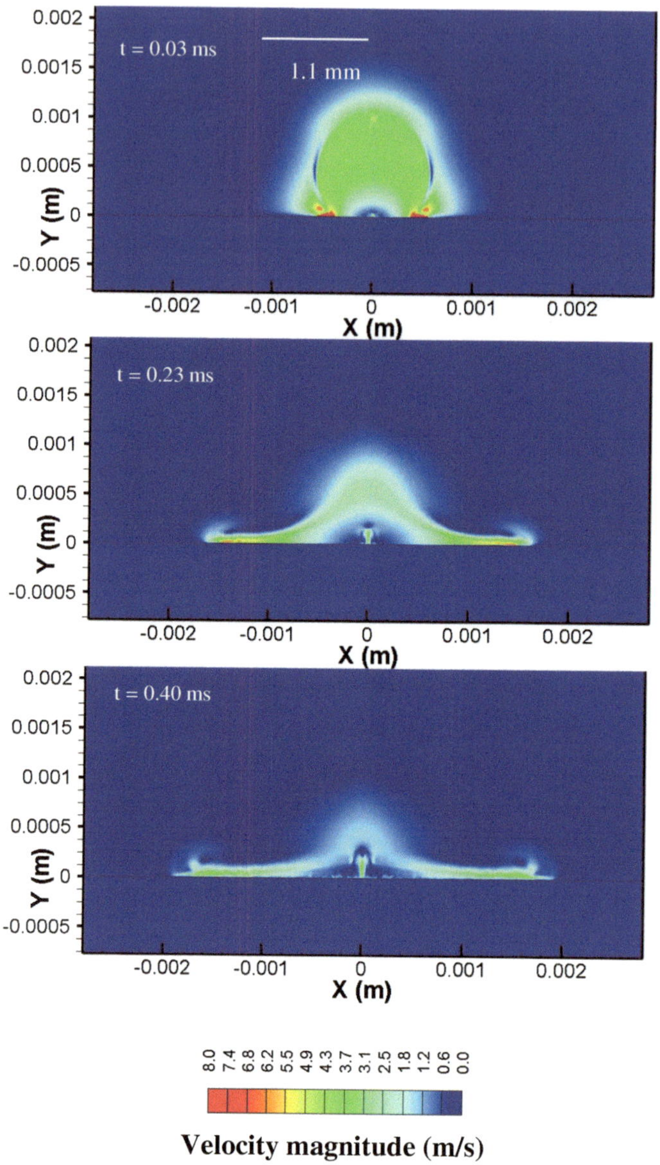

Fig. 7 Spreading velocity at different instants

equation that includes solidification kinetics and macrosegregation of a binary alloy. There is a built up of Pb concentration at the edge of the splat. This built up is caused by transport of solute toward the periphery by the high velocity at the periphery.

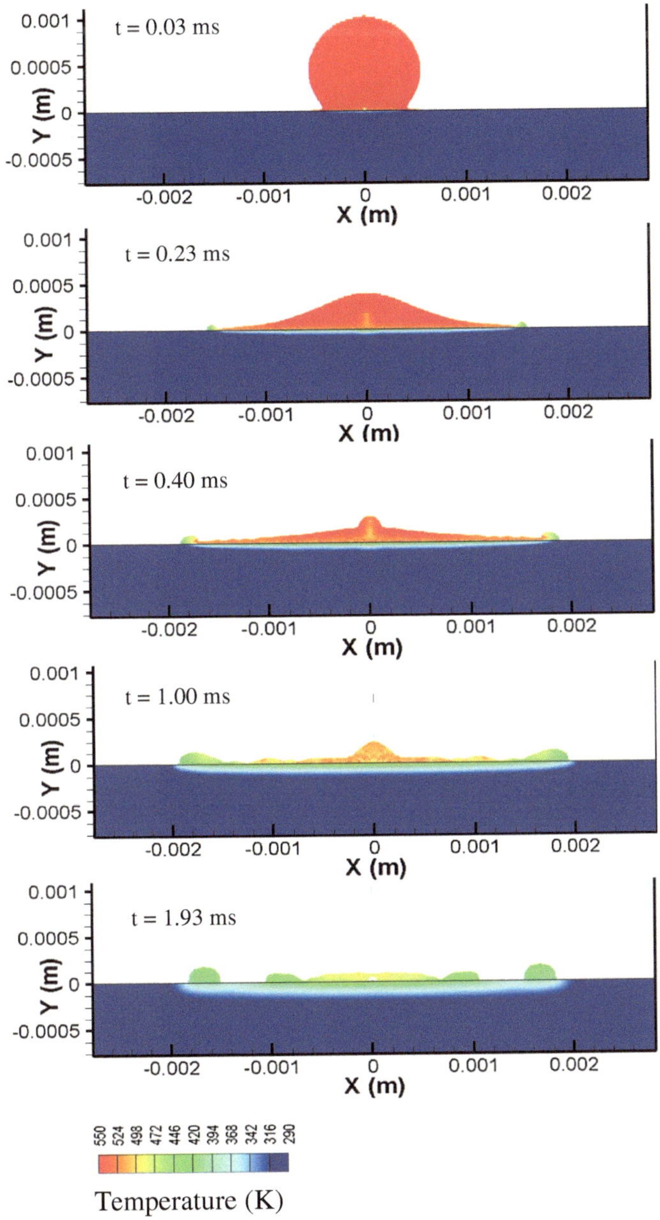

Fig. 8 Temperature distribution at different instants

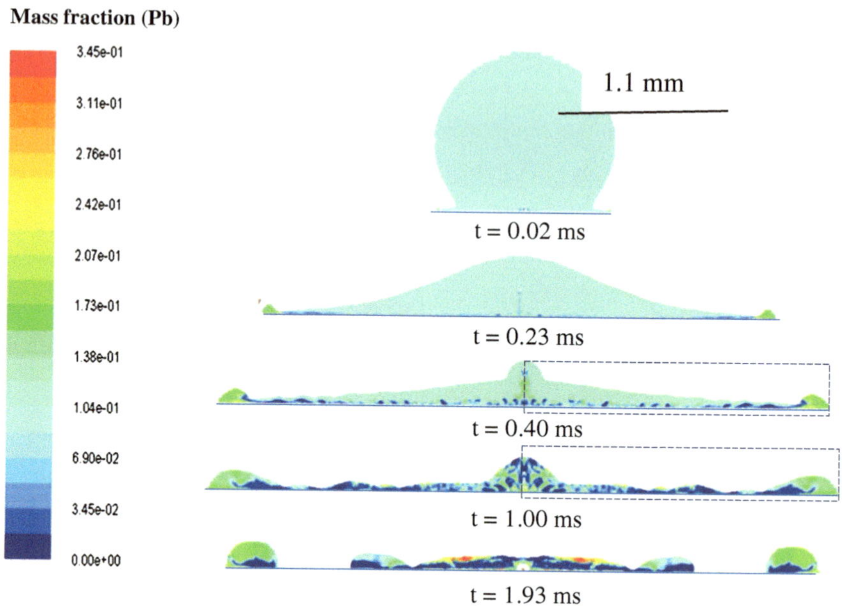

Fig. 9 Snapshots of species distribution (Pb)

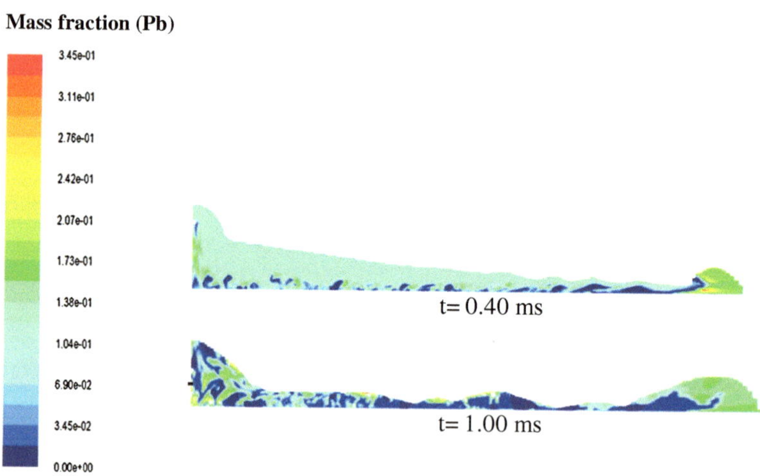

Fig. 10 Zoomed images of species (Pb) distribution in rectangular portions marked in Fig. 6

4 Conclusions

Pure substance solidifies at a constant temperature but in case of alloy it solidifies in a range of temperature with formation of mushy region (solid–liquid region). Comprehensive CFD-based model of alloy droplet deposition on a cold substrate is developed. Alloy droplet spreading and solidification is very different from a pure metal droplet spreading and solidification. The mushy zone formed in case of alloy droplet is disordered and wavy which is very different from the ordered solidification interface formed, as usually reported in the literature, in case of pure metal droplet. The presence of disordered and wavy mushy zone along with the high spreading velocity near the periphery causes breakup in the flattening droplet which finally produces a discontinuous coating/splat. Significant species variation is found to occur due to macrosegregation caused by fluid flow in the flattening droplet. Such inhomogeneous species composition in the splat can influence the mechanical properties of the coating.

The macrosegregation developed in the coating can significantly influence the coating strength and its properties. In future, the flow in the mushy zone, its shape, and relation with coating morphology are important issues to look upon.

References

Ahmed AM, Rangel RH (2002) Metal droplet deposition on non-flat surfaces: effect of substrate morphology. Int J Heat Mass Transf 45:1077–1091

Attinger D, Poulikakos D (2001) Melting and resolidification of a substrate caused by molten micro droplet impact. ASME J Heat Transf 123:1110–1122

Aziz SD, Chandra S (2000) Impact, recoil and splashing of molten metal droplets. Int J Heat Mass Transf 43:2841–2857

Bellet M, Combeau H, Fautrelle Y, Gobin D, Rady M, Arquis E, Budenkova O, Dussoubs B, Duterrail Y, Kumar A, Gandin CA, Goyeau B, Mosbah S, Zaloznik M et al (2009) Call for contributions to a numerical benchmark problem for 2D columnar solidification of binary alloys. Int J Therm Sci 48:2013–2016

Bennett T, Poulikakos D (1994) Heat transfer aspects of Splat-Quench solidification: modeling and experiment. J Mater Sci 29:2025–2039

Bussmann M, Chandra S, Mostaghimi J et al (2000) Modeling the splash of a droplet impacting a solid surface. Phys Fluids 12:3121–3132

Chandra S, Fauchais P (2009) Formation of solid splats during thermal spray deposition. J Therm Spray Technol 18:148–180

Fukumoto M, Huang Y (1999) Flattening mechanism in thermal sprayed Ni particles impinging on flat substrate surface. J Therm Spray Technol 8(3):427–432

Fukumoto M, Yang K, Tanaka K, Usami T, Yasui T, Yamada M et al (2011) Effect of substrate temperature and ambient pressure on heat transfer at interface between molten droplet and substrate surface. J Therm Spray Technol 20:48–58

Hong FJ, Qiu HH (2005) Modeling of substrate remelting, flow, and resolidification in microcasting. Numer Heat Transfer, Part A 48:987–1008

Inada S, Yang WJ (1994) Solidification of molten metal droplets impinging on a cold surface. Exp Heat Transfer 7:93–100

Kamnis S, Gu S (2005) Numerical modelling of droplet impingement. J Phys D Appl Phys 38:3664–3673

Kamnis S, Gu S, Lu TJ, Chen C et al (2008) Numerical modelling of sequential droplet impingements. J Phys D Appl Phys 41:165303 (7 pp)

Kang B, Zhao Z, Poulikakos D et al (1994) Solidification of liquid metal droplets impacting sequentially on a solid surface. ASME J Heat Transfer 116:436–445

Kumar A, Gu S, Tabbara H, Kamnis S et al (2012a) Study of impingement of hollow ZrO_2 droplets onto a substrate. Surf Coat Technol 220:164–169

Kumar A, Gu S, Kamnis S et al (2012b) Simulation of impact of a hollow droplet on a flat surface. Appl Phys A Mater Sci Process 109:101–109

Li L, Wang XY, Wei G, Vaida A, Zhang H, Sampath S et al (2004) Substrate melting during thermal spray splat quenching. Thin Solid Films 468:113–119

Liu H, Lavernia EJ, Rangel RH et al (1993) Numerical simulation of substrate impact and freezing of droplets in plasma spray processes. J Phys D Appl Phys 26:1900–1908

Madejski J (1976) Solidification of droplets on a cold surface. Int J Heat Mass Transf 19:1009–1013

Rangel RH, Bian X (1997) Metal-droplet deposition model including liquid deformation and substrate remelting. Int J Heat Mass Transf 40:2549–2564

Shakeri S, Chandra S (2002) Splashing of molten tin droplets on a rough steel surface. Int J Heat Mass Transf 45:4561–4575

Tabbara H, Gu S (2012) Modelling of impingement phenomena for molten metallic droplets with low to high velocities. Int J Heat Mass Transf 55:2081–2208

Waldvogel JM, Poulikakos D, Wallace DB, Marusak R et al (1996) Transport phenomena in picoliter size solder droplet dispersion. J Heat Transfer 118:148–156

Wang SP, Wang GX, Matthys EF et al (1999) Deposition of a molten layer of high melting point material: substrate melting and solidification. Mater Sci Eng A 262:25–32

Hybrid Friction Stir Welding of Similar and Dissimilar Materials

Deepak Kumar Yaduwanshi, Swarup Bag and Sukhomay Pal

Abstract Hybrid friction stir welding is an innovative solid-state joining technology which has great potential to produce effective and defect-free joint for similar materials and dissimilar materials irrespective of high chemical affinity and completely different physical and mechanical properties like aluminium and copper. Among the possible preheating source, plasma arc provides unique combination of high arc stability, concentrated energy density and low equipment cost. Plasma arc usually coupled with various manufacturing system in order to enhance the performance of conventional machining and bonding processes. Hence, plasma-assisted friction stir welding (P-FSW), as a hybrid system, is investigated in order to improve the weld joint quality and joint efficiency. Preheating effect using plasma arc, the P-FSW of aluminium alloy and other high-strength alloys enables to decrease the plunging force and enhance mechanical properties of welded joint. The integration of plasma arc on FSW tool also aids to decrease the probability of formation of welding defects. In present work, an overview of plasma-assisted friction stir welding (P-FSW) is presented by means of experimental investigation and prediction of it through numerical modelling. Finite element-based simulation using ABAQUS is carried out to evaluate the temperature profiles. A comparative study of mechanical and macro-microstructural characterizations of weld joint by conventional FSW and P-FSW processes has been conducted on similar (AA1030) and dissimilar (AA1100-pure copper) materials joining. Overall, the influence of preheating acts quite homogeneously throughout the structure as compared to conventional FSW. However, the results reveal that the development of P-FSW is still in initial stage and needs to improve in various aspects. Although P-FSW

D.K. Yaduwanshi · S. Bag (✉) · S. Pal
Department of Mechanical Engineering, IIT Guwahati, Guwahati 781039, Assam, India
e-mail: swarupbag@iitg.ernet.in

D.K. Yaduwanshi
e-mail: y.deepak@iitg.ernet.in

S. Pal
e-mail: spal@iitg.ernet.in

R.G. Narayanan and U.S. Dixit (eds.), *Advances in Material Forming and Joining*,
Topics in Mining, Metallurgy and Materials Engineering,
DOI 10.1007/978-81-322-2355-9_17

process is quite effective to improve mechanical properties and reduction of plunging forces, it needs to investigate the potentiality for relatively harder materials and dissimilar materials joining.

Keywords Hybrid friction stir welding · Preheating · Finite element method · Numerical model · Heat transfer · Friction model · Dissimilar material joining

1 Introduction

Friction stir welding (FSW) is one of the recently developed solid-state joining processes where similar or dissimilar type of materials is joined by plasticization. The process consists of a rotating tool which is plunged into the interface of two plates tightly abutted along a line, and the line is then traversed by the rotating tool (Thomas et al. 1991). During the process, heat is generated by plastic deformation as well as by the friction between the tool and the workpieces. The workpieces are ultimately joined by the stirring action of the softened material with excellent combination of mechanical properties (Rai et al. 2011; Kima et al. 2010; Grujicic et al. 2010). This method is often preferred due to low residual stress, low energy input and fine grain size compared to the conventional fusion welding process (Chao et al. 2003). Hence, the understanding of the heat transfer mechanism in the workpiece is helpful to predict thermo-mechanical interactive phenomena, thermal history during the welding process and the mechanical properties of the weld joint that finally evaluate the weld joint quality.

FSW technique has encountered a number of challenges in widespread application of similar and dissimilar joints. The relatively high downward force on the tool produces the heat required to initiate the weld. However, applications in high-strength alloys remain limited due to large welding force, consequent tool wear and low welding speed (Fernandez and Murr 2004). To overcome these drawbacks, the concept of preheating is stated by Sinclair et al. (2010). The authors reported that preheating the workpiece results in a significant (43 %) reduction in the axial force experienced by the tool for a wide range of welding parameters. On the other hand, the patent developed by Cao and Kou (2004) claims a preheating system with a tungsten inert gas (TIG) welding torch and extends the range of FSW to harder materials such as copper (Scutelnicu et al. 2008) and mild steel (Ferrando 2008). Oeystein et al. (1999) used a preheating source by applying a moving induction coil as the primary heat source to plasticize the material of the adjacent members. Palm (2001) developed a laser-supported FSW method to overcome the drawbacks of conventional FSW by using laser as preheating source. Due to preheating, friction stir weld joint comprises a refined grain structure having grain size of less than about 5 μm so that joint efficiency and joint quality are improved (Litwinski 2004; Murphy et al. 2008; Zaeh et al. 2008). Liu et al. (2012) develop hybrid FSW set-up assistance by microplasma arc. Preheating temperature is adjusted by changing arc current and tool travel speed.

In many emerging applications, the joining of dissimilar materials is required to take the advantages of their properties by different joining methods especially by FSW. Due to different chemical, mechanical and thermal properties of materials, dissimilar materials joining pose more challenging problems than similar materials joining (Kimapong and Watanabe 2004). However, when joining dissimilar materials by FSW, the problems not only arise from a material properties point of view, but also from the possibility of the formation of brittle inter-metallic and low melting point eutectics (Xue et al. 2010).

In this context, FSW is an impeccable method to weld materials that are hard to be joined by the conventional fusion welding methods such as Cu–Al with optimum combination of process parameters. Murr et al. (1998) first reported sound welds without any defect which were difficult to obtain using FSW. Afterwards, Ochi et al. (2004) investigated the formation of inter-metallic compounds (IMC) at interface of Cu–Al joint and effect of IMC on joint strength. Many authors have detected the existence of IMC compound layers in Cu/Al dissimilar metal with FSW process. Sahin (2010) determines the phases that occurred during welding as grey layer observed at the fracture surfaces of welded parts. However, this layer decreases the strength of the joints. Galvao et al. (2011) performed a thorough macro- and microstructural analysis of Al–Cu welded joint. It was concluded that the formation of inter-metallic phases deeply influences the weld crown morphology, which remains one of the main concerns in Al/Cu joining by FSW. These factors also contribute to asymmetry in both heat generation and material flow during FSW (DebRoy and Bhadeshia 2010). However, when FSW has been adopted to join dissimilar materials with a high difference in melting points, it is difficult to produce sound joints with satisfactory strength as applicable due to their significantly different plastic deformation and material properties (Murr et al. 1998; Liu et al. 2012).

To overcome above-mentioned problems, researchers investigated on the application of hybrid friction stir welding to join dissimilar materials with the aid of external heat source. Merklein and Giera (2008) studied laser-assisted friction stir welding of steel to aluminium alloy in butt joint configuration. The laser beam takes the role of preheating the steel blank to overcome asymmetry in heat generation and enhance the weldability. Bang and Bang (2010) and Bang et al. (2012, 2013) reported laser- or GTA-assisted friction stir welding of steel to aluminium alloy. The joint strength was achieved $\sim$90–93 % with reference to Al alloy as compared to $\sim$60–78 % without the effect of external heat source. Furthermore, it was demonstrated that tool life and decrement of applied load were expanded during joining of dissimilar materials with the aid of external heat source.

The objective of present work is to develop a plasma-assisted friction stir welding (P-FSW) system and demonstration of the process by means of experimental investigation and numerical modelling for joining similar and dissimilar materials. P-FSW as a hybrid system is developed to improve the joint efficiency and weld joint quality by integrating the preheating source with FSW process.

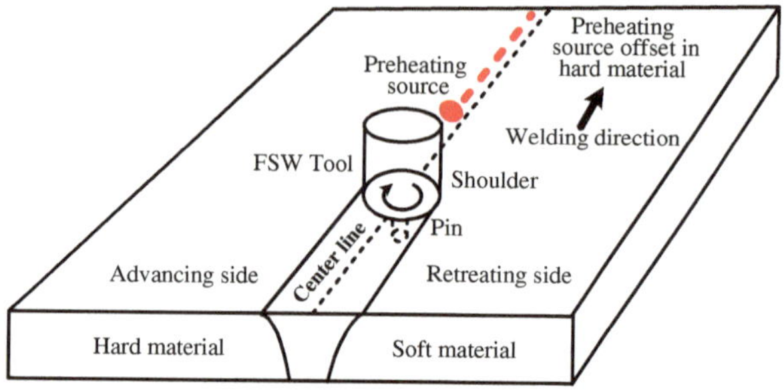

Fig. 1 Basic principle of plasma-assisted friction stir welding of dissimilar joint

A low-cost plasma arc is utilized to develop the hybrid system. Figure 1 schematically presents the P-FSW process. Adding an additional heat source, the base materials are preheated and some remarkable advantages are obtained such as faster and better plasticization of the base material, reduced FSW tool wear and clamping forces, faster welding speed and improved weld quality. In case of dissimilar joint, the additional heat source is located in the harder material side ahead of the tool in the welding direction, provides adequate metal flow around the tool and simultaneously prevents the formation of a large amount of brittle IMC. Although the preheating technique in FSW process is quite effective to improve mechanical properties and reduction of plunging forces, it needs to investigate from various directions.

2 Theoretical Background

FSW involves complex synergistic interaction of the physical phenomena during welding of various similar or dissimilar materials that are of practical significance by exploring and understanding the underlying physics of the process. Hence, this chapter aims at important physical insights into the mechanism of the process in terms of heat generation and heat transfer during FSW process. A three-dimensional heat transfer model for P-FSW is developed by considering temperature-dependent thermal properties. The heat input is linearly proportional to the distance from the centre of the tool, the contact condition between the rotating tool surfaces is considered, and the heat generation is due to work done by the friction force and plastic deformation.

2.1 Mathematical Model

In P-FSW, the heat generation in solution domain is due to plasma arc as well as friction and plastic deformation by FSW tool. Plasma arc preheats the workpiece ahead of FSW tool to influence the plasticization of material and improves material flow during welding. The heat generated by FSW tool is considered by means of friction and due to plastic deformation where heat is transferred into the workpiece following the Fourier's law of heat conduction. The transient temperature field T which is a function of time t and the spatial coordinates (x, y, z) is estimated by solving three-dimensional nonlinear heat transfer equation where the FSW tool moves with positive y-direction and is expressed as

$$\frac{\partial}{\partial x}\left\{k_x\frac{\partial T}{\partial x}\right\}+\frac{\partial}{\partial y}\left\{k_y\frac{\partial T}{\partial y}\right\}+\frac{\partial}{\partial z}\left\{k_z\frac{\partial T}{\partial z}\right\}+\dot{Q}=\rho C_p\left(\frac{\partial T}{\partial t}-V_{\mathrm{T}}\frac{\partial T}{\partial y}\right) \quad (1)$$

where ρ, C_p and k refer, respectively, to density, specific heat and thermal conductivity of the workpiece material; $\dot{Q}$ is the rate of heat generation; and V_{T} is the transverse speed of the tool. However, the interface thermal resistance between two plates is neglected here.

2.2 Heat Generation Model for FSW

Total heat generation involved in hybrid friction stir welding process (Q_{total}) is equivalent to the input by plasma arc (Q_{pre}) and FSW tool (Q_{FSW}).

$$Q_{\mathrm{total}} = Q_{\mathrm{pre}} + Q_{\mathrm{FSW}} \quad (2)$$

The heat that is transferred to the workpiece by preheating source is determined by

$$Q_{\mathrm{pre}} = \eta IV \quad (3)$$

where η is the thermal power efficiency, I is the welding current, and V is the welding voltage. The heat generated at the interface of tool and workpiece due to friction is Q_{f} and due to plastic deformation is Q_{p}. Therefore, total elemental heat due to friction and plastic deformation is given by

$$dQ_{\mathrm{FSW}} = dQ_{\mathrm{f}} + dQ_{\mathrm{p}} \quad (4)$$

The total amount of heat in FSW is Q_{FSW}, generated by the tool due to sliding and sticking friction conditions. Therefore, the heat energy generated at the contact interface between a rotating FSW tool and a stationary workpiece is subdivided into Q_1, Q_2 and Q_3, i.e. on the tool shoulder's surface, tool pin's side surface and tool pin's tip surface, respectively. The analytical estimation of heat generation from the shoulder is expressed as (Neto and Neto 2013),

$$Q_1 = \int_0^{2\pi} \int_{R_{\text{probe}}}^{R_{\text{shoulder}}} \omega \tau_{\text{contact}} r^2 (1 + \tan \alpha) \mathrm{d}r \mathrm{d}\theta \tag{5}$$

Heat generation from the probe is expressed as

$$Q_2 = \int_0^{2\pi} \int_0^{H_{\text{probe}}} \omega \tau_{\text{contact}} R_{\text{probe}}^2 \mathrm{d}r \mathrm{d}\theta = 2\pi \omega \tau_{\text{contact}} R_{\text{probe}}^2 H_{\text{probe}} \tag{6}$$

$$Q_3 = \int_0^{2\pi} \int_0^{R_{\text{probe}}} \omega \tau_{\text{contact}} r^2 \mathrm{d}r \mathrm{d}\theta = \frac{2}{3} \pi \omega \tau_{\text{contact}} R_{\text{probe}}^3 \tag{7}$$

Therefore, the total heat generation due to sliding friction is estimated as

$$Q = \frac{2}{3} \pi \omega \tau_{\text{contact}} \left(\left(R_{\text{shoulder}}^3 - R_{\text{probe}}^3 \right) (1 + \tan \alpha) + R_{\text{probe}}^3 + 3 R_{\text{probe}}^2 H_{\text{probe}} \right) \tag{8}$$

The material flow and heat generation are characterized by the contact conditions at the interface and are described as sliding, sticking or partial sliding/sticking (Askari et al. 2001). The contact condition under the shoulder can be described by sliding friction, using a friction coefficient μ and interfacial pressure p, or sticking friction, based on the interfacial shear strength at an appropriate temperature and strain rate. It is convenient to define a contact state variable δ, which relates the velocity of the contact workpiece surface with the velocity of the tool surface. The contact state variable (δ) is assumed to change linearly with the distance from the centre of the pin. Based on these assumptions and geometrical aspects of the deformation zone, the contact state variable can be expressed as

$$\delta = \frac{V_{\text{matrix}}}{V_{\text{tool}}} \tag{9}$$

where V_{matrix} is peripheral velocity of workpiece and V_{tool} is peripheral velocity of tool. A 100 % sticking condition at the pin/material interface is assumed, i.e. same as tool velocity. The velocity of the material is equal to zero at the outer edge of the deformation. The contact shear stress is then

$$\tau_{\text{contact}} = \tau_y = \frac{\sigma_y}{\sqrt{3}} \tag{10}$$

The expression of heat generation, assuming the sticking condition with convex shoulder, is

$$Q_{\text{FSW,sticking}} = \frac{2}{3} \pi \omega \frac{\sigma_y}{\sqrt{3}} \left\{ \left(R_{\text{shoulder}}^3 - R_{\text{probe}}^3 \right) (1 + \tan \alpha) + R_{\text{probe}}^3 + 3 R_{\text{probe}}^2 H_{\text{probe}} \right\} \tag{11}$$

where α is conical angle of shoulder surface in case of convex shoulder. Coulomb's friction law to describe the shear stress estimates the critical friction stress necessary for a sliding condition:

$$\tau_{\text{contact}} = \mu p \tag{12}$$

$$Q_{\text{FSW}} = \delta * Q_{\text{FSW,sticking}} + (1 - \delta) * Q_{\text{FSW,sliding}} \tag{13}$$

For a flat shoulder and straight cylindrical tool, the total heat generation is expressed as

$$Q_{\text{FSW}} = \frac{2}{3}\pi\omega\left[\delta\tau_y + (1 - \delta)\mu p\right]\left\{\left(R^3_{\text{shoulder}} - R^3_{\text{probe}}\right) + R^3_{\text{probe}} + 3R^2_{\text{probe}}H_{\text{probe}}\right\} \tag{14}$$

2.3 Initial and Boundary Conditions

Boundary condition for FSW thermal model is specified as surface interaction of the solution domain. At time zero, a uniform temperature in the workpiece is assumed. The initial condition for the calculation is expressed as

$$T(x, y, z, 0) = T_i \tag{15}$$

where T_i is the ambient temperature at time $t = 0$.

The boundary interaction of the solution domain is schematically represented in Fig. 2. At the free surface of the workpiece, natural convective boundary conditions

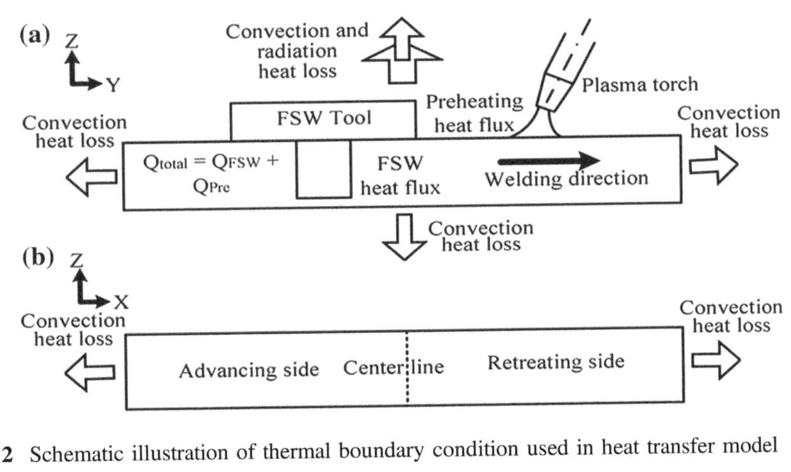

Fig. 2 Schematic illustration of thermal boundary condition used in heat transfer model

are considered except top surface with radiation. Mathematically, the boundary condition is expressed as

$$\text{P-FSW: } k\frac{\partial T}{\partial n} = h(T - T_0) + \delta\theta\left(T_0^4 - T^4\right) - q_s - q_a \tag{16}$$

where n is the normal direction vector of boundary, h is the convection coefficient, q_s is heat flux between tool and workpiece, q_a is the heat flux from plasma arc, k is the thermal conductivity of workpiece, T is temperature of workpiece, T_0 is initial temperature, δ is the emissivity of surface, and θ is the Stefan Boltzmann constant.

2.4 Hybrid Heat Source Model for P-FSW

A hybrid heat source model is developed by combining FSW tool and plasma arc. Preheating source is ahead of FSW tool with some lead distance on centre line in case of similar material joint. But in case of dissimilar joint, preheating source is offset towards hard material and FSW heat source is offset towards soft material from centre line. Lead distance and offset distance of heat source are not constant, but it depends on welding condition. Figure 3 illustrates the distribution of heat flux for FSW heat source over the plate surface due to the tool shoulder, on pin–workpiece interface due to tool pin, and due to tool pin bottom surface which is given as (Schmidt et al. 2004),

$$q_{\text{shoulder}} = \frac{3Q_1 r}{2\pi R_{\text{shoulder}}^3}; \tag{17}$$

$$q_{\text{pside}} = \frac{Q_2}{2\pi R_{\text{probe}} H_{\text{probe}}}; \tag{18}$$

$$q_{\text{pbottom}} = \frac{Q_3}{\pi R_{\text{probe}}^2} \tag{19}$$

where τ_{contact} and δ in the expression of Q_1, Q_2 and Q_3 are applied according to sliding or sticking friction conditions (Neto and Neto 2013). Therefore, total heat flux from the FSW tool is expressed as,

$$q_s = q_{\text{shoulder}} + q_{\text{pside}} + q_{\text{pbottom}} \tag{20}$$

To model the preheating source, a circular surface heat source with effective radius ($\bar{r}$) is considered. The effective radius of heat source depends on nozzle diameter of plasma torch and its stand-off distance from plate surface. However, the heat flux follows a Gaussian distribution and is expressed as

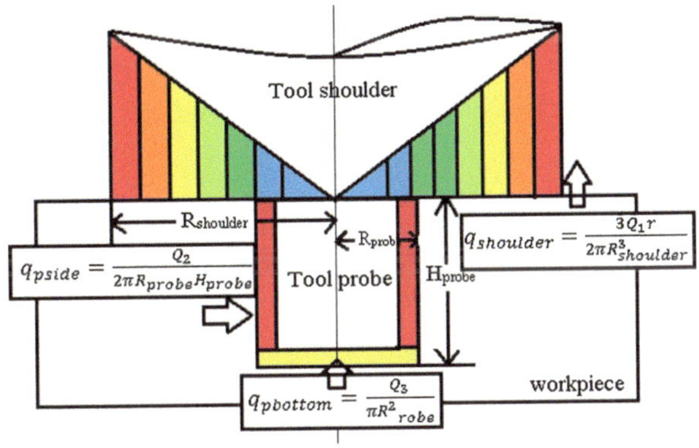

Fig. 3 Schematic illustration of heat flux distribution in moving heat source model for FSW

$$q_a(r) = \frac{3Q_{\text{pre}}}{\pi \bar{r}^2} e^{-3\left(\frac{r}{\bar{r}}\right)^2} \tag{21}$$

where $q_a(r)$ is the surface flux at radius r and $\bar{r}$ is the effective arc radius over which 95 % energy falls.

A moving heat source with heat distribution for P-FSW of similar and dissimilar joints is illustrated in Fig. 4a, b, respectively, and is used in the heat transfer analysis. Moving heat source is a combination of FSW tool and plasma arc as surface heat source. Heat distribution for plasma arc is exponential with maximum heat density in centre and radial heat distribution for FSW with maximum heat density at outer radius. Similarly, in P-FSW, heat source is aligned and moves on weld centre line in welding direction. But in dissimilar P-FSW, plasma arc heat source is offset from weld centre line in hard material (Cu) to improve plasticized state as compared to soft material (Al) and whether FSW tool is offset depends on process conditions.

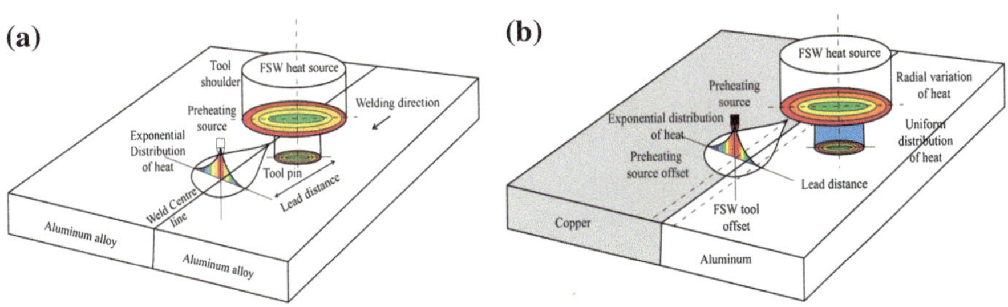

Fig. 4 Schematic illustration of moving heat source model for P-FSW **a** similar material and **b** dissimilar material joining

2.5 Material Properties for Dissimilar Joint

In hybrid friction stir welding of dissimilar joint, two different materials are plasticized and mixed together to form weld joint. Heat generated by tool due to friction and plastic deformation to plasticize the material and cause plastic flow of material at material–tool interface by stirring action. With forward motion of rotating tool on workpiece, two materials are mixed and form thick layer of inter-metallic compound which have mixed property of both material. There is a change in material properties of weld zone due to mixing which depends on forward motion of tool. For better estimation of thermal history of heat transfer model of the dissimilar materials, the concept of time-varying functionally graded material (FGM) is applied to define welding zone. In real FSW process, the weld nugget zone increases in size with time as the tool travels along weld line of the specimens (Li et al. 2009). This phenomenon is simulated by increasing the size of the FGM area, while the welding tool is moving forward. The rate of growth of the FGM region is estimated from the rotational and forward velocities of the tool illustrated in Figs. 5 and 6.

For joining copper with aluminium, workpiece is divided into three zones, i.e. advancing side (left) for copper, retreating side (right) for aluminium and middle trapezoidal zone for weld nugget as shown in Fig. 7. The time variation of material properties in weld zone depends on movement of tool axis, and these phenomena are implemented by "USDFLD" subroutine in ABAQUS. Upgrade material

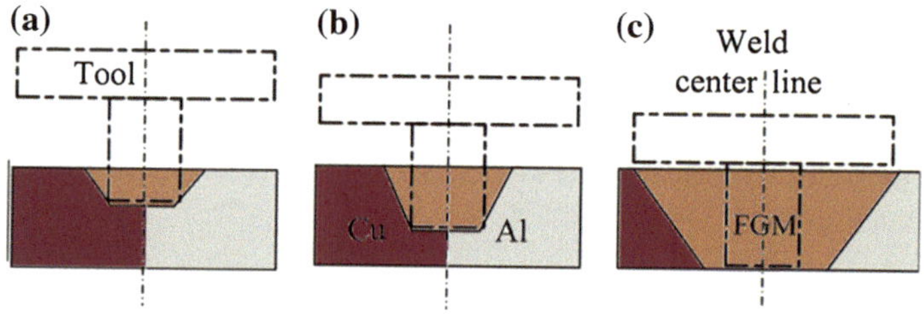

Fig. 5 Formation of functionally graded material during plunging of tool at **a** 2 mm, **b** 3 mm and **c** 5 mm plunging depth

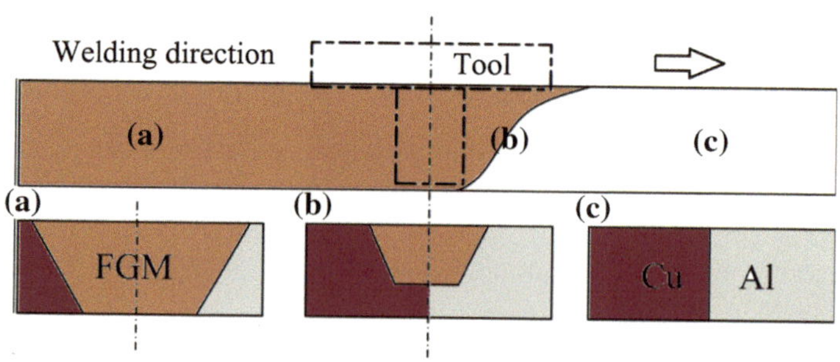

Fig. 6 Formation of functionally graded material during welding and forward motion of tool

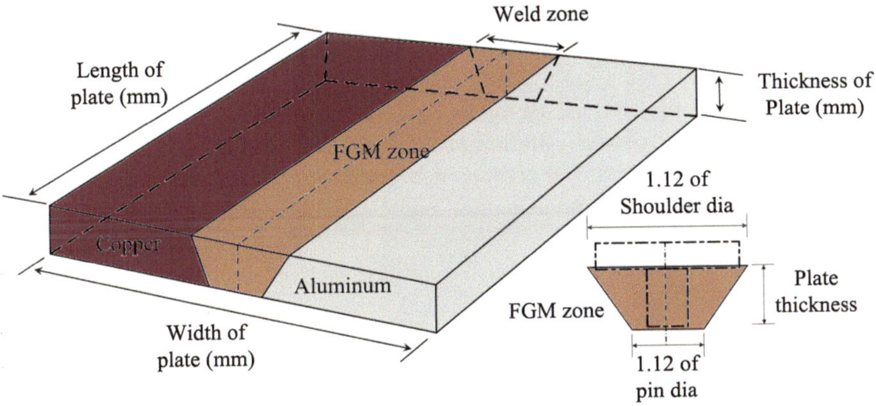

Fig. 7 Schematic illustration of applied functionally graded material concept in welding zone of dissimilar material joining

properties are assigned to location of this region at each time increment during simulation. In the no-weld region, two different material properties are assigned for different materials that do not change during the simulation.

According to the mixture rule of FGM (Kingery et al. 1976; Mahjmoud 2003), the FGM material properties such as thermal conductivity, specific heat and mass density can be estimated by using the following relations

$$k = k_1[1 + (3(k_1 - k_2)v_2)/(3k_1v_2 + (k_1 + 2k_2)(1 - v_2))] \tag{22}$$

$$c = (c_1\rho_1v_1 + c_2\rho_2v_2)/(\rho_1v_1 + \rho_2v_2) \tag{23}$$

$$\rho = \rho_1v_1 + \rho_2v_2 \tag{24}$$

Equation (22) is the expression of thermal conductivity of FGM area in terms of material 1 (Cu), material 2 (Al) and the volume fraction of material 2 (V_2), and Eqs. (23) and (24) are the specific heat and mass density, respectively. Figure 7 represents schematically the application of FGM concept in welding zone of dissimilar joint. The variation of material properties in FGM depends on the volume fraction of material 2 if $V_1 = 0$, then it indicate pure material 2. If the value of v_2 is between 0 and 1, it implies that the material is within FGM region. Thermal conductivity and the specific heat are not linear function of the volume fraction, while mass density is the linear function of the volume fraction.

2.6 Solution Methodology

The heat transfer analysis of the P-FSW process for establishing the numerical model is performed using commercial software ABAQUS. The solution domain of the model is assumed as rectangular plate. Solution domain has three partitions during dissimilar material joint, while middle section is weld zone and other two is base material. During simulation of similar materials, joining only two partitions of the solution domain with similar material properties is considered. The convective heat transfer coefficient is used for the top and sides of the workpiece, which is typical for natural convection between aluminium and air (Schmidt et al. 2004). Also, the higher convective heat transfer coefficient is used for the bottom surface of the plate which is typically considered as natural convection between steel and air.

The procedure of modelling is adopted based on the actual experimental procedure in order to obtain the numerical solution, such as temperature profile of the solution domain. The workpiece is meshed using eight nodded brick elements of type DCC3D8. The temperature is considered as single degree of freedom at each node with fine meshing in the weld zone. The heat source model is coded in "DFLUX" user subroutine. The time variation of material properties is applied by "USDFLD" subroutine in dissimilar joint to update the material properties of that location. The plunge-in velocity of 0.12 mm/s and plunge-out velocity of 1.16 mm/s are considered in present simulation. The temperature-dependent thermal properties of the aluminium alloy and pure copper are incorporated in the analysis that makes the analysis nonlinear iterative solution.

3 Experimental Investigation

Experiment is conducted under different welding conditions for joining of similar (AA1100) and dissimilar (AA1100-Pure Cu) materials in butt joint configuration. The chemical composition of AA1100 aluminium alloy (wt%) is reported in Table 1. Two rectangular plates of 200 mm × 100 mm size are butt joined by P-FSW process. The plate surface is cleaned by acetone prior to welding. In this study, a vertical milling machine with 7.5 hp motor capacity is used to perform experiment as described in Fig. 8. The welding tool is made of H13 with the shoulder and pin diameters of 20 and 6 mm, respectively. The rotating tool is perpendicular of the plate with 0° tilt angle with fate shoulder. The pin length is slightly less than the thickness of workpieces, with 5.65 mm tool pin length. The edges of the test pieces are machined to obtain a neat square butt. They are clamped to the horizontal bed with zero root opening. As shown in Fig. 8a, the preceding plasma torch is placed 13 mm away from the shoulder edge with an angle of 60° adjacent to the joint which ensures less effect of arc on tool and enhanced the material flow during welding. The aluminium alloy plate is located on the retreating side (RS) and the copper plate on the advancing side (AS). The melting temperature

Table 1 Chemical composition of aluminium alloy (wt%)

Cu	Fe	Si	Mn	Mg	Zn	Cr	Ti	Al
0.15–0.40	0.70	0.40–0.80	0.15	0.80–1.20	0.25	0.01–0.35	0.15	Bl

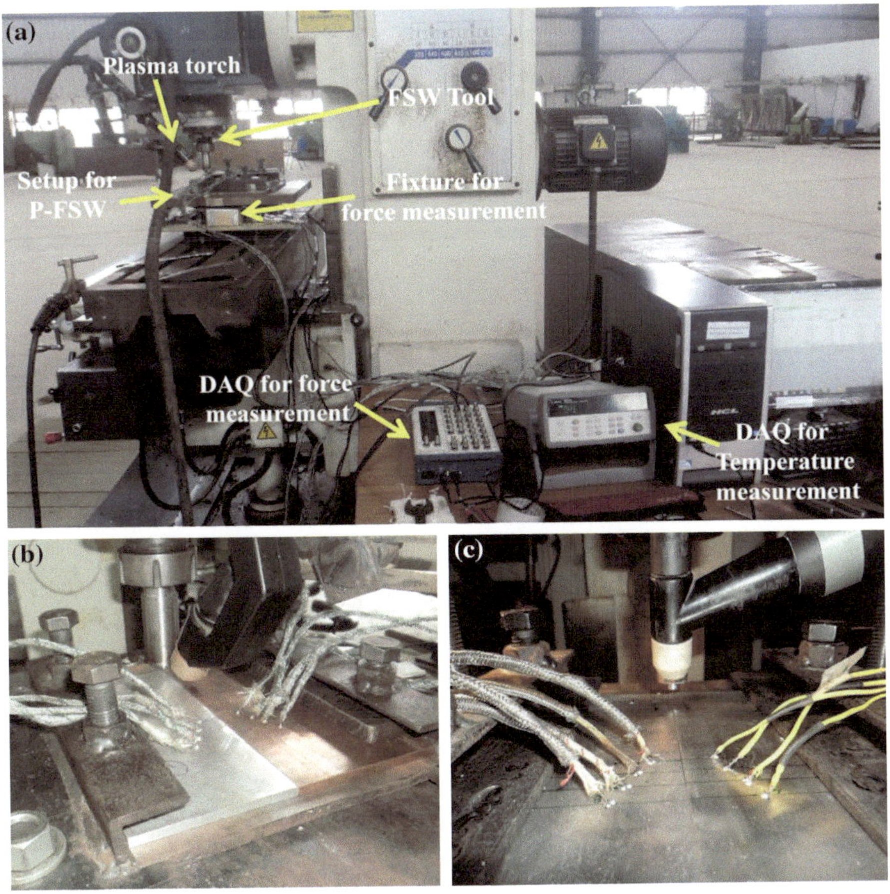

Fig. 8 **a** Experimental set-up for P-FSW; **b** closer view of fixture and set-up for dissimilar material joint; **c** closer view of fixture and set-up for similar material joint

of pure copper and AA1100 is considered as 1083 and 660 °C respectively. Five welding conditions are considered with variable preheating current, and the experimental conditions are described in Table 2.

Table 2 Process parameters of P-FSW

Exp. no.	Process parameter for FSW		Preheating current (amp)	
	Transverse speed (mm/s)	Rotational speed (RPM)	AA1100	AA1100-Cu
1	1.05	815	0	0
2	1.05	815	15	25
3	1.05	815	25	35
4	1.05	815	35	45
5	1.05	815	45	55

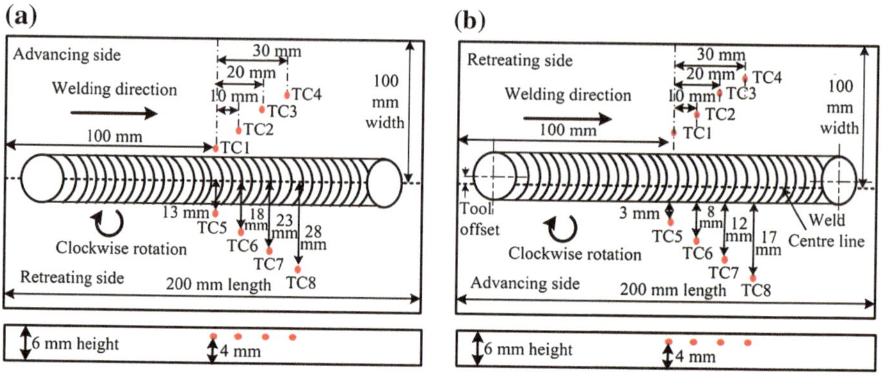

Fig. 9 Layout of thermocouple point during welding: **a** similar material joint and **b** dissimilar material joint

For measurement of tool plunging force, a fixture with four strain gauges at four hexagonal supporting members of backing plate as shown in Fig. 9a is used which convert force into corresponding voltage. This voltage is measured by data aquatint system with amplifier and multiplies it with calculated factor. Transient temperature is recorded at the regions near the top surface; eight different locations during P-FSW process using 36 grade K-type thermocouples are placed. All the temperature measurements are made in four on the AS as well as four on RS of welds. Thermocouples are used at the middle length of the plate along the longitudinal direction as shown in Fig. 9b. Vickers hardness test and tensile test are carried out to find effect of preheating on mechanical properties of weld joints. The Vickers hardness was measured along the transverse cross section of the welded specimen using a load of 0.1 Kgf and dwell time of 10 s (Fig. 10). Tensile test of the welded specimens was conducted based on ASTM E8 M-04 standard.

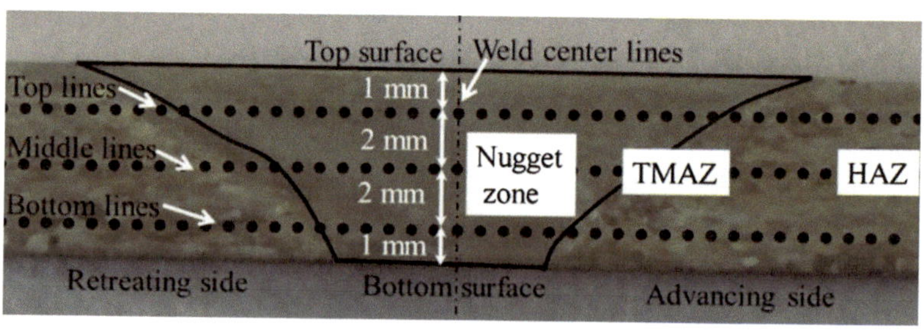

Fig. 10 Schematic illustrations of weld cross section for hardness measurement

4 Results and Discussion

P-FSW of similar and dissimilar materials is investigated on the aspects of thermal behaviour and mechanical properties of welded joints. In thermal analysis, preheating phenomena in FSW include thermal history and variation of cooling rate. In mechanical properties, the preheating effect is carried out on the basis of tensile strength, hardness and fractography analysis.

4.1 Thermal History and Cooling Rate

Comparative effect of preheating on temperature profile for similar material joining (Table 2) with time is shown in Fig. 11a, b, respectively, for AS and RS with preheating (Exp. 2) and without preheating (Exp. 1). It is observed that the peak temperature at the weld centre line is asymmetrical due to plastic flow of material. The area under the curve of thermal history represents total heat input during welding which is higher in case of preheating. This additional heat input due to plasma arc makes an effect on rate of change of temperature during welding that may improve the weld joint properties. This additional heat also increases the temperature of material around the weld zone so that material tends to lose some of its mechanical strength and facilities the ease of welding. The experimental method reveals that the peak temperature reached in the case of the hybrid welding is around 320–385 °C at a distance of 13 mm from weld centre line.

Preheating is more effective during FSW of dissimilar material than similar material; it minimizes the difference of thermo-mechanical properties of two different materials. As shown in Fig. 12, AS (Cu) temperature at TC 1 is 21 °C higher than RS (Al) in case of Exp. 1 without preheating, but temperature difference increases to 58 °C in case of Exp. 2. Elevated temperature difference in the weld

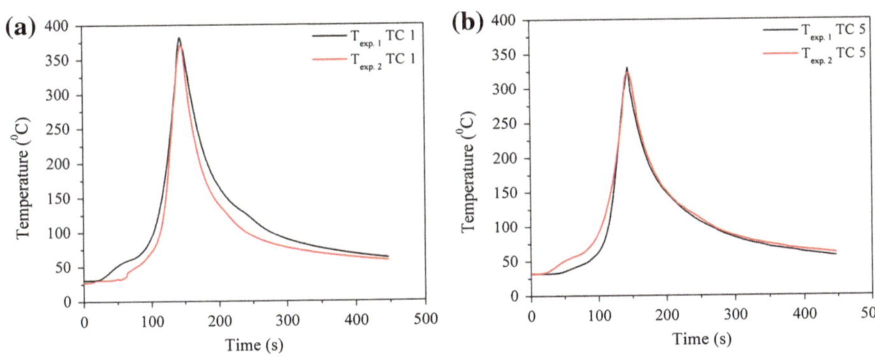

Fig. 11 Comparison between temperature profiles of Exp. 1 and 2 for joining similar materials at 13 mm away from weld centre line: **a** advancing side and **b** retreating side

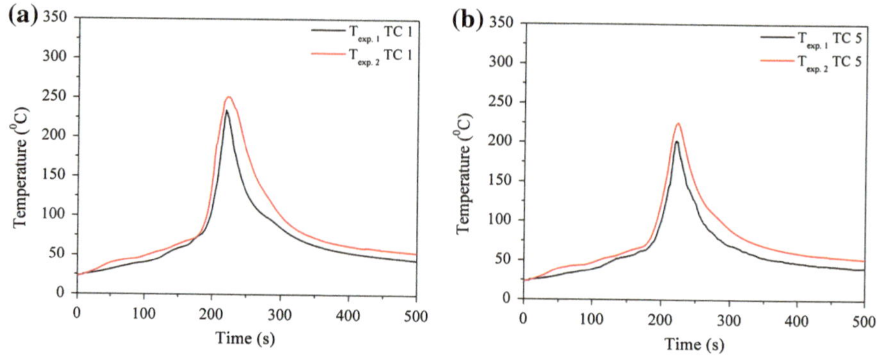

Fig. 12 Comparison between temperature profiles of Exp. 1 and 2 for dissimilar material joint at 15 mm away from weld centre line: **a** aluminium side and **b** copper side

reduces plasticized state temperature difference of two different materials. This result indicates that the sufficient plastic flow and partial annealing effect in weld joint by plasma preheating on the copper side result in significant increase in joint quality.

The cooling rate is estimated by change of temperature with respect to change of time ($\sim dT/dt$) for a fixed thermocouple location. It is obvious form Fig. 13 for similar joint and Fig. 14 for dissimilar joint that there is wide variation of cooling rate when the FSW tool is nearer to thermocouple location and asymptotically approaches to zero value when the tool is away from the fixed point. The preheating effect due to plasma arc decreases the variability and magnitude of cooling rate as compared to conventional FSW process. Preheating builds up heat in the metal, and it takes little longer time for it to cool down. Since the maximum cooling rate is less

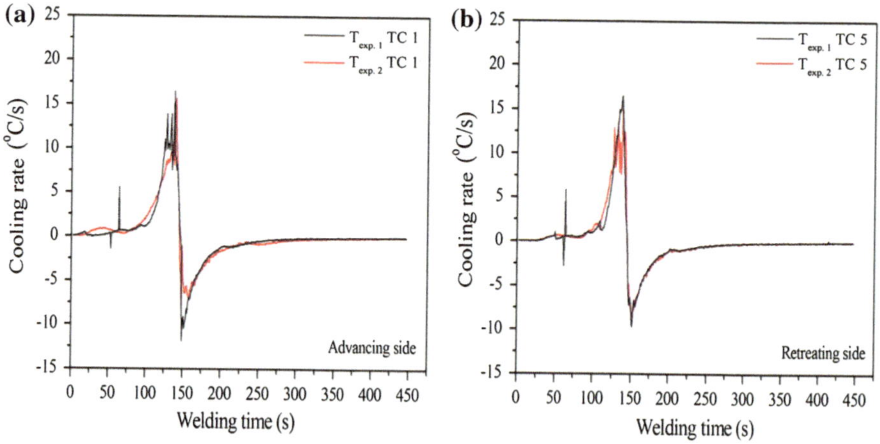

Fig. 13 Comparison of cooling rate of Exp. 1 and 2 for similar material joining: **a** advancing side and **b** retreating side

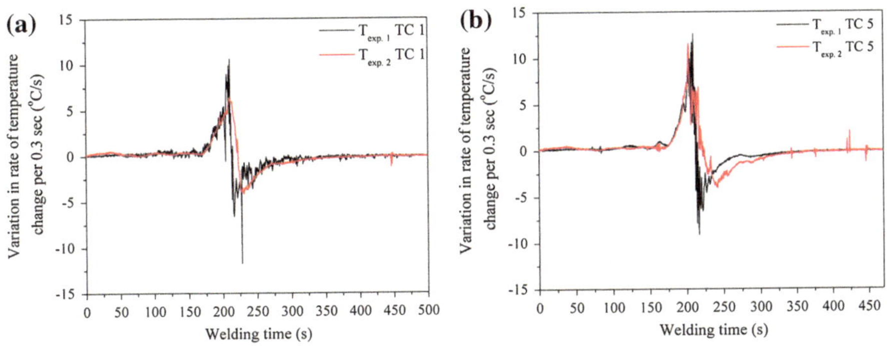

Fig. 14 Comparison of cooling rate of Exp. 1 and 2 for dissimilar material joining: **a** aluminium side and **b** copper side

than 17 K/s in case of similar joint and 13 K/s in case of dissimilar joint, the diffusional transformation will occur and this transformation is more stable for P-FSW process due to less variation of cooling rate (Sinclair et al. 2010). However, this much variability in cooling rate may influence the grain coarsening rate in this temperature range (Askari et al. 2001). Since the average cooling rate or temperature gradient ($\sim dT/dx$) from the top surface to the bottom surface decreases, it may influence the hardness distribution. Less variability of hardness distribution is expected for P-FSW process than FSW process. During dissimilar joint, cooling rate variation affected aluminium side, but the copper side not affected due to low operating temperature.

4.2 Characterization of Weld Joint

The effect of preheating is more prominent in characterization of mechanical properties such as microhardness, ultimate tensile strength, elongation and bending strength of the welded joints.

4.2.1 Micro Hardness

The hardness profiles measured along the transverse cross section of welded specimen at three different lines along centre, bottom and top of the weld zone corresponding to welding condition in Table 2. In the case of similar P-FSW welds of Exp. 1, hardness profile is strongly affected by work hardening rather than grain size in the weld and it is dominantly governed by the grain boundary strengthening and the dislocation strengthening. This hardening mechanism leads the case of irregular profile of hardness depicted in Fig. 15a. In the case of P-FSW welds (Exp. 2) as shown in Fig. 15b, the hardness profile is relatively uniform and

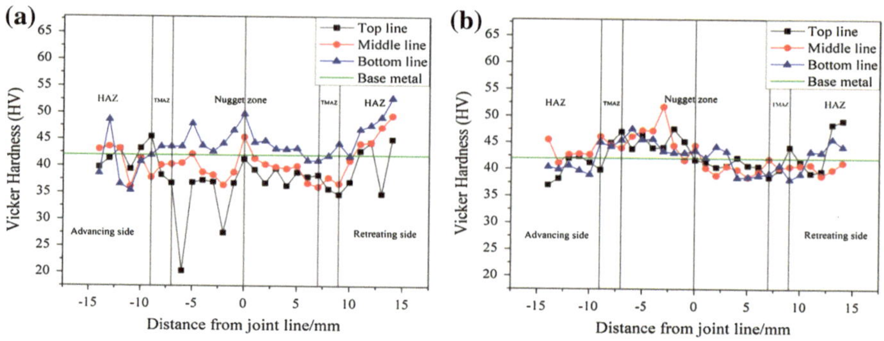

Fig. 15 Hardness distributions on the transverse cross section for **a** Exp. 1 and **b** Exp. 2 in Table 2 of similar material joining

comparatively less variation is observed. There are significant differences in the hardness distribution between RS and AS. The hardness within the stirring zone is higher on the AS than the RS, i.e. the hardness distribution in the FZ is asymmetric with respect to the tool rotation axis. It is attributed by the fact that the strain rate change due to the high-temperature deformation of pure aluminium depends on stress and temperature. This determines the steady-state deformation due to the effect of preheating and left more uniform finer grain throughout the structure.

The microhardness distribution profile on the transverse cross section of dissimilar joint for Exp. 1 and Exp. 2 of Table 2 is shown in Fig. 16a, b, respectively, as per hardness points illustrated in Fig. 10. The average hardness values of base metals Al and Cu are 46 and 130 HV, respectively. An inhomogeneous distribution of hardness values is observed in the nugget zone. The higher hardness value in the nugget zone relative to Al base metal is primarily associated with the formation of very fine recrystallized grains and the Cu-rich dispersed particles, corresponding to the effect of preheating observed in Fig. 16b. It is worth noticing that an abrupt

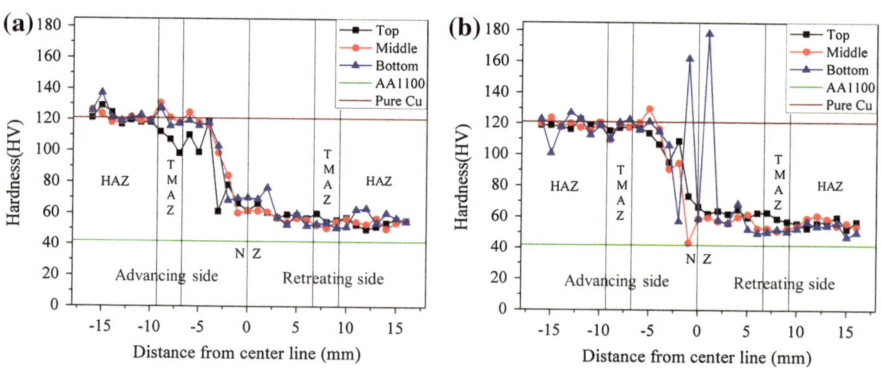

Fig. 16 Hardness distributions on the transverse cross section for **a** Exp. 1 and **b** Exp. 2 of dissimilar material joining

change in hardness value occurs adjacent to the interface. The values at the interface of top, middle and bottom are reached to 172.4 and 195.3 HV, which are far higher than the base metals.

4.2.2 Tensile Strength and Fracture Behaviour

The comparison of tensile strength of welded joints corresponding to welding condition of Table 2 is described by Fig. 17a. For these range of parameters, the tensile strength of Exp. 2–5 weld is ~5–8 % higher than that of Exp. 1 welds without preheating. The maximum tensile strength of P-FSW welds obtained at preheating current of 35 A is 134 MPa which is ~95 % of base metal tensile strength. This is mainly due to strain ageing by plastic deformation and dispersion strengthening by fine Al_2O_3 particles at stirred zone. Although with the Al_2O_3 particle addition, the ultimate tensile strength increases marginally, and further increase of tensile strength is attributed to the grain structure difference and the Orowan strengthening due to stirring of finely dispersed aluminium oxide particles. Overall, the recovery and continuous dynamic recrystallization for aluminium refine the grain size over the base material with the expense of heavily deformed work hardened material (Martin 1980).

Figure 17b shows the percentage of elongation for tensile test specimens. There is markedly difference in elongation of friction stir welded specimens as compared to base material. This is due to the fact that in P-FSW, the material undergoes continuous dynamic recrystallization, and as a result, a fine grain structure is produced with Al_2O_3 particles at stirred zone. The deformation in coarse-grained aluminium is known to have steep strain gradient in the vicinity of grain boundary and the centre of the grain. As the grain size decreases and strain increases, the deformation becomes more homogeneous. Therefore, the preheating does not improve its elongation much and the elongation is around ~9 ± 2 % for all cases.

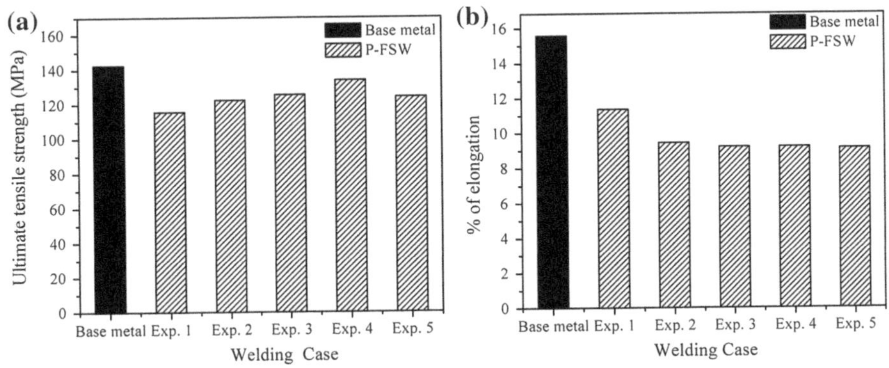

Fig. 17 Tensile strength and percentage of elongation at different welding conditions depicted in Table 2 for similar material joints

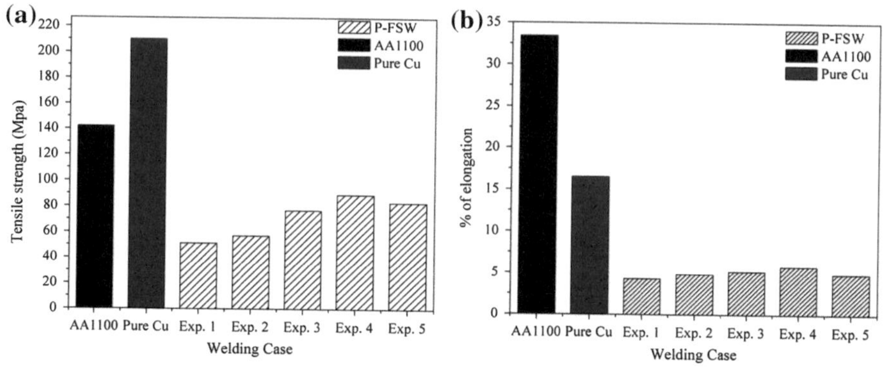

Fig. 18 Tensile strength and percentage of elongation at different welding conditions depicted in Table 2 for dissimilar material joints

This happens in case the base metal has high thermal conductivity and thermal diffusivity which leads the intuition that the effect of preheating may work quite homogeneously for highly thermal conductive materials.

The comparison of tensile strength and percentage elongation of joints welded at different preheating current corresponding to Table 2 is shown in Fig. 18. The tensile strength of joint produced at relatively low and without preheating current is rather weak with average value of 46–65 MPa. However, the maximum strength of joint fabricated at 45 amp preheating current reached 107.2 MPa, representing joint efficiency of 75.6 % with reference to Al alloy (Fig. 18a). The results indicate that the tensile strength of P-FSW welds is improved with increase in preheating current up to some optimum value. Figure 18b indicates that there is little improvement of ductile fracture behaviour in case of P-FSW. This result also indicates that the improvement of material flow and partial annealing effect in dissimilar materials by P-FSW significantly increases elongation and improves joint strength between aluminium alloy and pure copper when preheating occurs towards copper side.

It is noteworthy that without preheating, the weld joint shows brittle fracture behaviour. Also, a large number of voids are observed without preheating due to lack of proper material flow caused by insufficient stirring and heat generation. In case of P-FSW, increase of tensile strength and elongation is caused by additional heat input due to preheating towards copper side in case of dissimilar joint.

4.2.3 Bending Strength

Bending properties are tested to determine the flexural strength of similar and dissimilar joints which are always required in industrial application. The bending tests are carried out as per ASTM E08 for welded specimen. The dimensions for the bending specimen are 100 × 20 × 6 mm. The experimental bending results for different combinations of process parameters are listed in Tables 3 and 4 for similar and dissimilar joints, respectively.

Table 3 Bending strength for different welding conditions of Table 2 for similar material joints

Exp. no.	Ultimate load in N	Disp. at Failure in mm	Max. disp. in mm	Bend angle in deg.
1	1120.17	2.04	3.68	140° 50′
2	1365.11	2.31	4.85	140° 50′
3	1380.95	2.70	5.01	140° 50′
4	1770.28	3.21	7.42	140° 50′

Table 4 Bending strength for different welding conditions of Table 2 for dissimilar material joints

Exp. no.	Ultimate load in N	Disp. at failure in mm	Max. disp. in mm	Bend angle in deg.
1	812.93	1.84	3.51	5° 40′
2	818.31	1.91	4.64	12° 26′
3	829.45	2.10	3.56	13° 09′
4	1019.04	2.33	4.86	16° 56′

The appearance of the sample after the root surface bending test is shown in Fig. 19a, b for similar and dissimilar joints, respectively. The bending load is applied in the weld centre and is obvious in Fig. 19. The results indicate that the joint bending strength is increased with increase in preheating current. Crack is observed on the outer surface, and the bending strength and bending angle increase with the effect of preheating in both cases. Increase in preheating current allows weld joint to bend at higher bend angle which shows improvement in ductility of weld joints in dissimilar joint due to improvement in plastic flow. But temperature is not enough to soften the base material (Cu) at low current or without preheating. The materials are not sufficiently plasticized to be stirred and forged easily and resulting in lesser bending strength. Defect is in the root for all the joints that are observed, and this defect is known as the tunnel hole defect. Although the appearance of the welded surface seems to be good, tunnel defects can be observed at the AS of stirred zone in the weld.

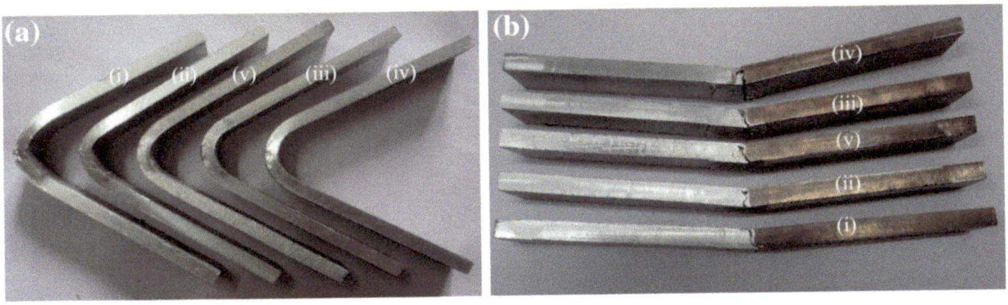

Fig. 19 Specimens after the bending test: **a** similar material joint and **b** dissimilar material joint

5 Validation of Numerical Model

Estimation of thermal history during P-FSW is based on analytical expressions of generated heat that provides the amount of heat generated on active surfaces of the welding tool. A comparison between the estimated maximum temperature obtained using the proposed model and the experimental results is depicted in Fig. 20. It is observed that the calculated and measured thermal histories for all thermocouple locations of advancing and retreating side of the weld show decent equivalent result. The results indicate that the calculated profile and the measured profile are of the same shape and that the maximum temperatures on both curves are of the approximately same value, which implies that the present model can accurately predict the heat transfer process in FSW.

A three-dimensional transient thermal analysis is performed using symmetric heat source profile to compare the results corresponding to Exp. No. 2 in Figs. 21 and 22 for both similar and dissimilar joints, respectively. Simulation result shows that the temperature distributions are not symmetric around welding line and maximum temperature occurs behind the tool pin. At the same time, the highest temperatures are always found in the workpiece material right below the tool shoulder and the temperatures progressively decrease from this region as a function of the distance in the radial and through-the-thickness directions. There is non-symmetric heat generation in stirring zone due to high thermal conductivity of copper which causes not as much of effective material flow during welding. To overcome heat loss due to high thermal conductivity of copper, a preheating source is used in copper side as shown in Fig. 22. Symmetric heat distribution in string zone is found in case of P-FSW with improvement in peak temperature about 37 °C.

Figure 23 shows the calculated isotherms of the peak temperatures in the welding zone compared to the experimentally obtained microstructure morphology

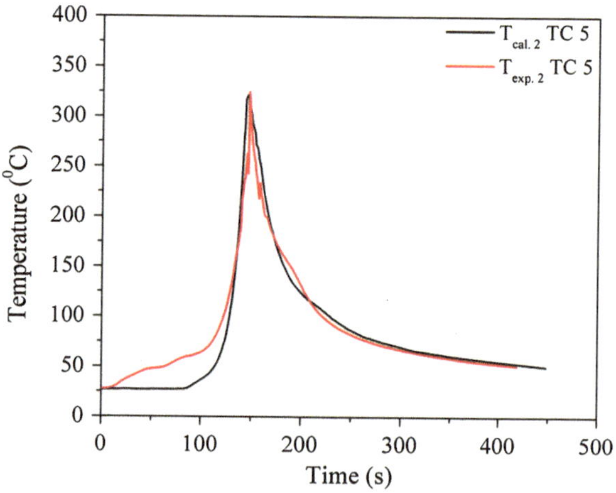

Fig. 20 Comparison between experimental and calculated time–temperature profile at TC 5 on the retreating side of Exp. 2 in Table 2 for similar joint

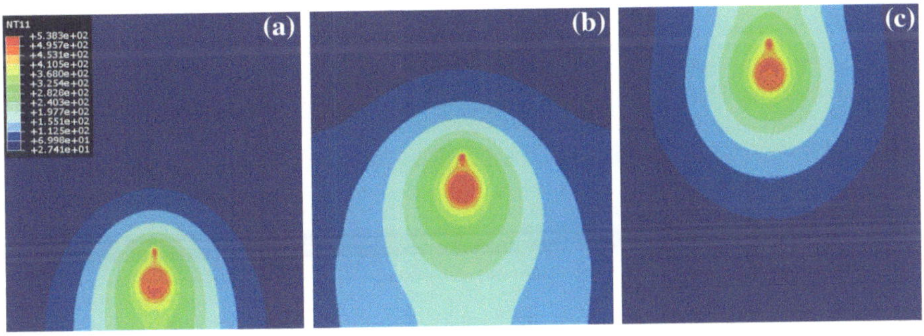

Fig. 21 Temperature distribution of similar material joining for Exp. 2 in Table 2 at welding time **a** 65 s, **b** 95 s and **c** 125 s

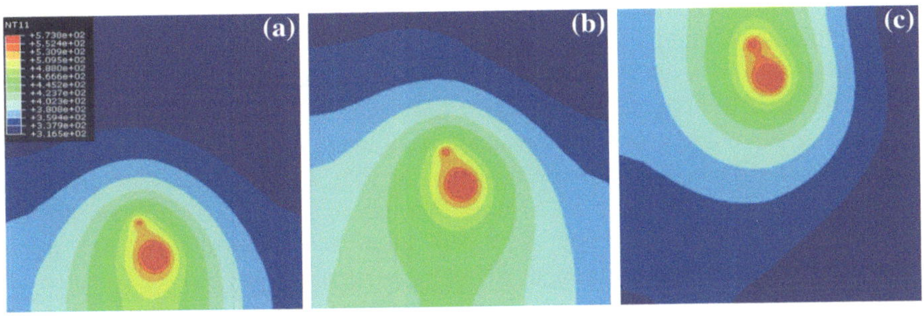

Fig. 22 Temperature distribution of dissimilar material joining for Exp. 2 in Table 2 at welding time **a** 65 s, **b** 95 s and **c** 125 s

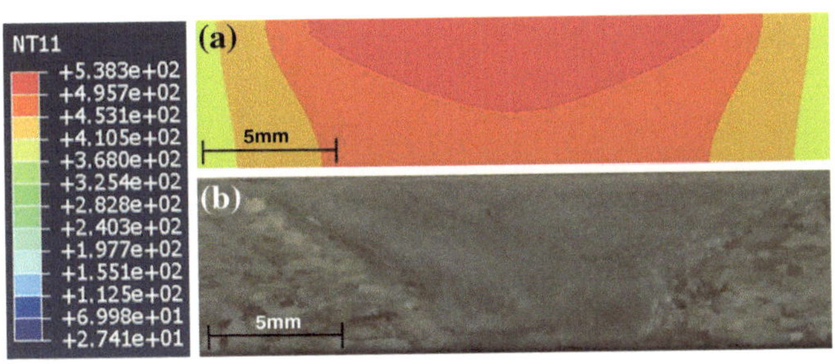

Fig. 23 Comparison of experimental and simulated isotherms for similar material joint of Exp. 2 in Table 2

of similar P-FSW joint. Weld zone demonstrated the grains within the stir zone which are roughly equiaxed and often an order of magnitude smaller than the grains in the parent material due to high temperature between 538 and 495 °C. It is obvious that the temperature isotherm provides useful information about various microstructural zones of weld cross section and this can be linked with dimensions of weld nugget.

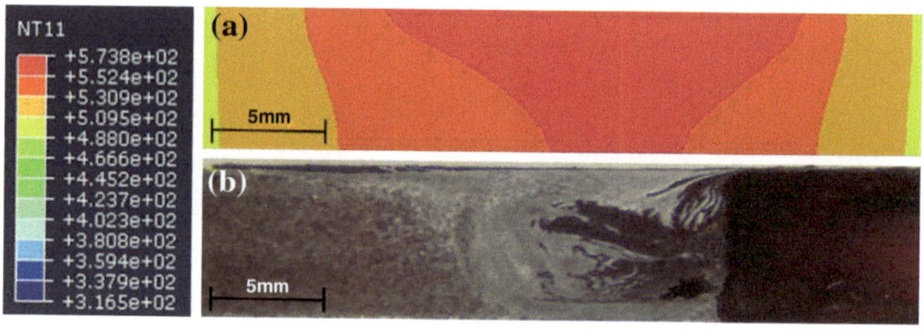

Fig. 24 Comparison of experimental and simulated isotherms for dissimilar joint of Exp. 2 in Table 2

Joining of dissimilar materials irrespective of high chemical affinity and completely different physical and mechanical properties like aluminium and copper is quite difficult. Although these factors contribute to the asymmetry in both heat generation and material flow, these drawbacks may be overcome by introducing additional local heating to preheat the copper side up to 200 °C and the optimum location of the tool to provide adequate metal flow around the tool. Figure 24 shows cross-sectional view in transverse direction of Exp. 2 in Table 2. The preheating source makes up heat loss due to high thermal conductivity in copper side and reduces the asymmetry in both heat generation and material flow. The microstructure morphology is related to the local thermal history. The calculated isotherms of the peak temperatures in the welding zone are compared with the experimentally obtained macrograph.

6 Conclusions

In this chapter, the P-FSW for similar and dissimilar materials is presented. The effects of preheating on the mechanical properties and the structure of a joint are described by means of experiments. The following conclusions can be derived from the present investigation:

- Plasma arc heat is used to preheat the workpiece at a localized area ahead of the rotating probe and thus to enhance the plasticization of local volume of material ahead of the probe. The preheating source basically enhances metal flow around the tool and reduces the plunging force over present range of process parameters.
- The high temperature ahead of the rotating tool softens the workpiece and enables joining without strong clamping fixtures. Less force is needed to move the welding tool forward; hence, wear is reduced. A further advantage of plasma arc energy for this process is the ability to weld at higher rates without causing excessive wear to the welding tool.

- In general, P-FSW of similar and dissimilar materials joining, UTS of welded joint increases with increases in heat input. For joining of similar materials, the microhardness in HAZ and nugget zone is approximately uniform throughout the weldment even with increase in heat input.
- An inhomogeneous distribution of hardness values is observed in the nugget zone during dissimilar joint. The values of hardness in weld zone in some point are far higher than Al and Cu due to formation of hard inter-metallic compounds.
- The joint bending strength is increased with increase in preheating current due to the enhanced plastic flow and partial annealing effect in both cases. But dissimilar joint bending strength is increased when the plasma arc is attached in harder material side.
- A finite element models are developed which pretend the welding of P-FSW process to join both similar and dissimilar materials and are validated with experimentally measured temperature distribution and macrograph of welded joint. Hence, the numerical model provides reasonably good prediction of temperature distribution using a hybrid heat generation model for P-FSW.

Although the preheating technique in FSW process is quite effective to improve mechanical properties and reduction of plunging forces, it needs to investigate from various aspects. The preheating in FSW (P-FSW) can be potentially used for relatively harder materials and joining of different dissimilar materials.

References

Askari A, Silling S, London B, Mahoney M (2001) Modeling and analysis of friction stir welding process. In: Friction stir welding and processing, pp 43–54

Bang HS, Bang HS (2010) A study on the weldability and mechanical characteristics of dissimilar materials butt joints by laser assisted friction stir welding. J Korean Weld Joining Soc 28: 678–683

Bang HS, Bang HS, Jeon GH, Oh IH, Ro CS (2012) Gas tungsten arc welding assisted hybrid friction stir welding of dissimilar materials Al6061-T6 aluminum alloy and STS304 stainless steel. Mater Des 37:48–55

Bang HS, Bang HS, Song H, Joo S (2013) Joint properties of dissimilar Al6061-T6 aluminum alloy/Ti–6 %Al–4 %V titanium alloy by gas tungsten arc welding assisted hybrid friction stir welding. Mater Des 51:544–551

Cao G, Kou K (2004) Arc-enhanced friction stir welding. GB Patent No. US7078647 B2, International patent application No. 10/970,058

Chao YJ, Qi X, Tang W (2003) Heat transfer in friction stir welding—experimental and numerical studies. Trans ASME 125:138–145

DebRoy T, Bhadeshia HKDH (2010) Friction stir welding of dissimilar alloys a perspective. Sci Technol Weld Joining 15(4):266–270

Fernandez GJ, Murr LE (2004) Characterization of tool wear and weld optimization in the friction-stir welding of cast aluminum 359+20 % SiC metal-matrix composite. Mater Charact 52(1): 65–75

Ferrando WA (2008) The concept of electrically assisted friction stir welding and application to the processing of various metals. Naval Surface Warfare Centre Carderock Division, pp 8–98
Galvao I, Oliveira JC, Loureiro A, Rodrigues DM (2011) Formation and distribution of brittle structures in friction stir welding of aluminium and copper: influence of process parameters. Sci Technol Weld Joining 16(8):681–689
Grujicic M, He T, Arakere G, Yalavarthy HV, Yen CF, Cheeseman BA (2010) Fully coupled thermomechanical finite element analysis of material evolution during friction-stir welding of AA5083. Proc Inst Mech Eng [B]: J Eng Manuf 224(4):609–625
Kima D, Badarinarayanb H, Kimc JH (2010) Numerical simulation of friction stir butt welding process for AA5083-H18 sheets. Eur J Mech A Solids 29:204–215
Kimapong K, Watanabe T (2004) Friction stir welding of aluminum alloy to steel. Weld J 83 (10):277–282
Kingery WD, Bowen H, Uhlmann DR (1976) Introduction of ceramics. Wiley, New York
Li K, Aidun D, Marzocca P (2009) Time-varying functionally graded material thermal modeling of friction stir welding joint of dissimilar metals. Trends in welding research 2008. ASM International, pp 731–735
Litwinski E (2004) High strength friction stir welding. GB Patent No. US6780525 B2, International patent application No. 10/035,865
Liu HJ, Shen JJ, Xie S, Huang YX, Cui F, Kuang LY (2012) Weld appearance and microstructural characteristics of friction stir butt barrier welded joints of aluminium alloy to copper. Sci Technol Weld Joining 17(2):104–110
Mahjmoud A (2003) Reduction of thermal stresses by developing two-dimensional functionally graded materials. Int J Solids Struct 40:7339–7356
Martin J (1980) Micromechanisms in particle hardened alloys. Cambridge University Press, Cambridge
Merklein M, Giera A (2008) Laser assisted friction stir welding of drawable steel-aluminium tailored hybrids. Int J Mater Form 1:1299–1302
Murphy A, Lynch F, Price M, Gibson A (2008) Modified stiffened panel analysis methods for laser beam and friction stir welded aircraft panels. Proc IMechE G: J Aerosp Eng 220:81–88
Murr LE, Li Y, Flores RD, Trillo EA, McClure JC (1998) Intercalation vortices and related microstructural features in the friction-stir welding of dissimilar metal. Mater Res Innovations 2(3):150–163
Neto DM, Neto P (2013) Numerical modelling of friction stir welding process: a literature review. Int J Adv Manuf Technol 65:115–126
Ochi H, Ogawa K, Yamamoto Y, Kawai G, Sawai T (2004) The formation of intermetallic compounds in aluminium alloy to copper friction-welded joints and their effect on joint efficiency. Weld J 18(7):516–523
Oeystein G, Arnt K, Midling O (1999) Modified friction stir welding. GB Patent No. WO1999039861 A1, International patent application No. PCT/NO1999/000042
Palm F (2001) Laser supported friction stir welding method. GB Patent No. US6793118 B2, International patent application No. 10/333,830
Rai R, De A, Bhadeshia HKDH, DebRoy T (2011) Review: friction stir welding tools. Sci Technol Weld Joining 16(4):325–342
Sahin M (2010) Joining of aluminium and copper materials with friction welding. Int J Adv Manuf Technol 49:527–534
Schmidt H, Hattel J, Wert J (2004) An analytical model for the heat generation in friction stir welding. Modell Simul Mater Sci Eng 12:143–157
Scutelnicu E, Birsan D, Cojocaru R (2008) Research on friction stir welding and tungsten inert gas assisted friction stir welding of copper. Recent Adv Manuf Eng 3:97–102
Sinclair P, Longhurst W, Cox C, Lammlein D, Strauss A, Cook M (2010) Heated friction stir welding: an experimental and theoretical investigation into how preheating influences on process forces. Mater Manuf Processes 25:1283–1291

Thomas WM, Nicholas ED, Needham JC (1991) Friction stir butt welding. International patent application No. PCT/GB2/02203. GB Patent application No. 9125978.8

Xue P, Xiao BP, Ni DR, Ma ZY (2010) Enhanced mechanical properties of friction stir welded dissimilar Al–Cu joint by intermetallic compounds. J Mater Sci Eng B 527:5723–5727

Zaeh MF, Gebhard P, Huber S, Ruhstorfer M (2008) Bifocal hybrid laser beam welding and friction stir welding of aluminum extrusion components. Adv Mater Res 43:69–80

Solid-State Joining by Roll Bonding and Accumulative Roll Bonding

Suprim Sardar, Atanu Mandal, Surjya K. Pal and Shiv Brat Singh

Abstract This chapter is intended to present the basic concept of roll bonding (RB) and accumulative roll bonding (ARB) of light-weight ductile metals. For this purpose, a discussion on previous research works on RB and ARB has been incorporated in this chapter. The general mechanisms of joining after rolling and several applications of RB and ARB have also been discussed. In the present study, an investigation on influence of various parameters on joint strength was conducted. Preheat temperature and percentages of thickness reduction are found to influence joint strength in a significant manner. The influence of annealing temperature and annealing time on mechanical properties of roll-bonded products in terms of controlling parameters has also been discussed here. The significant influence of preheat temperature and corresponding thickness reduction on the bond strength has been thoroughly discussed with a mechanism of peel testing. An attempt has been made to establish the mechanism behind failure along the joining line during peel testing of bond strength of solid-state welding through rolling.

Keywords Roll bonding · Accumulative roll bonding · Joint strength · Preheat · Peel test

S. Sardar (✉) · S.B. Singh
Department of Metallurgical and Materials Engineering,
Indian Institute of Technology Kharagpur, Kharagpur 721302, India
e-mail: suprim.sardar@gmail.com

S.B. Singh
e-mail: sbs22@metal.iitkgp.ernet.in

A. Mandal · S.K. Pal
Department of Mechanical Engineering, Indian Institute of Technology Kharagpur,
Kharagpur 721302, India
e-mail: atanuju88@gmail.com

S.K. Pal
e-mail: skpal@mech.iitkgp.ernet.in

R.G. Narayanan and U.S. Dixit (eds.), *Advances in Material Forming and Joining*,
Topics in Mining, Metallurgy and Materials Engineering,
DOI 10.1007/978-81-322-2355-9_18

1 Introduction

Previous research on mechanical properties and behavior of a material reveals that not only the chemical compositions and chemical bonding of materials but also the grain size have a significant influence on the overall mechanical behavior of materials. Most of the mechanical properties, viz. yield strength, hardness, ductility, malleability, formability, brittleness, etc., can be improved by refining the grain size. By the change of modern civilization, the demand of light-weight high-strength materials is increasing day to day. The ultra-fine-grained (UFG) materials have a large range of applications to date such as, in the field of aerospace, defence, automotive, interior decoration, household appliances, food processing, and chemical industries. As materials are processed through high-speed super-plastic deformation at elevated temperature, they change into UFG microstructure (Saito et al. 1998). Through these processing, fabricated products, having refined grain structure, exhibit excessive high strength and hardness. The modern non-traditional manufacturing processes such as cryogenic metal forming, chemical vapor deposition, mechanical alloying, rapid solidification, and intense plastic straining are used to produce UFG microstructure (Saito et al. 1998). Due to limited research works done on intense plastic straining out of other modern manufacturing processes, the current research work should investigate to find out the new techniques for industrial application. Through the intense plastic straining, macro-sized grained materials are transformed into UFG structured materials. As a result, refined microstructure exhibit and consequently high strength and toughness at ambient and elevated temperatures are achieved. The widely proposed methods of rolling were chosen for their superiority as intense plastic straining, due to their cost-effective mass production in nature. The RB and ARB are the solid-state welding processes, applied to bond similar or dissimilar metals by modified rolling; and these processes are the easiest and low-cost fabrication methods for the mass production of sheet metals. Through the various repetitive rolling processes, it can be identified that, the strength increases by 200–300 % for ultra-low carbon steel which are processed through ARB at 500 °C, while ductility decreases from 75 to 4 % after 32 passes (Krallics and Lenard 2004). At the surface region, the shear stress reaches the maximum value under compressive force. As a result under the action of high stress on the surface region, the grains are refined more (Costa et al. 2006). Following the subsequent passes of ARB cycle, ultra-grain refinement occurs along the thickness in the center region. On account of this action, the parabolic grain size distribution profile is seen along the thickness, due to development of shear strain. For every fabricated sample, the local grain size and hardness distribution follow the Hall–Petch relationship. It has been seen that grain refinement and hardness have direct relationship with shear strain of ARB-processed materials. Therefore, ARB is a very promising method to produce UFG. With the help of this process 80-nm grain can be produced with AA5083 (Toroghinejad et al. 2013). Milner et al. (2013) reached to the conclusion of transformation of micro-size of commercially pure Ti to the average grain size of 100 nm after seven cycle of processing at 450 °C.

Ultra-fine-grained 3003 alloy having mean grain size of 700–800 nm can be produced by ARB at 250 °C (Xing et al. 2002). The average grain sizes of samples were decreased very much from 10.2 μm to 700–800 nm after 6 cycles of ARB. In another experiment of ARB, the average grain size of pure copper (99.9 %) resolved into <100 nm after eight cycles of rotation (Shaarbaf and Toroghinejad 2008). By the cyclic action of accumulative rolling, the strength increased from 210 to 430 MPa; however, elongation decreased from 50 to 2 %. The average peel strength increases with the increase of annealing temperature from 200 to 400 °C (Jamaati and Toroghinejad 2010). As a result, entire material is contained with ultra-fine grains with high-angle boundaries. It has been clearly seen that, in accordance with tensile strength, the peel strength also increases with percentage of thickness reduction (Manesh and Taheri 2004). In general, the effects of annealing temperature and annealing time have been studied after rolling in various research works (Jamaati and Toroghinejad 2010; Yoda et al. 2011; Su et al. 2013).

Authors feel that there is a requirement of detailed presentation of these findings and further experiment for improve the readability of researches in this field. For this purpose, a detailed discussion containing *solid-state bonding, diffusion bonding, RB* and *ARB, controlling factors of RB* and *ARB, Advantages of RB* and *ARB, literature review*, and a *case study* on RB are made thoroughly in this chapter. The finding suggests that the pre-annealing increases the strength initially and the bond strength is very much dependent on preheating temperature and thickness reduction.

2 Solid-State Bonding

Welding is a widely used technology in metal fabrication field. The solid-state welding or solid-state bonding was introduced in the middle of twentieth century. The metals, which are difficult to weld with traditional fusion welding, tend to respond well in diffusion welding. Solid-state bonding is a diffusion welding or diffusion bonding process applied to join similar and dissimilar metals. Kazakov, one of the eminent soviet researchers, investigated and explained the basic method of diffusion bonding (Kazakov 1985a, b, c). According to his theory, the development of permanent joint can also be possible through interatomic bonds between workpiece under the action of localized or overall heating or plastic deformation or simultaneous action of both. On the accomplishment of above statement, the workpieces are definitely to be brought together to a close proximity. The surface geometry, chemical conditions, and physical state play an important role in assembling two surfaces at an atomic spacing. The interdiffused atoms travel across the interface of the weld zone on the action of applied bonding pressure. As a result, bonding takes place.

Diffusion bonding can be defined as adhesion and cohesion of atoms between two separate bodies due to short-range mobility of atoms through the surface area, by which two separate bodies can be transformed into a single entity. The migration of atoms across the interface of two closely attached bodies occur due to concentration gradients, which generates driving force of diffusion. The two separate materials are deformed under the action of compressive force together at an elevated temperature generally between half and three-fourth of the melting temperature for warm and hot working. The compressive force is applied to reduce the voids and air film, which may tap in joint surface interface. The tendency to cling toward similar or dissimilar type of atoms is activated among high thermal energy with high pressure. The diffusion bonding occurs only when minimum and essential energy of bonding overcomes the initial energy barrier of atoms. Lemus-Ruiz et al. (2011) explained the mechanism of diffusion bonding. Contemporary adhesion of close proximity between two regions of nearer surfaces produces interfacial joining. Klomp and Van De Ven (1980) asserted that temperature, pressure, and time are pivotal parameters affecting solid-state bonding process. It has been seen that adherence between lead and indium does not alter with increasing bonding temperature. On the other hand, the adherence of the aluminum and alumina bonds enhanced with increasing bonding temperature. Calvo et al. (1988) showed that the solid-state diffusion happened between iron and copper at elevated temperatures along with bonding pressure, which produced microstructural change in bond interface. A closer interlayer eutectoid substance was formed to the base iron, when bonding between iron and copper took place at a temperature higher than 900 °C. A narrow interlayer substance of precipitated copper ε-phase formed in the base iron in between 800 and 900 °C. Chen et al. (2012) applied low-temperature solid-state bonding techniques to fabricate electrodeposited Ni micro cones with an Au anti-oxidation layer with the application of pressure among Sn-capped Cu bumps. Interconnected bonding of Sn and electrodeposited Ni cone layer was developed as solid-state bonding with sufficient shear strength. In the solid-state welding, the bond strength was found to vary with respect to diffusion time (Cooper and Allwood 2014). The solid-state diffusion welding obeys the first and second Fick's laws of diffusion.

Therefore, temperature, pressure, and time are the three vital parameters, which govern the solid-state bonding of diffusion process. The diffusion of solid-state bonding produced interlayer bonding substances, which possesses same or different crystallographic structure with base metals. The joint interface of two surfaces exhibits adequate shear strength or bond strength in micro- and macro-level of bonding. There are numerous other solid-state welding processes in use, such as, projection welding, spot welding, seam welding, friction welding, and friction-stir welding.

3 RB and ARB

Roll Bonding is a plastic deformation process. Through this deformation process, solid-state welding is performed for similar and dissimilar types of metallic strips. Moreover, the essential five steps of the common RB process involves (i) dimensioning, (ii) surface preparation, (iii) riveting or clamping, (iv) preheatingin furnace (for hot working), and (v) rolling. All steps of RB are performed in rolling mill arrangement. A 2-HI rolling mill specified with alloy forged steel work roll of Ø320 mm × L300 mm barrel can be seen in Fig. 1. A brief clarification of RB steps has been discussed as follows. At first, dimensioning is the cutting of metallic sheets with required dimension, i.e., width × length (where as the thickness of the sheets should be specified, at the time of material selection). However, the final thickness of roll-bonded sheet should be chosen before the start of experiment. Surface preparation is the second stage. Before the bonding, surface preparation is an important and essential part of the process. Generally, surface roughness enhancement and contaminant layer removal from the surfaces are necessary two important steps of the surface preparation. Common processes to increase surface roughness are wire brushing, scratching, filing, or knurling. Both surface roughness increase and the removal of the contaminant layers take place at the same time. However, in knurling operation, the surface roughness can be increased as per requirement just by changing the knurling tools; but removal of contaminant layers is very less or sometimes not possible. The contaminant layers are mainly oxide and age-hardened strong layers, whereas loose layers are hydrocarbon. The hydrocarbon layers are oil, grease, and other lubricants which can be removed easily by application of chemicals such as weak acids, alcohols, and acetone. The strong bond layers like oxides can only be removed by wire brushing, scratching, filing, sand blasting or grid blasting, and some other metal removal processes such as turning and shaping with the special arrangement of sheet holding devices. In the third step, riveting is done just by folding the metallic sheets at both the sides or spot welding at one side. Except riveting, clamping may also be done, with the help of a wire or pin. The fourth and the most important step is the preheating(only for hot working process). The preheating is the heating operation, which is done

Fig. 1 2-HI rolling mill with capacity of maximum 75-ton roll separating force

immediately before the rolling, with the help of electric furnace. Sometimes automatic conveyers, attached with furnace, can carry the hot metallic sheets to the rolling mill. Following that, rolling is done. After proper adjustment of the roll gap and setting of the roll speed, the metallic sheets are passed through the rolls. The required pressure is produced by the rolls during the rolling process. The schematic representation of RB can be seen in Fig. 2 with the essential steps.

Accumulative Roll Bonding: Recently it has been revealed that ultra-fine grains of average diameter of less than 1 μm can be achieved with intense plastic straining process. The intense plastic straining processes technically belong to severe plastic deformation (SPD) process. Therefore, the UFGs can be obtained by intense plastic straining SPD, which is widely a proposed method to produce materials with UFG microstructure. In SPD, the most precise and determined study has been completed on ARB, throughout the last decade. Applying the roll bonding process, it is possible to manufacture the fine and ultra-fine-grained products. The ARB processes can accomplish both UFG and high-strained finished product simultaneously with or without preheating. Through the high straining, the manufactured materials exhibit 3–5 times more strength with respect to parent metals. ARB was first published by Saito et al. (1998). In ARB, the basic fundamental experimental steps are same as the RB process, except cyclic repetition. In addition to the experimental proceedings, the last step is the cyclic repetitions of the products. Those steps are practised with main common five steps of the RB process, i.e., from (i) to (v). Alternatively it can be stated that after the first cycle, all the steps are repeated in cyclic manner infinitely (theoretically). Nevertheless, the cyclic process is closed when surface cracks are found on the metallic surfaces. The schematic representation of general ARB process is shown in Fig. 3.

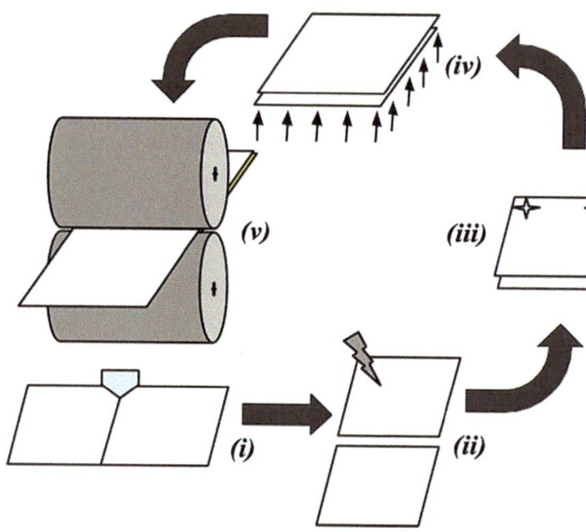

Fig. 2 The schematic of RB process with essential working steps, (*i*) dimensioning, (*ii*) surface preparation, (*iii*) riveting or clamping, (*iv*) preheating in furnace, and (*v*) rolling

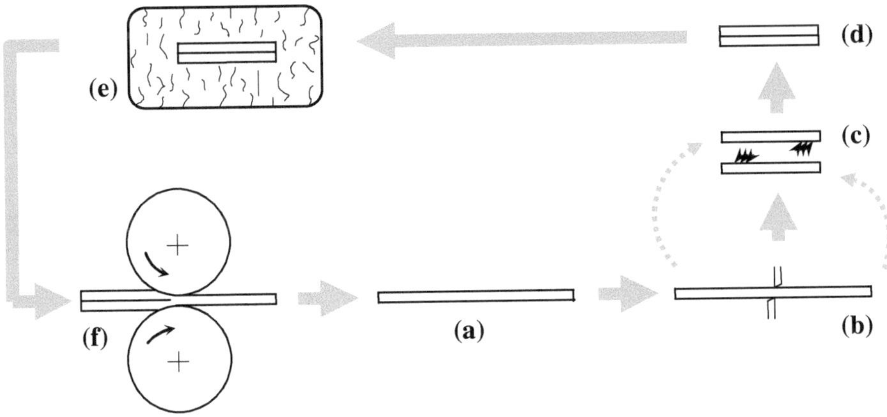

Fig. 3 Schematic diagram of various steps of accumulative roll bonding, **a** parent sheet metal, **b** cutting into two equal halves, **c** roughing and degreasing, **d** stacking of two strips, **e** preheating in furnace, **f** roll bonding

4 Controlling Factors of RB and ARB

The key controlling factors of RB and ARB are annealing temperature, annealing time, preheat temperature, rolling temperature, applied pressure, thickness reduction, rolling speed, surface roughness of parent metals, coefficient of friction between work piece and rollers, roll diameter, initial grain size of parent metal, ultimate strength, and ductility of parent metals. Recently, research works are being carried out controlling annealing temperature, annealing time, preheat temperature, percentage of thickness reduction, and number of cycles of ARB. Furthermore, it has already been revealed with various research works that the plastic deformation and severe plastic deformation performed in RB and ARB, respectively, for relatively light-weight ductile metals (Saito et al. 1998; Krallics and Lenard 2004; Costa et al. 2006; Toroghinejad et al. 2013; Milner et al. 2013; Xing et al. 2002; Shaarbaf and Toroghinejad 2008; Jamaati and Toroghinejad 2010; Saito et al. 1999; Tsuji et al. 1999).

4.1 Grain Size Factor

The scope of the ARB experimental work is to investigate and record the possibilities for involvements of SPD by producing of UFG with the help of ARB. The achievable range to manufacture fine grains (average diameter <10 μm), ultra-fine grain (average diameter <1 μm), sometimes nano grains (<100 nm in size) can be obtained through ARB (Toroghinejad et al. 2013; Milner et al. 2013; Shaarbaf and Toroghinejad 2008).

4.2 Strength and Hardness Improvement

The main intention of RB and ARB is to join the materials, and consequently, grains are refined. Grain refinement causes strengthening and enriches other mechanical properties during ARB for each cycle of material processing. According to literature, the refinement and strengthening occur simultaneously through high straining with SPD techniques. Saito et al. (1999) have conducted the experiment on AA1100, AA5083, and IF steel. The results indicate that the strength of AA1100 increased from 84 to 304 MPa after 8 cycles, and the elongation was decreased from 42 to 8 %. The strength of AA5083 increased from 319 to 551 MPa after 7 cycles, and the elongation reduced from 25 to 6 %; but for IF steel, the strength increased from 274 to 751 MPa after 5 cycles, and the elongation decreased from 57 to 6 %. Mean grain sizes of 1100 pure aluminum, 5083 Al–Mg alloy, and IF-steel materials before rolling were 37, 18, and 27 μm, respectively, and it got transformed to metals with final grain sizes of 670, 280, and 420 nm, respectively (Tsuji et al. 1999). The metals and alloys having UFG exhibit exceptional mechanical properties, such as high strength, high toughness, and super-plasticity at low temperatures, etc. The increment in hardness through RB happened not only due to strain hardening but also for grain refinement. This is the usual theory behind transformation of mechanical properties during traditional rolling process with respect to accumulative rolling process applied with preheat temperatures, on light metal such as commercially pure aluminum and other aluminum alloys.

4.3 The Mechanism Involved in RB

To discuss the mechanism of RB, the main focus has been given to *mechanical interlocking, interfacial bonding, captivating of energy barrier,* and *recrystallization during annealing* of the roll-bonded sheet metals which are the leading causes of RB (Manesh and Taheri 2004). Manesh and Taheri proposed a model (Manesh and Taheri 2004) for enlightenment on mechanism for cold roll welding which revealed that contact surfaces consisted of hard and relatively brittle layers. During rolling, the relative movement between two layers faced high friction. After that, scratch-brushed layers were fractured and it formed debris of scattered blocks. The mechanical interlocking happened when parent metal went through the roller. The work-hardened scratched and brushed layers split up and the same to form coherent blocks along the interfacial surface. While the rolling pass goes on, the underlying base metal is exposed in between two harder blocks; as a result, the base metal clings to the hardened layer of blocks (Wright et al. 1978). Eizadjou et al. (2009) conducted experiment and commented that with scratch-brushing, the covering surfaces of parent metals were transformed to work-hardened and brittle ones. During rolling, by the action of roll force, the scratch-brushed brittle surface layer breaks up coherently, due to surface expansion. As a result of increasing surface area, the surface cracks and fissures are generated on the contact surface. It has been found that the direction of these cracks and fissures is transverse to the rolling direction. As the rolling deformation increases over the deformation zone, generated surface cracks and fissures become wider, and therefore, the distance between each coherent block increases. Simultaneously, the virgin metal, staying under the hardened layer, extrudes through the coherent broken blocks and meets with opposite surface of hardened blocks or with virgin base metals, by which bonding is done.

5 Advantages and Disadvantages of RB and ARB

Merits of RB—RB is a cost-effective solid-state fabrication method, by which two separate sheet metals are transformed into a single body. Through the RB process, materials are strained well and the strength of the material increases. RB process can join similar and dissimilar materials with the formation of reduced grain size.

Demerits of RB—Joint failure rate is more in RB with respect to other welding processes. Since materials get strain-hardened, ductility decreases. Annealing is needed to recover the ductility after rolling.

Merits of ARB—ARB can produce UFG materials. High straining is possible to be achieved through ARB process. Strength of the materials increases enough. ARB can produce micro-grains and nano-grains through the entire volume of bulk material. This process requires relatively lower force and cheaper forming tools such as rollers, with respective to the other SPD processes, e.g., *equal channel angular extrusion, continuous confined strip shearing, continuous cyclic bending, cyclic extrusion compression, repetitive corrugation and straightening, high pressure torsion,* and *mechanical milling*. Composite materials can be easily created with the help of ARB processing.

Demerits of ARB—This process is applicable only to light-weight ductile materials. Through the ARB, strength of processed materials increases over the cycles; however, ductility decreases drastically with number of cycles. ARB shows better result only for the thin sheets.

6 Literature Review

To facilitate the present research work, a detailed literature study has been made on RB and ARB processes. The evaluation of RB processes applied on bi-layered aluminum–aluminum strips and copper strips, bi-layered bi-metallic alloys, bi-layered and tri-layered composites has been discussed in this section.

6.1 RB on Bi-Layered with Same Metals

In the characteristic examination with commercial pure aluminum AA1100, the experiment was conducted to find out the threshold deformation and bond strength of material having surface roughness of 4.2 μm through cold roll bonding (Shaarbaf and Toroghinejad 2008). The effects of annealing time, annealing temperature, hardness, and coefficient of friction on bond strength have been examined thoroughly. It has been clearly observed that bond strength increases with the increase in annealing time, annealing temperature, and coefficient of friction. However, bond strength decreases with the increase in hardness of parent metals.

With the aid of simulation and rolling test of 6111 aluminum alloy, Hong-zhi found that the heat generation increases with the increase in rolling (Hong-zhi 2006). Contact time decreases with the speed, as a result bond strength decreases. Bond strength will increase as the entry temperature increases. The three important parameters such as entry temperature, annealing temperature, and rolling speed control the bond strength.

In an extensive study on bond strength in cold pressure welding, Wright et al. (1978) conducted experiment on commercially pure aluminum composites to achieve 75 % deformations in rolling to measure the ultimate shear strength of the welded samples. Metallographic observations revealed that below the threshold deformation, the non-welded surfaces consisted of the actual scratch-brushed areas and the regions of the underlying base metal, due to the presence of contaminants. A theoretical relationship between weld strength and rolling deformation had been developed. The weld strength is a function of interfacial area bonded and the ultimate shear strength of each bonded junction.

Eizadjou et al. (2009) conducted experiment with commercial pure aluminum for 60 % deformation and reached to the conclusion that scratch-brushed layers are fractured coherently after entering roll gap. In that experiment, 21 % deformation was set as the threshold deformation by adjusting roll gap at ambient temperature. Hosseini and Kokabi performed peel testingon 5754 aluminum alloy to calculate the weld efficiency (Madaah-Hosseini and Kokabi 2002). They established a relation between the surface expansion and rolling pressure on the weld efficiency.

6.2 RB on Bi-Layered Bi-Metallic Metals

With the experiment on chromized interlayer of MS90 (CuZn10) alloy strips and steel sheets, Tolaminejad and Arabi (2008) commented that strength and the area fraction of the blank bond are two controlling factors of bond strength. It concludes that the bond strength of block bond is eight times less than the bond strength of the blank bond. This indicates that the overall bond strength depends on area fraction of the blank bond and its strength.

6.3 Bi-Metallic Composite Through RB

Zhang and Bay (1997) experimentally developed a shear testing system for measuring weld strength. Here, five different metal combinations such as Al–Al, Cu–Cu, steel–steel, Al–Cu, and Al–steel; and seven different surface preparation methods such as anodizing of Al, semi-bright electrolytic Ni plating, electroless Ni plating, heat-treated electroless Ni plating, electrolytic hard Cr plating, scratch brushing, and degreasing with acetone were used to cold-rolled welding tests. It was found that different surface preparation methods had a significant effect on weld strengths.

Roll bonding techniques were applied to form metal–metal composites. Roll-bonded laminates, hot-pressed sheet, and roll-bonded rod were produced with the bi-metallic composite of Al–steel and Ti–steel sheet (Moore et al. 1981). To produce these bi-metal composites, the optimum fabrication processes were developed. The bond strength was measured in between matrix and reinforcement carefully. Conversely, the produced composites were controlled through the processing of another plastic deformation process. The production cost of metal–metal composite was relatively low, and it was ready to further forming operation with high strength but low weight in terms of mechanical properties.

6.4 Tri-Layered Composite Through RB

The research work carried out by Manesh and Taheri (2003) explored the effect of annealing heat treatment on mechanical properties of aluminum-claded steel sheet. The main focus of that research work was to find the effects of annealing temperature and time on mechanical properties, bond strength and threshold deformation percentage of aluminum-claded steel sheet. The results revealed that bonding strength between the clad layer and base metal, and formability attained their maximum values at an optimum annealing temperature and optimum annealing time. The brittle intermetallic layer at the intimate interface of the claded layers decreased on the application of optimum annealing temperature and annealing time. It has been observed that bond strength is directly proportional to the total thickness reduction of Al-claded steel sheets. The annealing temperature up to 500 °C at constant annealing time promoted the formation of intermetallic phase in between intimate surfaces of aluminum and steel at high level. The research work suggested that the optimum annealing condition was at 450 °C for 16 h, which was able to decrease the work-hardening effect under cold rolling and reduce intermetallic phase thickness.

In an experimental study, aluminum-based metal matrix composites were performed at different thicknesses on the bond strength of Al–Ni–Cu layers

(Shabani et al. 2012). Additionally, to identify the effect of reduction in thickness on the bond strength and optimum parameter in rolling to form Al/Cu/Ni composites, a post-rolling annealing was evaluated. Surprisingly, it had been observed that bond strength was reduced due to the presence of Ni-coated Cu in Al–Cu strips, during peel testing. On the other hand, the bond strength increased with increase in thickness of coating, and then, bond strength reached to its high value in comparison with uncoated Al strips, although, the post-rolling annealing heat treatment reduced the bond strength. In addition to this, the bond strength increased with the decrease of thickness during rolling.

Heat treatment process has been performed on the Al/Cu roll-bonded material, and the interfacial bond strength study has been done (Sheng et al. 2011). The results signify that low-temperature heat treatment is able to develop better bond strength and Al/Cu interfacial morphology. Nevertheless, in long time heat treatment and higher temperature produce Al_2Cu intermetallic compound, which is one of the main causes of decreasing bond strength. In addition to this, Al_2O_3 particle precipitated along the interface, thereafter. The experiment suggests that to get better result, heat treatment should be limited to 200 °C at 20 h for claded sandwich sheet of Al/Cu/Al.

Wire brushing has a significant effect on roll bonding for cladding the aluminum sheets (Kim et al. 2012). The waviness and crack formation lead to an incompatible deformation between the clad and core which may strengthen the bonding in non-brushed sheets.

Laminated metal composites of Mg and Al prepared by roll bondingwere regularly used in various applications such as in aerospace (Liu et al. 2012). In both the laminated metal composites, Mg shows a complete dynamic recrystallization and the Al layers show a decrease in grain size, the increment of dislocation volume fraction with increase in roll bonding reduction. No intermetallic compound was found in Mg/Al interface. In comparison with two types of laminated metal composites such as Mg/Mg and Mg/Al/Mg, both the laminated metal composites showed an elongation of 35 %. Nevertheless, Mg/Al/Mg shows higher elongation at every reduction ratio and Mg/Mg shows higher bonding toughness at lower rolling ratio due to delamination mechanism. The properties of a laminated composite prepared by roll bonding had significantly affected by rolling reduction and reinforce particles (Masoumi and Emadoddinm 2013). Lap shear and Erichsen cup test showed that formability and bond strength of two- and three-layer Al/SiCp sheets decreased with the increase in number of layers. The reinforce particle on each layer and the number of layers depict the formability property. A bond criterion was developed for laminated composites and simulated by using finite element method. 7075 Al/AZ31B Mg/7075 Al-laminated composites were prepared by hot rolling with various reduction ratios and temperatures (Zhang et al. 2011).

7 Case Study

A detailed two-stage experiment is conducted, the first one is, to study the effect of annealing on the strength of roll-bonded strips, and the second, the effect on the bond strength of the rolled strip by changing preheatingtemperature and thickness reduction.

7.1 *Materials and Rolling Parameter Selection*

Commercially pure aluminum AA1100 is soft and ductile, and it has the property of excellent workability, after annealing. It is better for complicated forming operation, because it work-hardens more slowly than other alloys. It is the most weldable among aluminum alloys; therefore, solid-state bonding is possible. It has excellent properties for corrosion resistance and is commonly used in the field of food processing and chemical industries. It also produces super-finished products. The aluminum AA1100 which has thickness (t_0) = 1.2 mm, surface area of length (l) = 150 mm × width (w) = 25 mm, was taken as the parent metal. The temperatures 360, 400, 440, and 480 °C were chosen as annealing temperatures, and annealing time was varying from 1- to 7-h experiment. In second part, RB fabrication was done in the temperature range from 200, 300, and 400 °C with AA1100 for understanding the effect of preheat temperature and thickness reduction on peel strength. The metal samples were heated up to the above-mentioned temperature so that easy investigation could be done on the variety of fabrication having improved bond strength.

7.2 *Experimental Setup*

In the present work, the common RB process was practised with the help of 2-HI laboratory-scale rolling mill with rollers having 320 mm diameter and 300 mm roll length, respectively. At first, two sheets with the same dimensions and materials were taken. After that, the cleaning, roughing, and degreasing were done on surfaces, which were to be joined. The rolling temperature has been taken same as the preheat temperature, because the time elapsed between preheated metallic sheets taken out from the furnace and kept into roll gap was less than 2 s. For that reason, no heat loss was considered. Instantly, sheets were passed through the roller and roll bonding was done.

7.3 Part-I: Effect of Pre-Annealing Temperature and Time on RB Test

The experiment conducted on the annealing effect after rolling has been discussed. In this part of case study, the study of annealing before rolling is made. For this reason, the annealing can be termed as pre-annealing. The strips of AA1100 were prepared with proper dimension for this test. For the pre-annealing purpose, the samples were heated in furnace separately up to 360, 400, 440, and 480 °C and kept for 30 min each for soaking and then cooled for 3 h in furnace. To study the annealing time effect, the annealing time was selected as 1, 2, 3, 4, 5, 6, and 7 h only for 440 °C annealing temperature. By varying the annealing temperature and annealing time, annealed samples were prepared. RB was then conducted at 300 °C rolling temperature with 50 % thickness reduction. After completing the RB, the tensile test samples were prepared according to ASTM specifications. The tensile test was carried out in 100-kN electro-mechanical controlled universal testing machine (INSTRON 8862) and the samples were loaded at a crosshead velocity of 0.1 mm/min.

7.3.1 Outcome of Annealing Effect on Strength and Ductility

In Figs. 4, 5, and 6, a graphical representation of the effects of annealing temperatures on the mechanical properties of Al–Al sheet is presented. The annealing temperature is held constant.

- UTS and YS both increase with the increase in annealing temperature initially 400 °C and decrease beyond 440 °C.
- The % elongation of the Al–Al roll-bonded sheet decreases with the increase in annealing temperature till 440 °C, and increases beyond 480 °C.

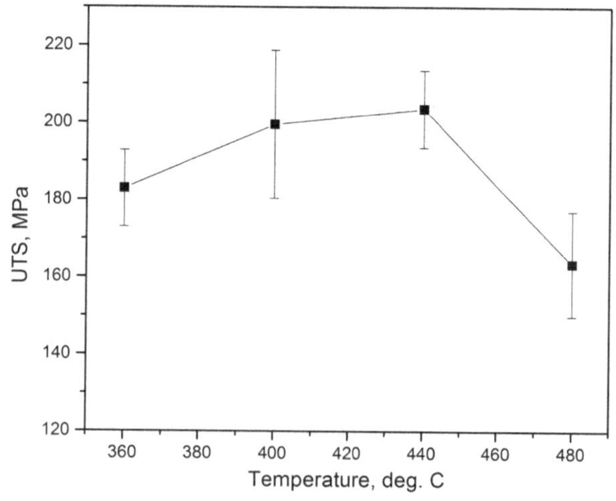

Fig. 4 Effect of pre-annealing temperature on UTS of Al–Al sheet at constant annealing time of 3 h

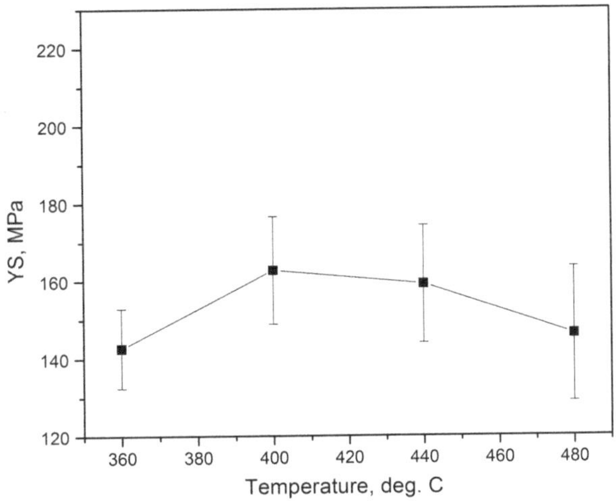

Fig. 5 Effect of pre-annealing temperature on YS of Al–Al sheet at constant annealing time of 3 h

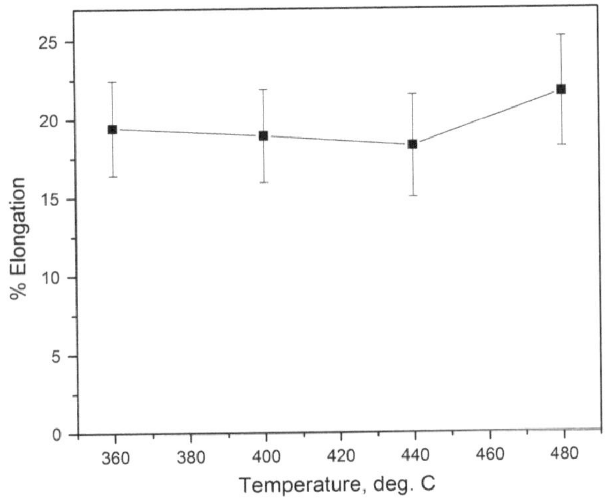

Fig. 6 Effect of pre-annealing temperature on % elongation of Al–Al sheet at constant annealing time of 3 h

Static and dynamic strain aging of Al–Al sheet at annealing temperature of less than 400 °C and overaging of Al–Al sheet with increasing the annealing temperature to 480 °C is observed (Manesh and Taheri 2003; Taheri et al. 1995).

The variation of UTS, yield strength, and % elongation of Al–Al annealed at a temperature of 440 °C with different time duration between 1 and 7 h are plotted in Figs. 7, 8, and 9, respectively.

The observable fact is that the static and dynamic strain-aging processes of Al–Al sheets have been seen at annealing time between 1 and 3 h. Overaging also takes place, during the annealing time between 3 and 7 h.

Fig. 7 Effect of pre-annealing time on UTS of Al–Al sheets that were pre-annealed at 440 °C

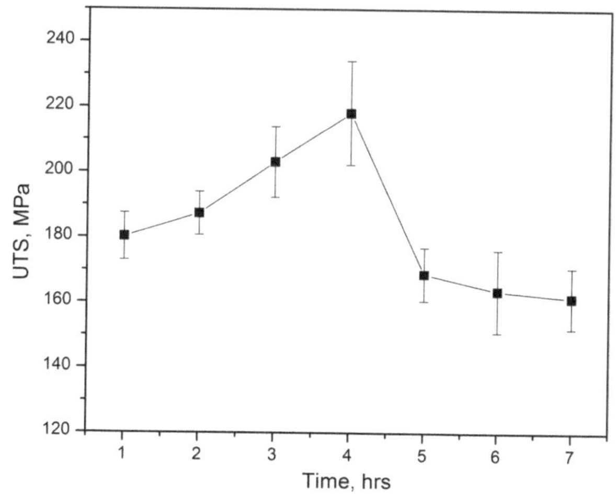

Fig. 8 Effect of pre-annealing time on YS of Al–Al sheets that were pre-annealed at 440 °C

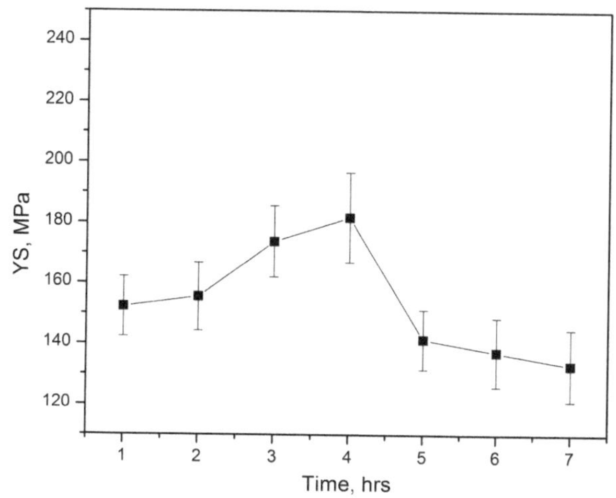

7.4 Part-II: Effect of Preheat Temperature and Thickness Reduction on Peel Strength Test

For the peel test experiment, some portions (50 mm × 25 mm) were smeared with grease AP3 in one side along length of the both metallic strips. The smeared surfaces of two sheets were placed one upon another, and then, riveting was done with tearing and folding only once in the opposite side along length. Next the stacked sheets were preheated and rolled. The schematic representation of RB is shown in Fig. 10.

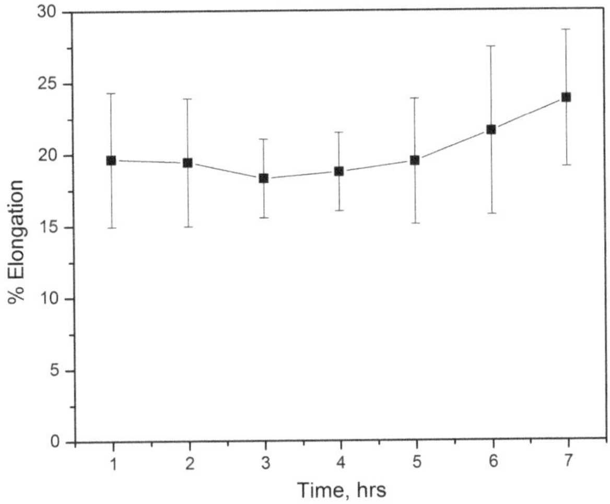

Fig. 9 Effect of pre-annealing time on % elongation of Al–Al sheets that were pre-annealed at 440 °C

In this study, 20, 50, and 80 % reductions on thickness were done for samples with the preheat temperatures of 200, 300, and 400 °C, respectively. Preheated sheets were then rolled separately in a 2-HI rolling mill having rolling speed of 10 m/min in non-lubricating condition of rollers. Subsequently, in the same manner roll bonding was done at other preheat temperatures. Therefore, eighteen samples were prepared for bond test experiment on roll bonding.

Peel testing is a method to measure the bond strength. It is a suitable characterization technique of bonding strength or shear bond strength of tape, adhesives, and flexible substrates and metallic materials where diffusion can occur. The peel test is generally done on rubber band, biomaterials, dental materials, medical packaging organic films and inorganic films, adhesive joint with wood or metals etc.

With the help of a strip cutter, all successful roll-bonded sheets were cut to a length of 80 and 25 mm in width. Out of 80, 40 mm was gauge length; 35 mm on each side was outstretched in a same straight line which was perpendicular to the gauge length, and the radius of curvature in between gauge and grip was 5 mm. The schematic representation of peel testing has been shown in Fig. 11. Along the gauge length, the point where grips were spread out is called grip-side or arm-side; and the reverse side is called end-side; this grip-side and end-side can be seen in Fig. 11. The real specimen peel test has been shown in Fig. 12a. Figure 12b is clearly representing gauge length and grip lengths of 40 and 35 mm, respectively. The bond strength test has been performed in 100-kN electro-mechanical controlled universal testing machine, INSTRON 8862. Crosshead speed of the machine was set as 0.1 mm/min.

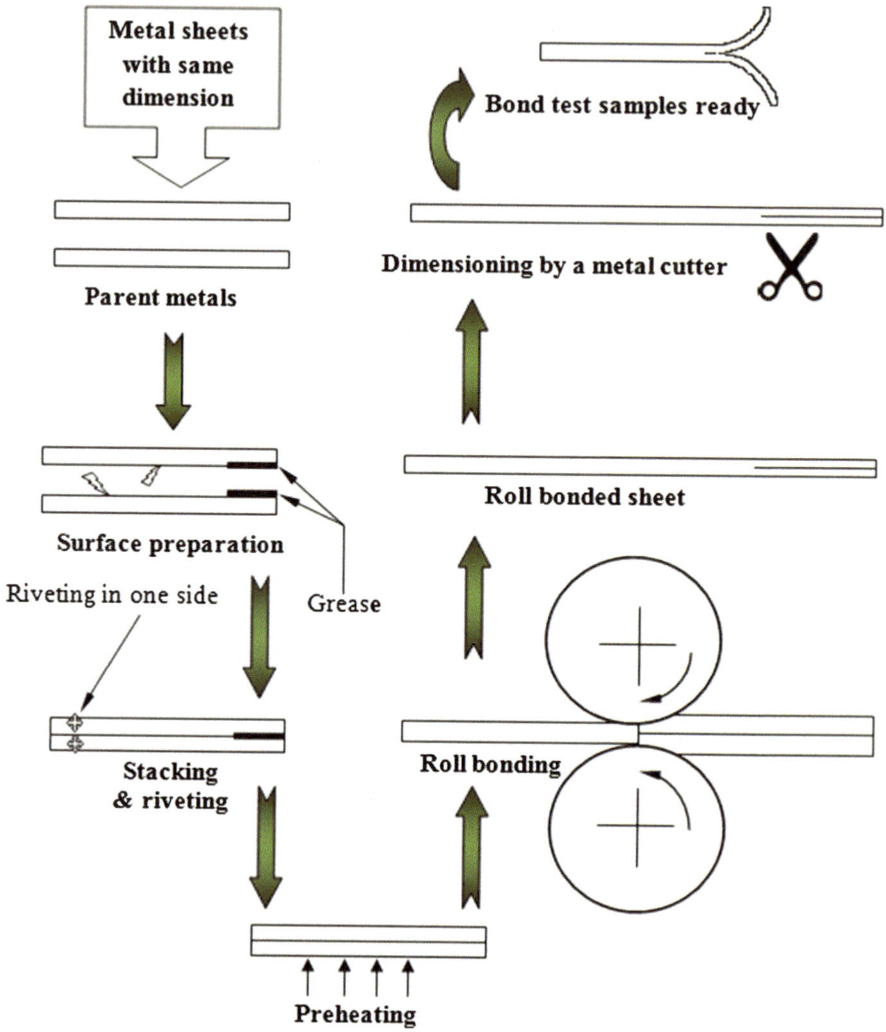

Fig. 10 Schematic diagram of roll bonding for joining strength measurement

7.4.1 Outcome of Bond Strength Study

In the present experiment, the bond or joining strength of roll-bonded AA1100 aluminum sheets with different parametric combinations has been shown in Table 1. Except the first combination, i.e., rolling at 200 °C with 20 % thickness reduction, all combinations produced successful joining. From Table 1, it is seen that

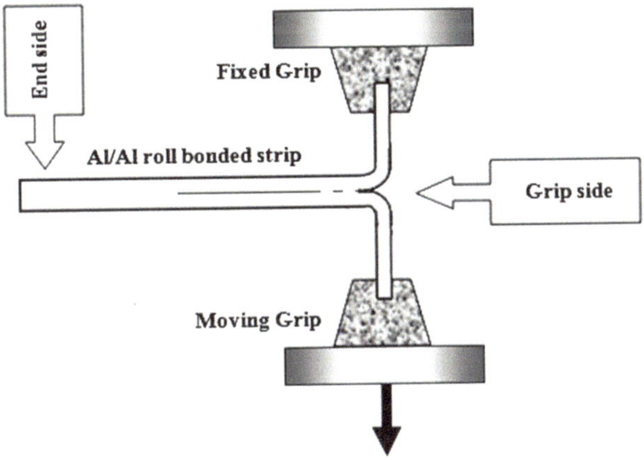

Fig. 11 Schematic diagram of peel testing for bond strength measurement

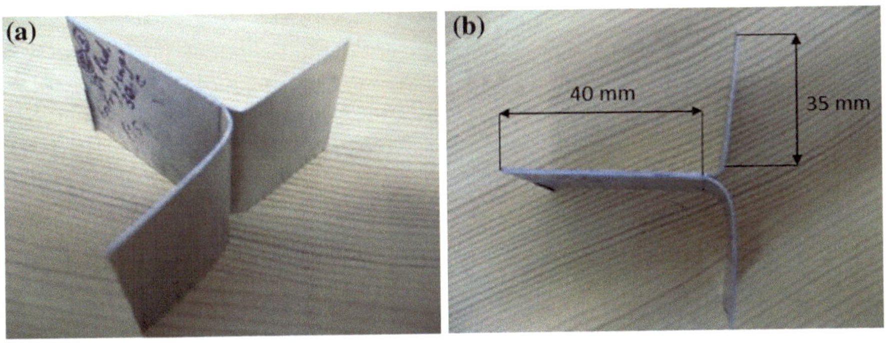

Fig. 12 Peeling test samples with dimensions, **a** isometric view of a sample before peeling, **b** view of an actual sample, which indicates gauge and grip lengths

presenting minimum bond strength is 21 N/mm, for the combination of 20 % thickness reduction at 300 °C, and the maximum bond strength is 92 N/mm for 80 % thickness reduction at 400 °C combination.

Microstructural study has been conducted on different samples to clarify the bonding phenomenon. In this microstructural study, a very thick line of trapped air, along the rolling direction (RD), is depicted in Fig. 13a, which represents the microstructure of roll-bonded joint produced at 200 °C with 50 % thickness reduction. Whereas Fig. 13b shows the microstructure of roll-bonded joint produced at 300 °C with 50 % thickness reduction, more bonding is found along the joining line which is comparatively less thick. Positions of trapped air are shown here by three white arrows. However, in the upper portion along the RD, this line is

Table 1 Bond strength of roll-bonded AA1100 sheets

Bond strength (N/mm)			
Thickness reduction (%)	Rolling at 200 °C	Rolling at 300 °C	Rolling at 400 °C
20	Not joined	21	32
50	46	50	59
80	65	73	92

not found. It may happen due to the increase in preheat temperature from 200 to 300 °C. Consequently, an increment in the bond strength is also found from 46 N/mm to 50 N/mm, according Table 1. A very thin line along RD is found in Fig. 13c, which represents the microstructure of roll-bonded joint produced at 400 °C with 50 % thickness reduction. In the similar way, the bond strength increases to 59 N/mm according to Table 1. Therefore, it may be concluded that the bond strength increases with an increase in preheat temperature for a constant thickness reduction.

Figure 14a–c represent the microstructures of roll-bonded joints produced at 20, 50, and 80 % thickness reduction and a constant 300 °C preheat temperature, respectively. The joining lines are found to be thin, thinner, and thinnest for Fig. 14a–c, respectively. The joining lines are pointed with black arrows in these figures. The bond strengths for roll-bonded joints produced with 20, 50, and 80 % thickness reductions with 300 °C preheat temperature are 21, 50, and 73 N/mm, respectively, and presented in Table 1. It is also found from Fig. 15a–c that the bonding separation decreases with the increase in percentage of thickness reduction by keeping the same preheat temperature. Consequently, the bond strength increases with an increase in percentage reduction for a constant preheat temperature.

Similarly, the microstructures, acquired at 500× magnification, of roll-bonded joints produced with 20, 50, and 80 % thickness reductions at a constant preheat temperature of 400 °C are shown in Fig. 15a–c, respectively. The similar phenomenon, such as 300 °C preheat temperature, can also be observed here.

7.4.2 Discussion

Through the present study, it can be observed that the bond strength is increasing with the increase in preheat temperature by keeping percentage of thickness reduction the same and with the increase in percentage of thickness reduction by keeping preheat temperature the same. These phenomena are explained as follows. After the application of external tearing force on two arms of peel test samples, the grains along the common surface line present in bending portion of the inner and outer surfaces of the arm root are under tensile force, compressive force,

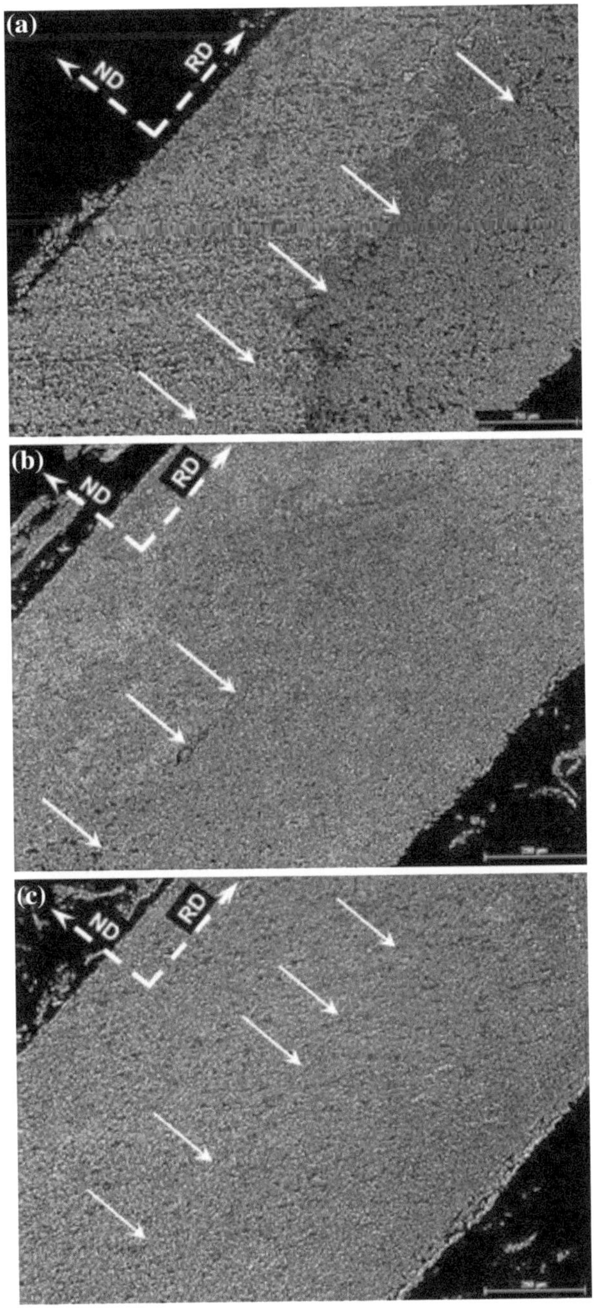

Fig. 13 **a** Optical micrograph of RB at preheat temperature 200 °C and 50 % reduction, 100× magnification. **b** Optical micrograph of RB at preheat temperature 300 °C and 50 % reduction, 100× magnification. **c** Optical micrograph of RB at preheat temperature 400 °C and 50 % reduction, 100× magnification

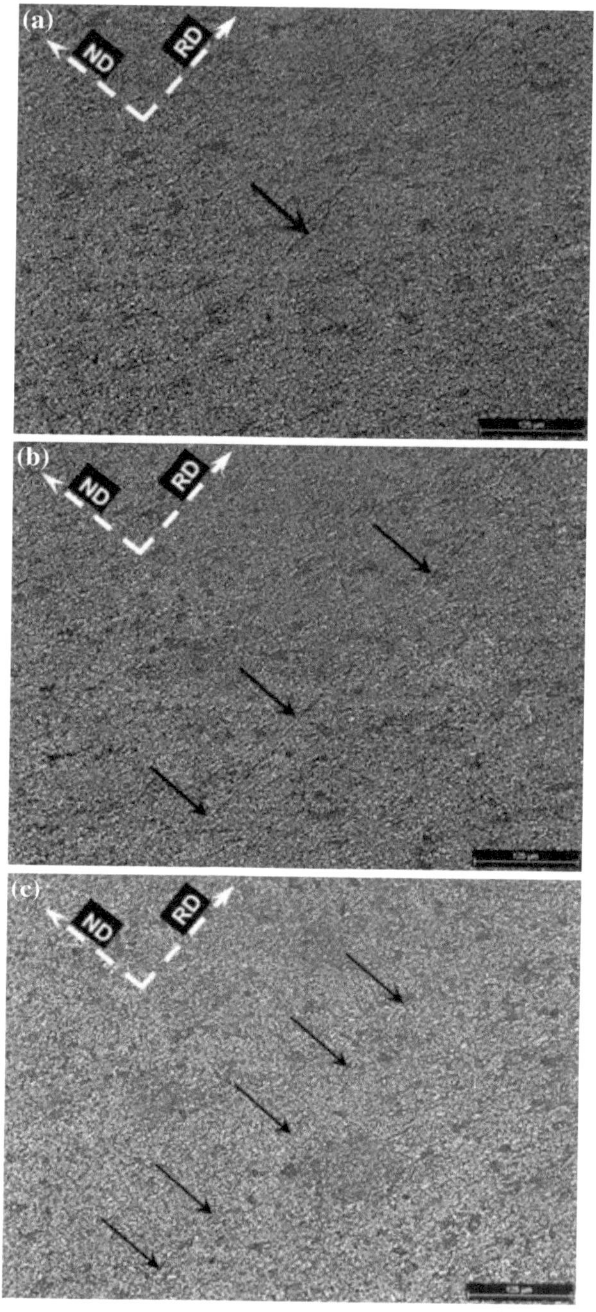

Fig. 14 **a** Rolling with 20 % reduction at preheat temperature 300 °C, 200× magnification. **b** Rolling with 50 % reduction at preheat temperature 300 °C, 200× magnification. **c** Rolling with 80 % reduction at preheat temperature 300 °C, 200× magnification

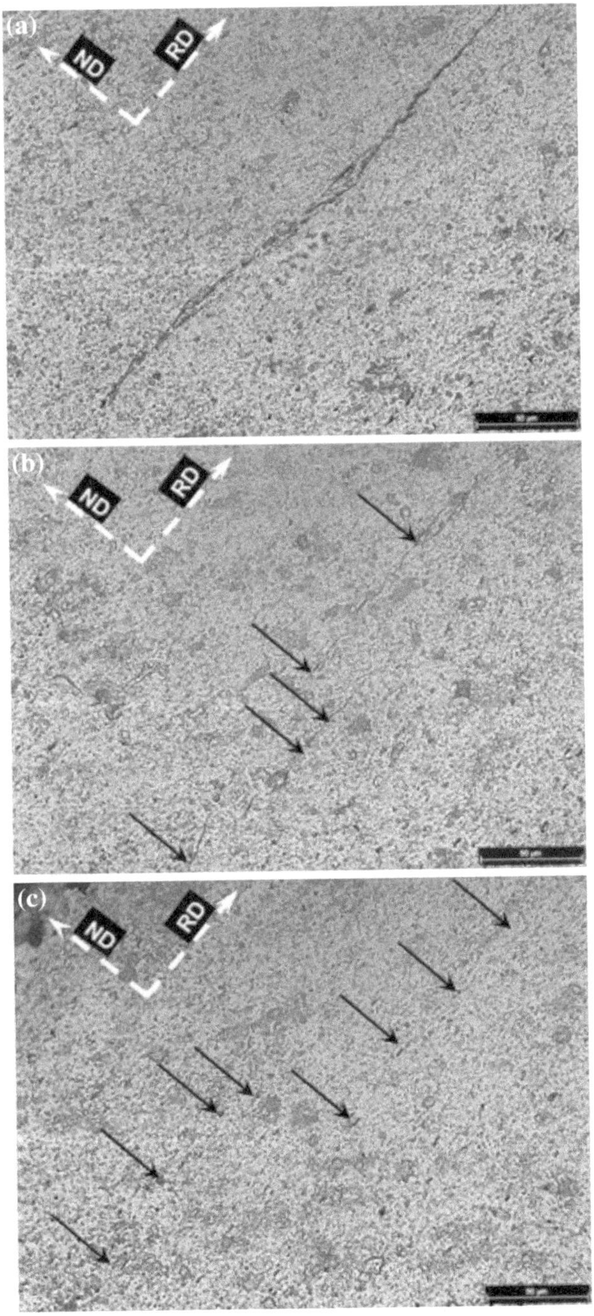

Fig. 15 **a** Optical micrograph of RB with 20 % reduction at preheat temperature 400 °C, 500× magnification. **b** Optical micrograph of RB with 50 % reduction at preheat temperature 400 °C, 500× magnification. **c** Optical micrograph of RB with 80 % reduction at preheat temperature 400 °C, 500× magnification

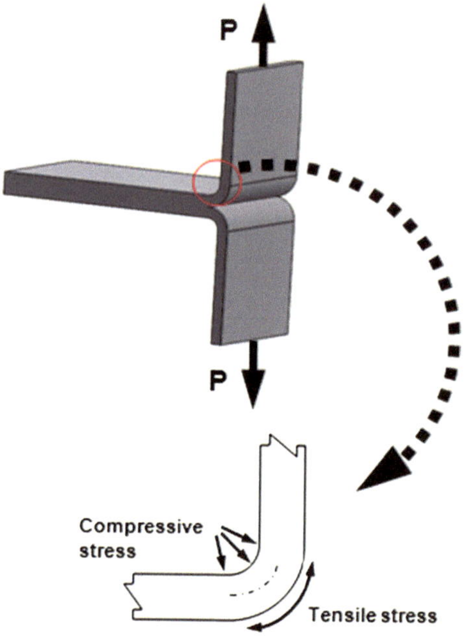

Fig. 16 Schematic of instantaneous stresses at bending corner of a peel test sample along RD throughout duration of testing

respectively, which is explained in Fig. 16. The upper half segment of the neutral fiber, which is closer to concave curvature, is under compressive stress and the other segment, which is closer to convex curvature, is under tensile stress.

High-stress field is generated around the cleavage of lips of air-trapped flaw. With the simultaneous action of tensile stress, crack spreads and it produces an increase in the surface area of the side of the flaw. By this action of crack propagation along the joining line, each crack is conglomerated with the adjacent enlarging cracks, as shown in Fig. 17. Therefore, the specimen having more tiny air-trapped flaws produces low peel strength. Tiny air-trapped flaws are less in number for the roll-bonded joints produced with more percentage of thickness reduction and more preheat temperature. Consequently, the peel strength is higher due to the less flaw propagation along the joining line of two metallic strips for these roll-bonded joints.

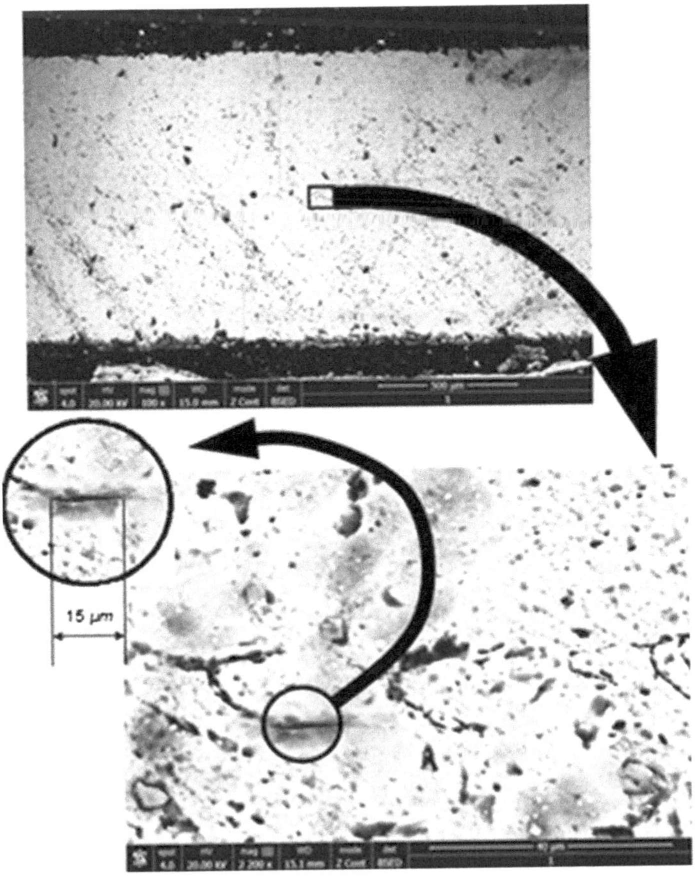

Fig. 17 Electron micrograph of a sample; roll-bonded at 200 °C with 50 % reduction

8 Concluding Remarks

This chapter discusses about the important aspects of the solid-state joining by using roll bonding and accumulative roll bonding processes. The mechanisms involved in RB and ARB are thoroughly discussed here. Merits and demerits of RB and ARB are also presented in this chapter.

An experimental study about the influences of process parameters on bond strengths of roll-bonded joints has been discussed in a systematic manner. From this study, it can be concluded that the bond strength or peel strength increases with an increase in percentage of thickness reduction and preheat temperature.

References

Calvo FA, Urena A, Gomez De Salazar JM, Molleda F, Criado AJ (1988) Solid-state transformations during diffusion bonding of copper to iron. J Mater Sci 23:1231–1236

Chen Z, Luo T, Hang T, Li M, Hu A (2012) Low-temperature solid state bonding of Sn and Nickel micro cones for micro interconnection. ECS Solid State Lett 1:7–12

Cooper DR, Allwood JM (2014) The influence of diffusion mechanisms in aluminium solid-state welding processes. In: Procedia Engineering 11th International Conference on Technology of Plasticity, ICTP 2014

Costa ALM, Reis ACC, Kestens L, Andrade MS (2006) Ultra grain refinement and hardening of IF-steel during accumulative roll-bonding. Mater Sci Eng A 406:279–285

Eizadjou M, Danesh Manesh H, Janghorban K (2009) Mechanism of warm and cold roll bonding of aluminum alloy strips. Mater Des 30:4156–4161

Hong-zhi Y (2006) Key factors for warm rolled bond of 6111-aluminium strip. Trans Nonferrous Met Soc China 16:84–90

Jamaati R, Toroghinejad MR (2010) Effect of friction, annealing conditions and hardness on the bond strength of Al/Al strips produced by cold roll bonding process. Mater Des 31:4508–4513

Kazakov NF (1985a) An outline of diffusion bonding in vacuum. Diffus Bond Mater 10–16

Kazakov NF (1985b) A theory of diffusion bonding. Diffus Bond Mater 17–48

Kazakov NF (1985c) Principal bonding variables and recommended. Procedures for diffusion bonding in vacuum. Diffus Bond Mater 49–70

Kim SH, Kim HW, Euh K, Kang JH, Cho JH (2012) Effect of wire brushing on warm roll bonding of 6XXX/5XXX/6XXX aluminum alloy clad sheets. Mater Des 35:290–295

Klomp JT, Van De Ven AJC (1980) Parameters in solid-state bonding of metals to oxide materials and the adherence of bonds. J Mater Sci 15:2483–2488

Krallics G, Lenard JG (2004) An examination of the accumulative roll-bonding process. J Mater Process Technol 152:154–161

Lemus-Ruiz J, Ceja-Cardenas L, Bedolla-Becerril E, Lopez-Morelos VH (2011) Production, characterization, and mechanical evaluation of dissimilar metal/ceramic joints. Nanocomposites Unique Prop Appl Med Ind. InTech 205–224

Liu CY, Wang Q, Jia YZ, Jing R, Zhang B, Ma MZ, Liu RP (2012) Microstructures and mechanical properties of Mg/Mg and Mg/Al/Mg laminated composites prepared via warm roll bonding. Mater Sci Eng A 556:1–8

Madaah-Hosseini HR, Kokabi AH (2002) Cold roll bonding of 5754-aluminum strips. Mater Sci Eng A 335:186–190

Manesh DH, Taheri AK (2003) The effect of annealing treatment on mechanical properties of aluminum clad steel sheet. Mater Des 24:617–622

Manesh HD, Taheri AK (2004) Study of mechanisms of cold roll welding of aluminum alloy to steel strip. Mater Sci Technol 20:1064–1068

Masoumi M, Emadoddinm E (2013) Interface characterization and formability of two and three-layer composite sheets manufactured by roll bonding. Mater Des 44:392–396

Milner JL, Abu-Farha F, Bunget C, Kurfess T, Hammond VH (2013) Grain refinement and mechanical properties of CP-Ti processed by warm accumulative roll bonding. Mater Sci Eng A 56:109–117

Moore JJ, Wilson DV, Roberts WT (1981) Evaluation of formable light metal-alloy steel composites. Mater Sci Eng 48:123–136

Saito Y, Tsuji N, Utsunomiya H, Sakai T, Hong RG (1998) Ultra-fine grained bulk aluminum produced by accumulative roll-bonding (ARB) process. Scripta Mater 39:1221–1227

Saito Y, Utsunomiya H, Tsuji N, Sakai T (1999) Novel ultra-high straining process for bulk materials development of the accumulative roll-bonding (ARB) process. Acta Mater 47: 579–583

Shaarbaf M, Toroghinejad MR (2008) Nano-grained copper strip produced by accumulative roll bonding process. Mater Sci Eng A 473:28–33

Shabani A, Toroghinejad MR, Shafyei A (2012) Effect of post-rolling annealing treatment and thickness of nickel coating on the bond strength of Al–Cu strips in cold roll bonding process. Mater Des 40:212–220

Sheng LY, Yang F, Xi TF, Lai C, Ye HQ (2011) Influence of heat treatment on interface of Cu/Al bimetal composite fabricated by cold rolling. Compos B 42:1468–1473

Su I, Lu C, Tieu K, Deng G (2013) Annealing behavior of accumulative roll bonding processed aluminum composites. Steel Res Int 84:1241–1245

Taheri AK, Maccagno TM, Jonas JJ (1995) Dynamic strain aging and the wire drawing of low carbon steel rods. ISIJ Int 35:1532–1540

Tolaminejad B, Arabi H (2008) A study of roll-bonding MS90 alloy to steel utilizing chromized interlayer. Iran J Sci Technol Trans B Eng 32:631–640

Toroghinejad MR, Ashrafizadeh F, Jamaati R (2013) On the use of accumulative roll bonding process to develop nanostructured aluminum alloy 5083. Mater Sci Eng A 561:145–151

Tsuji N, Saito Y, Utsunomiya H, Tanigawa S (1999) Ultra-fine grained bulk steel produced by accumulative roll-bonding (ARB) process. Scripta Mater 40:795–800

Wright PK, Snow DA, Tay CK (1978) Interfacial conditions and bond strength in cold pressure welding by rolling. Met Technol 444:24–31

Xing ZP, kang SB, Kim HW (2002) Structure and properties of AA3003 alloy produced by accumulative roll bonding process. J Mater Sci 37:717–722

Yoda R, Shibata K, Morimitsu T, Terada D, Tsuji N (2011) Formability of ultrafine-grained interstitial-free steel fabricated by accumulative roll-bonding and subsequent annealing. Scripta Mater 65:175–178

Zhang BW, Bay N (1997) Cold welding experimental the surface preparation investigation methods. Weld Res Suppl 326s–330s

Zhang XP, Yang TH, Castagne S, Gu CF, Wang JT (2011) Proposal of bond criterion for hot roll bonding and its application. Mater Des 32:2239–2245

Author Index

R.G. Narayanan and U.S. Dixit (eds.), *Advances in Material Forming and Joining*,
Topics in Mining, Metallurgy and Materials Engineering,
DOI 10.1007/978-81-322-2355-9

Subject Index

R.G. Narayanan and U.S. Dixit (eds.), *Advances in Material Forming and Joining*,
Topics in Mining, Metallurgy and Materials Engineering,
DOI 10.1007/978-81-322-2355-9